Air Conditioning Applications and Design

Air Conditioning Applications and Design

Second Edition

W. P. Jones
MSc, CEng, FInstE, FCIBSE, MASHRAE

Routledge
Taylor & Francis Group
LONDON AND NEW YORK

First published by Butterworth-Heinemann

This edition published 2011 by Routledge
2 Park Square, Milton Park, Abingdon, Oxon OX14 4RN
711 Third Avenue, New York, NY 10017, USA

Routledge is an imprint of the Taylor & Francis Group, an informa business

First published in Great Britain by Arnold 1980
Second edition 1997

British Library Cataloguing in Publication Data
A catalogue record for this book is available from the British Library

Library of Congress Cataloguing in Publication Data
A catalog record for this book is available from the Library of Congress

ISBN 978-0-415-50293-1

Typeset in Times and Univers. Produced by Gray Publishing, Tunbridge Wells

Contents

Preface to the First Edition

This book is essentially a text on system design and application, primarily intended for the use of the more advanced student of building services at a university or technical college but with its content also offered for the practising engineer. A knowledge of the basic principles of air conditioning is therefore presupposed, notably in the topics of climate, comfort, psychrometry, fluid flow in ducts, fans, refrigeration, automatic controls, heat gains, the determination of supply air quantity and simple system design. In this respect it is, therefore, a sequel and complement to *Air Conditioning Engineering* and is the outcome of my experience over many years, not only in lecturing but also in the practical design and installation of air conditioning systems. Despite this presupposition and the fact that reference to the psychrometric data published by the Chartered Institution of Building Services may be necessary, the book is, nonetheless, self-contained. Upon this foundation a more advanced study of air conditioning is established, theoretical considerations being used wherever possible to justify the choice and design of particular systems for their correct applications. The practical consequences of design are dealt with in so far as they affect performance, system space requirements, commissioning, the diagnosis and solution of the problems that inevitably arise, energy conservation, and comparative capital and running costs.

The evolution of the Institution of Heating and Ventilating Engineers into the Chartered Institution of Building Services, upon the acquisition of a royal charter, has been accompanied by a raising to degree standard of the level of technical qualification acceptable for admission to corporate membership – a move in line with other chartered institutions and professional bodies. Further, many academic centres now offer courses of study leading to the award of higher degrees in environmental engineering and building services. This elevation of standards has meant that a knowledge of design applications and system characteristics, extending beyond an appreciation of first principles, is increasingly necessary for aspirants to corporate membership. Although simplicity is a desirable feature of all systems there has been a growing trend to complication and the designer may now have more options available for an application than hitherto. This, coupled with advances in technique and the proliferation of packaged plant, makes mid-career training an increasing necessity among professional engineers in building services. Furthermore, because of the widening recognition in the UK of the need for successful air conditioning design and installation overseas, guidance on system performance at altitudes significantly above sea level has been included and, where apt, the consequences of system operation in hot climates mentioned in the text.

A chapter on economics has been provided, in the face of continuing inflation, aiming to allow the capital costs of various systems to be established at a budget level for an historical date, the contemporary costs then being evaluated by the application of an appropriate inflation index, obtainable from official sources. As a guide, inflation indices for the building services industry in the United Kingdom are given up to 1978.

I am grateful to Haden Young Ltd for kind permission to reproduce some of the data in the Appendix and elsewhere in the text, as indicated.

W.P.J.
1979

Preface to the Second Edition

Since the first edition was published in 1980 there have been some significant changes in the building services industry. Notable among these are the rapid increase in the use of computers for engineering calculations and the way this has broadened the horizons of engineers by offering them a choice of solutions to design problems that would not have otherwise been possible without tedious calculations. Nevertheless, an understanding of the practical realities of plant and system choice and performance remain essential and this edition addresses the need. The increasing importance of energy conservation and the effect that engineering activities have on the environment, coupled with the continuing modification to the Building Regulations in the UK, have also had an influence on design attitudes and on the systems and plant that result from them.

The induction system has become obsolescent and less space is allotted to it in this edition, In parallel with this, more space has been allowed for the fan coil, variable air volume and chilled ceiling systems. An essential section has been added on the atrium in buildings and a new section has also been provided that discusses the basic issues related to the choice of combined heat and power installations, More extensive coverage has been given to the services for swimming pools and the use of heat pumps for this application. In particular, the section on clean rooms for the semi-conductor industry has been brought up to date and greatly expanded. In the chapter on air distribution sections have been introduced on variable volume air distribution, the use of swirl diffusers and permeable textile ducting. Displacement ventilation is also dealt with. The chapter on acoustics has been updated, particularly by the introduction of the concept of room criterion and by the addition of a section on the noise radiated by plant to areas outside a building. A few changes have been made in the section on economics, principally by taking advantage of the work done in the preparation of the CIBSE Energy Code for air conditioned buildings to address the problem of estimating the thermal and electrical energy consumptions of VAV and fan coil systems. By kind permission of the CIBSE, tables dealing with the solar gain through windows (for lightweight buildings with internal blinds) and sol-air temperatures for walls and roofs, have been added to the Appendix.

W P Jones
1996

1

Practical Load Assessment

1.1 The aims of load assessment

Most air conditioning systems operate at their design loads for only a small part of their life and it follows, therefore, that the designer should be concerned not only with the maximum heat gains and cooling loads but also with the way these change throughout the day and over the year. Establishing the pattern of such variations will be of help in choosing the correct system and in selecting the best form of automatic control. Applications lie in the commercial, industrial, institutional and domestic sectors for the climates of the United Kingdom, Europe and the rest of the world. It must therefore be expected that the size of the contribution made by each of the principal elements in the heat gain will not be constant but, nonetheless, the approach to the calculation will be essentially the same in all instances, although the same importance will not be attached to each element. Consequently, as a starting point, we shall examine the practical assessment of loads for one particular application – an office block in London This will provide a theme for later development.

1.2 A hypothetical office block

Figure 1.1 illustrates a notional office block of simple design, with details of a typical module for an intermediate floor. The areas to accommodate lifts, escape staircases, builders' voids for ducts and pipes, lavatories, and so on, are assumed to be in a pair of relatively small service blocks, one at each end of the building, adjoining the two short walls where there are no windows. For simplicity in the calculations, and probably without introducing significant error, it is further assumed that the presence of these two blocks does not influence the U-value of the two end walls or the heat flow through them. The other basic assumptions for the building are:

Windows
Area of glass (A_g): 40% of the outer facade, on the two long faces only.
Type: openable, double, clear, panes of 6 mm float or plate glass, with metal frames over not more than 10% of the gross window area.
Shading: internal, white Venetian blinds, to be drawn by the occupants as necessary to exclude the entry of the direct rays of the sun.
Thermal transmittance (U_g): 3.3 W m^{-2} K^{-1}.

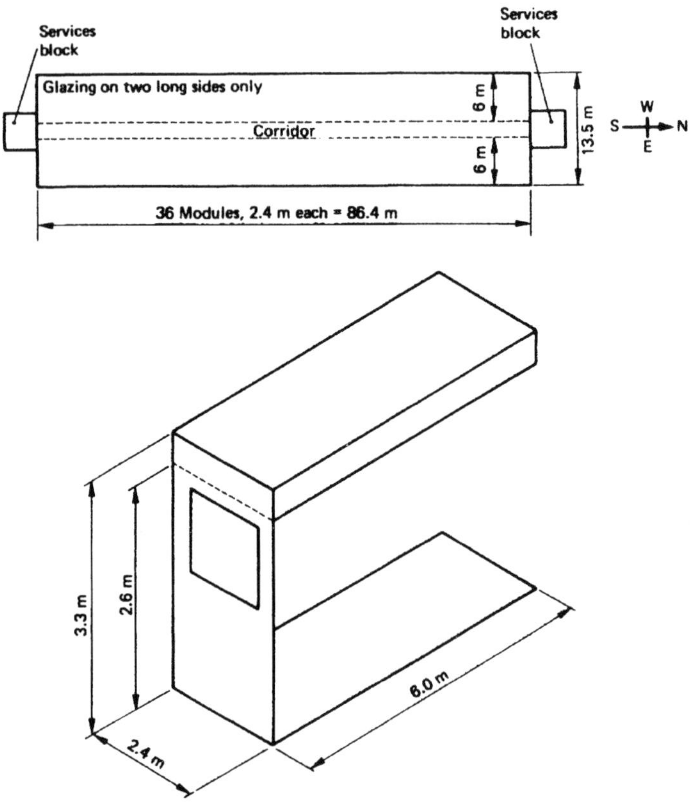

Figure 1.1 Plan and typical module of a hypothetical office block.

Roof
Construction: 19 mm asphalt, 150 mm aerated concrete slab, 50 mm UF foam, 16 mm plasterboard ceiling.
Thermal transmittance (U_r): 0.45 W^{-2} K^{-1}.
Surface density: 108 kg m^{-2}.
Decrement factor (f): 0.77.
Time lag (ϕ): 5 h.

Intermediate floors and ceilings
Construction: hollow pots in concrete, 50 mm cement screed, carpet, overall thickness 200 mm.
Surface density: 300 kg m^{-2}.
Suspended ceiling: proprietary acoustic panels, approximate surface density 7 kg m^{-2}.

External walls
Construction: 105 mm brick, 75 mm UF foam, 100 mm heavy-weight concrete blocks, 13 mm light-weight plaster, overall thickness 293 mm.

Thermal transmittance (U_w): 0.45 W m^{-2} K^{-1}.
Surface density: 417 kg m^{-2}.
Decrement factor (f): 0.2.
Time lag (ϕ): 9 h.
Area of wall (A_w): 60% of the outer facade on the two long faces, 100% of the outer facade on the two short faces.

Internal partitions
Construction: 2 mm × 12 mm perlite plasterboard sheets on timber studs, continued up to the soffit of the slab.
Surface density: 30 kg m^{-2}.

Doors opening onto central corridor
Construction: 50 mm deal, 800 mm wide × 2000 mm high.
Surface density: 30 kg m^{-2}.

Natural infiltration rates (n)
Summer: 0.5 h^{-1}; winter: 1.0 h^{-1}.

It is customary to suppose that the occupied area, and hence the area for which the loads are calculated, is the pair of peripheral strips, each 6 m wide. The central corridor is not subjected to the same loads because its population is transient, it is shielded from climatic effects and its lighting may well be at a lower level than elsewhere. The treated floor area, on the other hand, is often taken to include both the corridor and the peripheries.

In order to determine the air conditioning loads for the model building, the additional design assumptions listed below must be made:

- Outside states: 28°C dry-bulb, 19.5°C wet-bulb (sling), regarded as occurring at 15.00 hours sun time in July, and −2°C saturated as the design winter condition.
- Room states: 22°C dry-bulb, 50% saturation in summer with 20°C dry-bulb, 36% saturation in winter.
- Population density: 9 m^2 per person but two people assumed for an office comprising only one module of 14.4 m^2 floor area.
- Fresh air allowance. 1.4 l s^{-1} m^{-2}, over the total floor area of 13 997 m^2. (After allowing a diversity factor of 0.75 for the occupancy in summer design conditions this corresponds to about 16 l s^{-1} for each person, which is the fresh air allowance recommended by the CIBSE for premises where there is some smoking.)
- Metabolic rate: for sedentary workers, 90 W sensible and 50 W latent, per person.
- Power dissipated by electric lights: 17 W m^{-2}, including control gear
- Power dissipated by business machines: 20 W m^{-2}.

There is general agreement on the practical assessment of sensible heat gains, except in the estimation of solar gains through glass and the determination of the heat flow through walls and roofs; the first issue being of far greater importance than the second. Opinion on the better approach to both calculations is divided between methods advocated by American authorities and those used in the United Kingdom.

1.3 Solar heat gain through glass

Solar heat gains through glass may be calculated from first principles using data published by the CIBSE (1986a) and by ASHRAE (1993), but it is much more convenient to use tabulated results for a particular window, defined by its orientation, the time of the day, the month of the year, the latitude of the place, etc. In the CIBSE guide the cooling load arising from solar gain through vertical glass, i.e. the sensible heat gain from this source with due allowance for the storage effect of the building, is tabulated for lightweight and heavyweight structures, with and without internal shading on the windows, for latitude 51.7°N (approximately that of north London), assuming that the air conditioning system runs for 10 h a day to maintain either a constant dry resultant or a constant air temperature inside. Current engineering practice does not attempt to use dry resultant temperature in practical air conditioning and so only air temperature (dry-bulb) is relevant for determining heat gains. Further, maintaining a constant dry resultant temperature in a room imposes a bigger cooling load than does keeping the air temperature constant. Tables A.1–A.4, in the Appendix, provide tabulated solar loads based on the American method for latitude 51.5°N (approximately that of central London) and for typical building construction. Table A.5 reproduces CIBSE data.

Although the influence of the storage effect of a building upon the solar gain through glass is principally exercised by the floor slab, the other room surfaces also play a part and are often taken into account when estimating the average surface density of a room, per unit area of floor, prior to determining the solar load according to the American data (Tables A.1–A.4). The procedure is best illustrated by an example.

EXAMPLE 1.1

Estimate the mean surface density of a typical module on an intermediate floor of the hypothetical office block shown in Figure 1.1. Note that it is customary to halve the density of a floor slab if it is covered by a carpet, as in this case, since this is considered to insulate partially the mass of the floor slab from the solar radiation. The mass of the glass is insignificantly small. Only half the mass of the side partitions contributes to the mean surface density of the module because the adjoining offices are also air conditioned. On the other hand, the corridor and the door opening onto the corridor are fully effective if the corridor is not air conditioned, as is assumed in this case. The suspended ceiling is taken as fully effective because it is separated from the slab above by an air gap.

Answer

Floor: $0.5 \times (2.4 \times 6.0) \times 300$	2160 kg
Door: $(0.8 \times 2.0 \times 30$	48 kg
Corridor partition: $(2.6 \times 2.4) - (0.8 \times 2.0) \times 30$	139 kg
Side partitions: $0.5 \times 2.6 \times 6.0 \times 2 \times 30$	468 kg
Exterior wall: $(2.6 \times 2.4 - 0.4 \times 3.3 \times 2.4) \times 417$	1281 kg
Suspended ceiling: $(2.4 \times 6.0) \times 7$	101 kg
Total:	4197 kg

Over a floor area of $2.4 \times 6.0 = 14.4\ m^2$ this gives a mean surface density of 291 kg m^{-2}. Clearly some of the assumptions made regarding the relevant thickness of a partition or the effect of a carpet (Carrier, 1965) on the floor slab may be arbitrary, to some extent.

The tables based on the Carrier method (A.1–A.4 in the Appendix) that give direct values for the solar loads through windows are related to mean surface densities of 500 kg m^{-2} for heavy-weight buildings and 150 kg m^{-2} for light-weight buildings. Interpolation between the tables is probably not reliable, although the design engineer must exercise judgement. It is suggested that for the calculated value of 291 kg m^{-2} the tables for 150 kg m^{-2} should be used. If the floor slab had not been carpeted the mean surface density would have been (2160 + 4197)/14.4 = 441 kg m^{-2} and the use of the tables for 500 kg m^{-2} would be appropriate.

EXAMPLE 1.2

Using the appropriate tables, compare the solar heat gain through the windows of a typical module (Figure 1.1) by the CIBSE (1986a) and Carrier (1965) methods, for the month of July. Assume the floor is fitted with a carpet and that the steel-framed windows are virtually flush with the outer facade, i.e. ignore the shadow cast by any reveal.

Answer

Reference to Example 1.1 shows that the appropriate, modular, surface density is 150 kg m^{-2} and so, for glass internally shaded by Venetian blinds, Tables A.1(b) and A.2 (in the Appendix) yield the answers directly by the Carrier method. These are plotted as full lines in Figure 1.2 and, assuming office hours of 08.00 to 16.00, sun time (09.00 to 17.00, clock time in the UK), we see that the peak loads are 253 W m^{-2} at 08.00 h on the east face and 268 W m^{-2} at 17.00 h on the west.

The CIBSE (1986b) interprets a light-weight building as one having demountable partitions, suspended ceilings and either supported false floors or solid floors with a carpet

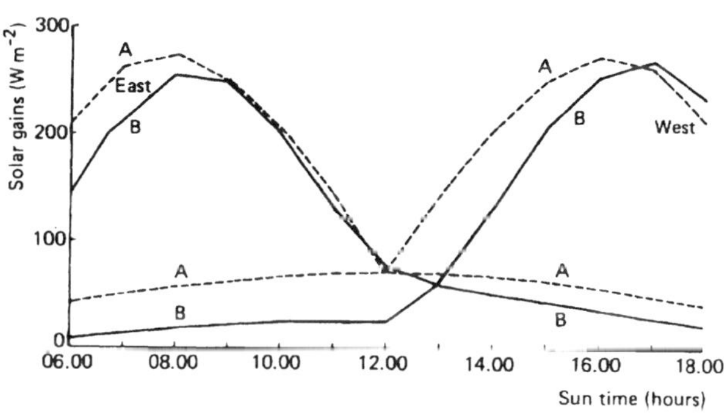

Figure 1.2 Cooling loads arising from solar gain through west-facing and east-facing single clear glass, shaded internally by white Venetian blinds. It is assumed that the blinds are closed by the occupants when direct solar radiation falls on the glass and are opened by them when the windows are in shade. Curves labelled A (dashed line) have been determined by the CIBSE method for a light-weight building, using data from Table A.5 in the Appendix. Curves labelled B (continuous line) have been calculated according to the Carrier method, using data from Tables A.1 and A.2 in the Appendix for a surface density of 150 kg m^{-2}.

or a wood-block surface. Note that the concept of the response factor, according to the CIBSE (1986c), is correctly applied for the calculating of heating duties.

The CIBSE cooling load by solar radiation is obtained directly from Table A.5 in the Appendix. This refers to light-weight buildings at latitude 51.7°N (London), fitted with internal shades and conditioned by a plant operating 10 h a day to maintain a constant internal dry-resultant temperature. The loads peak in July at 08.00 h sun time (09.00 h clock time) for the east-facing glass and at 16.00 h sun time (17.00 h clock time) for the west, as 306 and 293 W m^{-2}, respectively.

The corrections given in the footnote to the table must be applied. First, shading factors (denoted here by f_s) must be determined. Since it is a light-weight building having double-glazed windows fitted with internal light slatted blinds, the factors are 0.95 for the east windows at 15.00 h sun time but 0.74 for the west windows at the same time. Secondly, since the air conditioning system controls room temperature (air dry-bulb), not dry resultant temperature, an air point control factor (denoted here by f_c) must be determined. For a light-weight building having double glazing fitted with light slatted internal blinds, f_c is 0.91, whether the blinds are open or closed. There is no correction for the hours of plant operation, if these are different from 10 h per day. Hence the solar load is calculated by

$$Q_{sg} = f_s f_c q_{sg} A_g \tag{1.1}$$

where Q_{sg} is the cooling load due to solar gain through vertical glazing (W), q_{sg} is the specific cooling load due to solar gain through vertical glazing, read from Table A.5 (W m^{-2}) and A_g is the area of glass (wooden frame) or opening in the wall (metal frame) (m^2).

When comparing the CIBSE solar load with that determined by the Carrier method the different haze factors adopted must be taken into account. Carried values are based on a haze factor 0.9 whereas the CIBSE figures are based on 0.95. Hence the CIBSE values must be multiplied by 0.9/0.95 to put them on the same footing as the Carrier values, when comparison is made. This has been done when producing Figure 1.2.

It is possible to use the tabulated cooling loads according to the CIBSE (1986a) for other places in the world than London. Tabulations are also given for 10°N, 20°N, 30°N, 35°N, 40°N, 45°N, 50°N, 55°N and 60°N. For southern latitudes the values tabulated for the summer months (September to March, inclusive) must be multiplied by a factor of 1.07 because the earth is 3.5% closer to the sun in January than it is in July and because the intensity of radiation is inversely proportional to the square of the distance between the earth and the sun. Hence a value for a northern latitude from Table A.5 can be multiplied by 1.07 and used for the same numerical value of a southern latitude.

EXAMPLE 1.3

Calculate the cooling load by solar radiation through glass for a west window in the building shown in Figure 1.1, assuming it to be for the month of January in Perth, Western Australia, at 15.00 h sun time. Take the latitude of Perth to be 32°S.

Answer

Referring to the CIBSE guide (1986a) it is found that the solar cooling loads in January at 15.00 h sun time are 240 and 220 W m^{-2} for latitudes 30°N and 35°N, respectively. At 32°N the cooling load is 232 W m^{-2}, by interpolation. Hence the solar cooling load at 32°S is 1.07

$\times 232 \times 0.74 \times 0.91 = 167 \text{ W m}^{-2}$, where 0.74 and 0.91 are the shading and airport control factors, respectively.

The results using Table A.1 are generally a little less than those from the CIBSE guide, particularly the peak values. It is not easy to say which are the more correct but it must be pointed out that the American-based answers (Table A.1) are proved in use over a longer period than are the CIBSE values and are related to a continental-type (American) climate with longer stretches of continuous sunshine than are experienced in the UK. For these reasons, the Carrier figures are commonly in use throughout the world.

A difficulty arises when dealing with windows of heat-reflecting or heat-absorbing glass. Table A.2 lists factors for various glass types, to be applied to the loads given in Table A.1. Note that the factors in Table A.2 are not applied to the loads for bare, clear glass, quoted in Table A.3. This is because windows or window and shade combinations that absorb a lot of solar heat transmit this to the interior, and exterior, virtually instantaneously by convection and long-wave radiation and this form of heat transfer is not influenced by the thermal inertia of the building. Only the directly transmitted shortwave solar radiation is so affected. Therefore, the storage factors for windows shaded internally with Venetian blinds are larger than those for bare, clear glass. A proprietary type of bare glass that is strongly heat-absorbing therefore corresponds more closely to clear glass with internal blinds than it does to bare, clear glass. So the factors in Table A.2 are applied to the gains through shaded windows, in Table A.1, yielding answers that are approximately correct.

It should be noted that heat-absorbing glass invariably requires internal shades as well, if people within the room are not to feel uncomfortable when subjected to the direct solar radiation that is still transmitted through the glass. This is because the solar radiation is of high intensity, from a surface at 6000°C, whereas energy radiated from internal blinds is low intensity, from a surface at about 40 or 50°C. Blinds may be omitted with certain types of proprietary glass that are strongly heat-reflective, provided that the shading coefficient of such glass is low enough. There is no absolute yardstick for this but a tentative suggestion is that the shading coefficient should not exceed 0.27.

EXAMPLE 1.4

Calculate the shading coefficient for a proprietary brand of single, heat-absorbing glass, denoted as 49/66 bronze, with the following properties in comparison with 4 mm, single, clear glass.

1	2	3	4	5	6
Glass type	Absorbed heat	Fraction of the absorbed heat convected and reradiated to the room	Direct transmit-tance	Total transmit-tance	Shading coefficient
4 mm clear	0.08	0.03	0.84	0.87	1.00
49/66 bronze	0.34	0.10	0.56	0.66	0.76

The figures in column 3 are obtained by assuming that approximately 30% of the absorbed heat (column 2) enters the room, 70% being lost to the outside. Adding the value in column 3 to that in column 4 yields the total transmittance in column 5.

Answer

The shading coefficient is defined as the ratio of the total thermal transmittance of a particular glass, or glass and shade combination, to that of single, clear, 4 mm sheet. For 49/66 bronze it is thus 0.66/0.87 = 0.76 and internal Venetian blinds will certainly be needed if people in the room are to feel comfortable when the sun shines on them through the windows. Although adding blinds will not reduce the shading coefficient to as low as 0.27, the short-wave direct solar radiation that causes the discomfort can be excluded.

However, it is dangerous to fit reflective material, such as metallic foil or paper, or even Venetian blinds, on the inner surface of heat-absorbing glass. The risk is that the reflected ray from the foil, paper or blinds, is absorbed by the glass as it passes to outside, increasing its temperature and causing thermal expansion and stress. The glass can then crack, shatter, or even be ejected from its frame and fall into the street outside. The glass manufacturer must be consulted before attempting any such internal treatment.

EXAMPLE 1.5

Calculate the solar cooling load at 15.00 h sun time in July through a west-facing proprietary brand of heat-absorbing single glass with the characteristics listed in Example 1.4, (a) using CIBSE data in Table A.5 assuming a light-weight building, given that the total shading coefficient of 4 mm clear glass fitted with internal white Venetian blinds is 0.53 and (b) using Carrier data in Table A.1, assuming a surface density of 150 kg m^{-2}.

Answers

(a) From Table A.5 the load through clear single glass fitted with internal Venetian blinds is $0.77 \times 0.91 \times 270 = 189$ W m^{-2}. The shading coefficient of clear glass fitted with internal blinds is 0.53 and that of the heat-absorbing glass is 0.76. Hence the cooling load is $189 \times 0.76/0.53 = 271$ W m^{-2}. (b) From Table A.1 the load through clear single glass fitted with internal Venetian blinds is 205 W m^{-2}. From Table A.2 the factor for heat-absorbing glass with a shading coefficient of 0.76 is 1.43, as the footnote to the table explains. Hence the cooling load is $205 \times 1.43 = 293$ W m^{-2}.

Answers obtained by either method are approximately correct for peak values but for lesser loads, at other times, the accuracy of the methods is in doubt.

1.4 Variations in outside air temperature

It is seldom obvious initially at which time of the day a maximum heat gain will occur, so it is useful to have a means of estimating outside air temperature, t_o, at various times. A reasonable assumption is that temperature varies sinusoidally against time, θ, the peak, t_{15}, occurring at 15.00 h sun time. The difference between this and the minimum value equals the diurnal range, D. Then

$$t_o = t_{15} - \frac{D}{2}\left[1 - \sin\frac{(\theta\pi - 9\pi)}{12}\right] \qquad (1.2)$$

Meterological records quote mean monthly maximum dry-bulb temperatures, corresponding to t_{15}, and mean daily maximum and minimum values whose difference yields D. Table 1.1 shows the monthly variation in t_{15} and D obtained from records in Kew.

Table 1.1 Monthly variation in t_{15} and D obtained from records at Kew (in °C)

	March	April	May	June	July	August	September
t_{15}	15.5	18.7	23.3	25.9	26.9	26.2	23.4
D	6.8	7.8	8.4	8.8	8.4	8.2	7.2

The use of Equation (1.2) is relevant to the calculation of air-to-air transmission gains through glass, the determination of heat gain by infiltration and the estimation of sol-air temperatures. Two problems sometimes arise in the choice of a value for t_{15}. First, although it is customary to take the summer design value of the outside air temperature as that prevailing at 15.00 h sun time, this does not always equal the value of t_{15} given in tabulated meteorological data. For example, the design brief in Section 1.2 for the hypothetical office block quotes 28°C, which can be interpreted as t_{15} but Table 1.1 gives a value of 26.9°C for July. In such a case, 28°C is adopted for t_{15} in July but the tabulated value of 8.4°C for D is associated with it. This is because the diurnal range is typical of the month and is not tied to a particular maximum value. Secondly, we may wish to determine heat gains for a month other than that having the peak value of t_{15}, taken usually as July in the northern hemisphere. If so, although the choice is open to the designer, it is suggested that if the design value chosen for t_{15} in July is not the same as that in the meteorological tables, the values of t_{15} for adjacent months be altered by the amount of the difference. Thus, if 28°C were the design value for t_{15} in July, at Kew, 1.1°C would be added to the tabulated values for June and August to give 27°C and 27.3°C for t_{15} in those months.

1.5 Heat gain through walls and roofs

Although sol-air temperatures offer the only practical way of dealing with unusual structures, the heat gain through walls and roofs is sometimes conveniently calculated by using equivalent temperature differences developed by Stewart (1948), that take account of the diurnal variations in air temperature and solar radiation plus the time lag and decrement factor of the wall or roof. Equivalent temperature differences are given for some typical walls and roofs at a latitude of 51.5°N in the UK in Table A.8. The table is based on an air-to-air temperature difference of 6°C at 15.00 h sun time; if the difference is otherwise at 15.00 h, for a particular case, then the difference from 6°C must be applied as a correction to the tabulated values. No allowance should be made for any hourly variation of air temperature each side of 15.00 h since this has already been taken into consideration in the table.

The heat gain through a wall or roof, Q, is then easily calculated by

$$Q = AU(\Delta_e) \tag{1.3}$$

where Δ_e is the equivalent temperature difference, in K.

EXAMPLE 1.6

Determine the equivalent temperature difference for an east wall of 300 kg m^{-2} surface density at 09.00 h sun time in June at latitude 51.5°N, given that the outside air temperature is 27°C at 15.00 h sun time and the room is held at a constant value of 22°C.

Answer
Table A.8 quotes a value of 9.5 K. Since the room temperature is 22°C and that outside at 15.00 h is 27°C, the air-to-air difference is only 5°C and a correction of –1°C must be applied, yielding 8.5 K as the required answer. The fact that the outside air temperature at 09.00 h is less than 27°C plays no part in the use of the table, having already been allowed for. The equivalent temperature difference is then multiplied by the *U*-value of the wall and its area, to give the heat gain to the room at 09.00 h.

Using sol-air temperature, t_{eo} is a more tedious process because of the way it is defined which is, in approximate terms

$$t_{eo} = t_o + a(I_\delta + I_s)/h_{so} \tag{1.4}$$

where longwave radiant exchanges between the wall or roof and its surroundings are ignored, α is the absorption coefficient for solar radiation, I_δ is the intensity of direct radiation normally incident on the surface, tilted at an angle δ to the horizontal, I_s is the intensity of scattered solar radiation (sky plus ground) normally incident on the surface and h_{so} is its outside surface film coefficient of heat transfer. The heat gain, Q, through the wall or roof is then expressed by

$$Q = AU[(t_{em} - t_r) + f(t_{eo} - t_{em})] \tag{1.5}$$

where A is the area of the wall or roof, U is its thermal transmittance, t_{em} is the 24-h mean value of the sol-air temperature, t_r is the room air temperature (assumed to be constant over 24 h), f is the decrement factor for the wall or roof and t_{eo} is the sol-air temperature at the time the heat entered the outside surface.

Sol-air temperatures may be calculated from Equation (1.4) or obtained directly from Table A.9, for south-east England. Note that if the outside air dry-bulb temperature used for the particular calculation is not the same as that given in the second column in Table A.9, the difference must be applied as a correction to the tabulated values, as Equation (1.4) clearly justifies. Tables A.6 and A.7 give very approximate values of time lag and decrement factor in terms of building density and thermal insulation. It is much better to adopt the time lags and decrement factors given by the CIBSE (1986d), which are clearly quoted in terms of specific and practical building construction in everyday use. Values of direct and scattered solar radiation may be obtained from CIBSE (1986a), ASHRAE (1993), Carrier (1965) and Jones (1994c).

EXAMPLE 1.7

Using sol-air temperatures and the relevant data, calculate the heat gains through the east and west walls of modules in the hypothetical office building (Figure 1.1) at both 08.00 and 15.00 h sun time in July.

Answer
The design brief for the hypothetical office building gives a time lag of 9 h and a decrement factor of 0.2 for the walls. The colour of the external brickwork is interpreted as light. The following preliminary tabulation sorts out the related temperatures, prior to calculating the actual heat gain.

Design outside temperature at 15.00 h	28°C			
Tabulated outside temperature at 15.00 h	24.5°C			
Correction (see Equation (1.4))	+3.5 K			
Orientation	East	East	West	West
Time of heat gain to room (h)	08.00	15.00	08.00	15.00
Time lag of wall (h)	9	9	9	9
Time of relevant sol-air temperature (h)	23.00	06.00	23.00	06.00
Tabulated sol-air temperature (°C)	16.5	24.5	16.5	14.5
Correction (K)	+3.5	+3.5	+3.5	+3.5
Actual sol-air temperature (°C)	20.0	28.0	20.0	18.0
Tabulated 24 h mean sol-air temperature (°C)		22.5	22.5	
Correction (K)		+3.5	+3.5	
Actual 24 h mean sol-air temperature (°C)		26.0	26.0	

The area of a modular wall is $A_w = 0.6 \times 2.4 \times 3.3 = 4.752\ \mathrm{m}^2$ and its thermal transmittance is $U_w = 0.45\ \mathrm{W\ m^{-2}\ K^{-1}}$.

The heat gains to the rooms through the wall are then calculated by means of Equation (1.5).

East wall:
08.00 h $Q_w = 4.752 \times 0.45\ [(26 - 22) + 0.2\ (20 - 26)] = 6.0\ \mathrm{W}$
15.00 h $Q_w = 4.752 \times 0.45\ [(26 - 22) + 0.2\ (28 - 26)] = 9.4\ \mathrm{W}$

West wall:
08.00 h $Q_w = 4.752 \times 0.45\ [(26 - 22) + 0.2\ (20 - 26)] = 6.0\ \mathrm{W}$
15.00 h $Q_w = 4.752 \times 0.45\ [(26 - 22) + 0.2\ (18 - 26)] = 5.1\ \mathrm{W}$

It is seen that the heat gain through a wall appears to be insignificant and this is often so. The exception is for the case of heat gains through roofs, particularly for low-rise buildings of large plan area, such as hypermarkets or air conditioned warehouses. Such buildings often have roofs of low thermal transmittance but of small mass. They have virtually no thermal inertia and consequently the decrement factor is best taken as 1.0 and the time lag as zero. Equation (1.5) then simplifies to

$$Q_r = A_r U_r [t_{eo} - t_r] \tag{1.6}$$

Any discrepancy between answers obtained by the use of equivalent temperature differences and those obtained by using sol-air temperatures is of no great concern; in most instances the heat gain through the wall is only about 1% of the total sensible heat gain.

The advantage of sol-air temperatures in dealing with complicated structures can be seen in the following example.

EXAMPLE 1.8

Determine the sensible heat gain to a room, at 15.00 h sun time in July, through a horizontal flat roof consisting of 19 mm black felt-bitumen layers on 25 mm of expanded polystyrene fixed to metal decking and provided with a vapour seal. There is then a substantial air space and a suspended ceiling constructed from 100 mm thick gypsum plasterboard. Recessed light fittings liberate 10 W m^{-2} into the ceiling void and the temperatures of the outside air and the room air are 28 and 22°C, respectively. The U-value for the roof above the void is

1.1 W m^{-2} K^{-1}, the inside surface film resistances on each side of the suspended ceiling, for heat flow downward, are both 0.15 m^2 K W^{-1}, the inside surface film resistances on each side of the suspended ceiling, for heat flow downward, are both 0.15 m^2 K W^{-1} and the thermal conductivity of plasterboard is 0.16 W m^{-1} K^{-1}.

Answer

Assuming a time lag of 0 and a decrement factor of 1.0, a thermal balance is struck between heat flow from outside into the void, and heat flow from the void into the room. Denote the void temperature by t_v, the thermal transmittance of the roof by U_r and that of the ceiling by U_c. The heat balance is

$$U_f\,[t_{eo} - t_v] + 10 = U_c(t_v - t_r)$$

From Table A.9 the sol-air temperature at 15.00 h sun time in July is 45.5°C for a horizontal, black roof. The table is based on an outside air temperature of 24.5°C at 15.00 h and a correction of +3.5 K must be applied because the outside air temperature assumed at 15.00 h is 28°C. Hence the sol-air temperature to be used is 45.5°C + 3.5 K = 49°C.

The thermal transmittance must be calculated for the ceiling

$$U_c = 1/[0.15 + 0.01/0.16 + 0.15] = 2.76 \text{ W m}^{-2}\text{ K}^{-1}$$

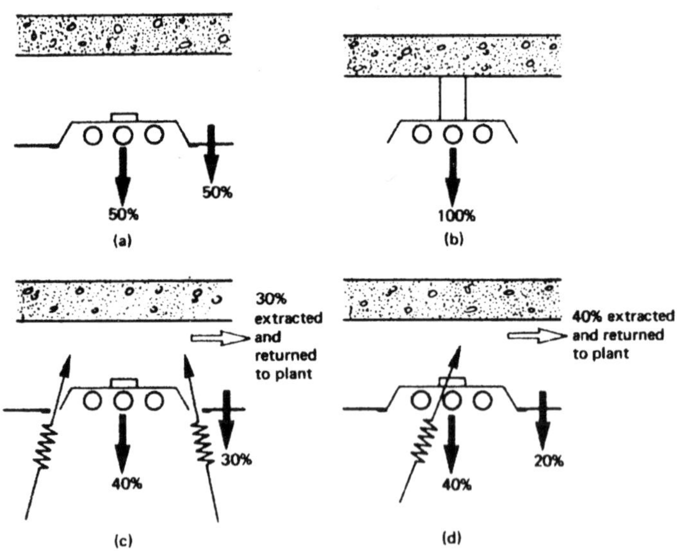

Figure 1.3 (a) Unventilated, flush-mounted luminaire. All the heat enters the treated space, mostly to the room below although some goes to the room above. (b) Unventilated surface-mounted luminaire. As (a). (c) Air extracted into the ceiling void. Heat goes to the room directly as well as indirectly through the ceiling and through the slab to the room above. Total gain to both rooms is 70%. (d) As (c) except that the luminaire is directly extracted-ventilated and the total gain to both rooms is 60%.

The thermal balance is then

$$1.1(49 - t_v) + 10 = 2.76(t_v - 22)$$

whence $t_v = 124.62/3.86 = 32.3$°C. Hence the heat gain to the room, from the ceiling, is 2.76 × (32.3 – 22) = 28.4 W m^{-2}.

1.6 Heat gain from electric lights

In the past, air conditioning systems in commercial premises have relied on the use of extract ventilated light fittings to remove a significantly large part of the heat liberated by the luminaires and their control gear. This extracted heat then largely becomes a load on the cooler coil in the central air-handling plant, instead of on the conditioned rooms. As a result, the air quantity that must be supplied is reduced and the capital and running costs of the installation made cheaper. Figure 1.3 illustrates the possibilities, and it can be seen that with an unducted extract light fitting, about 40% of the heat is transferred to the central plant. It is not worthwhile, technically or economically, to make duct connexions directly to the light fittings. The most effective method is to permit free airflow through the fitting into the ceiling void, from where it ultimately enters a rudimentary system of horizontal extract ducting connected at one or two places to vertical extract ducts, which may be in builder's work, leading to the plant. Fire dampers are located in the duct system, as required by the local authority. Such a system of horizontal ducting is as simple as possible and limited in its extent by the position of the dampered spigots in its walls. These spigots allow the air to flow from the ceiling void into the duct system and no extract luminaire should be further than 18 m from an extract spigot. The reason for this restriction is that otherwise air will prefer to leave the rooms through ill-fitting ceiling panels rather than through the lights and those luminaires remote from the spigots will liberate more heat into the conditioned space than was intended. There may also be some variation in the colour rendering. The IES Code (1973) recommends that between 15 and 30 l s^{-1} be extracted through a ventilated luminaire for best results. More than 30 l s^{-1} is not recommended as there is a fall in light output from conventional fluorescent tubes when the lamps are overcooled. It is generally thought that there is no problem with dust deposits on extract-ventilated light fittings.

It is important to note that the introduction of the smaller diameter, polyphosphor, fluorescent tubes has significantly reduced the consumption of electrical power for lighting and the emission of heat from luminaires, in commercial buildings. Table 1.2 gives some typical total heat emissions. However, before using extract-ventilated luminaires it is essential to consult the manufacturers of the fluorescent tubes. If polyphosphor tubes are overcooled by extract-ventilated luminaires the illuminance diminishes and the use of such fittings with polyphosphor tubes is not recommended.

1.7 Heat gains from people and business machines

PEOPLE

The sensible and latent heat emissions from people have been well researched and are given in Table 1.3. The total emission depends on the activity but the split between sensible and latent heat emission depends on the dry-bulb temperature: as the dry-bulb increases the body finds it more difficult to lose sensible heat by convection and radiation and so the proportion lost by evaporation, and hence the latent heat gain, rises.

Table 1.2 Typical total heat emissions for various illuminances and luminaires

Illuminance in lux	Watts liberated per m^2 of floor area, including power for control gear							
	Filament lamps		Discharge lighting		65 W white fluorescent			Polyphosphor fluorescent tube 58 W (1.5 m)
			MBF	SON				
	Open industrial reflector	General diffusing fitting	Open industrial reflector		Enamel plastic trough	Enclosed diffusing fitting	Louvred ceiling panel	
150	19–28	28–36	4–7	2–4	4–5	6–8	6–8	4–8
200	28–36	36–50	–	–	6–7	8–11	9–11	6–10
300	38–55	50–69	7–14	4–8	9–11	12–16	12–17	10–16
500	66–88	–	13–25	7–14	15–25	24–27	20–27	14–26
750	–	–	18–35	10–20	–	–	–	–
1000	–	–	–	–	32–38	48–54	43–57	30–58

Notes: The larger figure in the range quoted is for small rooms which normally need from one-third to one-half more energy because of losses in reflection. The heat liberated by polyphosphor tubes depends on the type of fitting used. Gaps in the illuminances may be covered by interpolation but extrapolation is more risky.

MBF, Mercury fluorescent high pressure.

SON, Sodium high pressure.

Table 1.3 Heat emissions from people

Activity	Heat liberated (W); room dry-bulb temperature								
	Metabolic rate (W)	20°C S	20°C L	22°C S	22°C L	24°C S	24°C L	26°C S	26°C L
Seated at rest	115	90	25	80	35	75	40	65	50
Office work	140	100	40	90	50	80	60	70	70
Standing	150	105	50	95	55	82	68	72	78
Eating in a restaurant	160	110	50	100	60	85	75	75	85
Light work in a factory	235	130	105	115	120	100	135	80	155
Dancing	265	140	125	125	140	105	160	90	175

Emissions in a restaurant include the heat given off by the food.
Population densities relate to the application and more guidance is given in Chapter 3. However, a typical allowance for offices is 9 m^2 per person, over an entire building.

BUSINESS MACHINES

The allowance for small power has been greatly overestimated in recent years. The result has been that systems have been installed with too much sensible cooling capacity. As a consequence, systems have controlled poorly, particularly variable air volume installations which have been restricted in the amount of turn-down possible, with inevitable complaints of draught at low duties.

The CIBSE (1992) give extensive and realistic information on nameplate powers and half-hour average powers. A reasonable allowance for design purposes in a speculative office block is 20 W m^{-2}, referred to the treated floor area.

1.8 Practical heat gains

A designer may have to calculate three sets of heat gains:

1. The maximum sensible heat gain for the room or module so that the necessary supply airflow rate can be determined, or the correct size of terminal unit selected. For rooms with external glazing, the dominant part of the sensible gain is the cooling load caused by solar heat gain through glass and this guides the designer when deciding on the time of the day and the month of the year for the calculation of maximum modular gain
2. The sensible heat gains coincident with the maximum cooling load for the entire building, so that the refrigeration plant can be chosen. A major component of the refrigeration duty is the fresh air load and this is usually a maximum at about 15.00 h sun time in July, when the enthalpy of the outside air is greatest.
3. The maximum simultaneous sensible heat gain for that part of the building dealt with by each air handling plant, so that the air handling plant can be sized, for the case of variable air volume systems. See Chapter 2.

Transmission gains through the walls and roof are complicated and have been dealt with (see Equations (1.3), (1.5) and (1.6)). Transmission gains through opaque parts of the envelope that are not subject to direct solar radiation, such as floors over open car parks that are exposed to the outside air, are easily determined

$$Q_f = A_f U_f (t_o - t_r) \tag{1.7}$$

where Q_f is the transmission gain through the floor (W), A_f is the area of the floor (m^2), U_f is its thermal transmittance coefficient (W m^{-2} K^{-1}), t_o is the outside air temperature (°C) and t_r is the room air temperature (°C).

In a similar way, the sensible heat transmission through glass, Q_g, is calculated by

$$Q_g = A_g U_g (t_o - t_r) \tag{1.8}$$

with obvious meanings for the symbols.

The component of the sensible gain due to the natural infiltration of air from outside, Q_{si}, is easily established by taking the density of air to be 1.2 kg m^{-3} and its specific heat capacity to be 1000 J kg^{-1} K^{-1}

$$Q_{si} = (nV/3)(t_o - t_r) \tag{1.9}$$

where n is the number of air changes per hour due to natural infiltration and V is the room volume (m^3).

EXAMPLE 1.9

Calculate the maximum sensible heat gains to a west-facing and an east-facing module on an intermediate floor of the hypothetical office building shown in Figure 1.1. Use CIBSE data (Tables A.5 and A.9), assume the building is light-weight and take relevant data from Example 1.7 as necessary.

The design conditions are as follows: outside summer design state (at 15.00 h sun time in July): 28°C dry-bulb, 19.5° wet-bulb (sling), 10.65 g kg^{-1} moisture content and 55.36 kJ kg^{-1} enthalpy.

Summer design state in the room: 22°C dry-bulb, 50% saturation, 8.366 g kg^{-1} moisture content, 43.39 kJ kg^{-1} enthalpy.

Natural infiltration rate in summer: 0.5 h^{-1}.

Population: two people, sedentary activity, 90 W sensible each, 50 W latent each.

Lighting: 500 lux, liberating a total of 17 W m^{-2}, referred to the floor area, with unventilated luminaires.

Business machines: 20 W m^{-2}, referred to the floor area.

Answer

(1) West-facing modules. Table A.5 shows that the maximum solar gain through glass occurs at 16.00 h sun time in June. Outside air temperatures in June are about two or three degrees less than those in July, prompting the designer to choose July. It must also be observed that 16.00 h sun time is 17.00 h clock time, in the summer in the UK, when the majority of the occupants will have left the office building and the air conditioning plant may have been switched off. This suggests that the maximum heat gain calculation should be done for 15.00 h sun time in July. On this basis the following sensible heat gains are calculated.

From Example 1.7, for a time lag of 9 h and a decrement factor of 0.2, the relevant sol-air temperatures (Table A.9 plus a correction of 3.5 K) are t_{eo} = 18°C and t_{em} = 26°C.

The specific solar gain from Table A.5 is 270 W m^{-2}, the air-point control factor, f_c, is 0.91 and the shading factor, f_s, is 0.74. Hence the sensible heat gains are calculated as follows:

			W	%
Glass:	$Q_g = 3.168 \times 3.3 \times (28 - 22)$	=	63	4.5
Wall:	$Q_w = 4.752 \times 0.45\ [(26 - 22) + 0.2(18 - 26)]$	=	5	0.4
Infiltration:	$Q_{si} = (0.5 \times 37.44/3)(28 - 22)$	=	37	2.6
Solar:	$Q_{sg} = 0.74 \times 0.91 \times 270 \times 3.168$	=	576	41.3
People: 2 × 90		=	180	12.9
Lights: 17 × 14.4		=	245	17.6
Business machines: 20 × 14.4		=	288	20.7
Total:		=	1394	100

The specific heat gain, referred to the floor area of 14.4 m^2, is 96.8 W m^{-1}. If the heat gain had been worked out for 15.00 h in June, the heat gain by transmission through glass and the wall, and by infiltration, would be about 60 W less, based on CIBSE (1986e) data for June, but the solar gain would have gone up by 30 W, giving a net reduction of 30 W.

(2) East-facing modules. The peak solar gain through glass in Table A.5 is 328 W m^{-2} at 08.00 h sun time in June but at 08.00 h sun time in July it is 306 W m^{-2}. Arguing that the lower outside air temperature in June is likely to compensate for the smaller specific solar gain through glass in July, the maximum sensible heat gain is calculated for 08.00 h sun time in July, as follows. However, it is first necessary to estimate the outside air temperature at 08.00 h sun time in July. This can be done by means of Equation (1.2), assuming a diurnal range of 11.5 K, according to CIBSE (1986e) or to some other source of meteorological data.

Alternatively, a value of 16°C may be read directly from Table A.9, and corrected by 3.5 K to 19.5°C. This is adopted here. From Example 1.7 the relevant sol-air temperatures are: t_{eo} = 20°C and t_{em} = 26°C.

			W	%
Glass:	$Q_g = 3.168 \times 3.3 \times (19.5 - 22)$	=	−26	−2.0
Wall:	$Q_w = 4.752 \times 0.45\ [(26 - 22) + 0.2(20 - 26)]$	=	6	0.5
Infiltration:	$Q_{si} = (0.5 \times 37.44/3)(19.5 - 22)$	=	−16	−1.2
Solar:	$Q_{sg} = 0.74 \times 0.91 \times 306 \times 3.168$	=	653	49.1
People: 2 × 90		=	180	13.5
Lights: 17 × 14.4		=	245	18.4
Business machines: 20 × 14.4		=	288	21.7
Total:		=	1330	100

The specific heat gain, referred to the floor area of 14.4 m^2, is 92.4 W m^{-2}.

Assuming an air density of 1.2 kg m^{-3} and a latent heat of evaporation of 2454 kJ kg^{-1} for water, the following simple equation can be derived for the latent heat gain to a room, Q_{li}, in watts, resulting from the natural infiltration of n air changes per hour

$$Q_{li} = 0.8nV(g_o - g_r) \tag{1.10}$$

where g_o and g_r are the moisture contents of the outside and room air, respectively (expressed in g kg^{-1}).

The latent heat given off by people, in watts, is simply the product of the number of people, n_p, and the appropriate latent emission per person, q_{lp}, from Table 1.3

$$Q_{lp} = n_p q_{lp} \tag{1.11}$$

The total latent heat gain, Q_l (in kW) is then

$$Q_l = [0.8nV(g_o - g_r) + n_p q_{lp})]/10^3 \tag{1.12}$$

The latent heat gains are the same for the west as for the east, in this case, because both are for the month of July, when the outside moisture content is virtually constant throughout the day at 10.65 g kg^{-1}, for 28°C dry-bulb and 19.5°C wet-bulb, Similarly, the room moisture content is fixed at 8.366 g kg^{-1}, for 22°C and 50% saturation.

EXAMPLE 1.10

Calculate the latent heat gains to a module in the hypothetical building, coincident with maximum sensible gains.

Answer

Infiltration: 0.5 × 37.44 × 0.8 × (10.65 – 8.366)	=	34
People: 2 × 50	=	100
Total:		134 W

1.9 Heat gains and refrigeration load

Refrigeration load is the sum of sensible and latent heat gains, fresh air load, fan power, duct heat gain and, in the case of water chillers, a small allowance for pump power and heat gains to pipes. In the case of poorly designed systems there may also be an additional item for wasteful heat that cancels part of the refrigeration capacity. Because the fresh air load is a significantly large proportion of the total, the refrigeration load is often calculated for the time when this will be greatest, for example 15.00 h in July or August (in the UK). However, this does not always follow, as Table A.10 (Appendix) shows. At such a time, the simultaneous sum of the sensible heat gains in the various treated areas in a building will be less than the sum of their maximum individual gains, because diversity factors may be applied to the gains from lights, people and machines and also because the solar load through the glass varies with time. However, it is not always the case that the maximum refrigeration load will occur at the times mentioned and the designer must use his or her common sense. For example, the peak load for a dining room may occur at 13.00 h sun time (14.00 h clock time) when the occupancy is most dense. With office blocks it is reasonable to suppose that some of the people will be on holiday or absent because of illness and that a proportion of the lights will be switched off or need replacing. Business machines may have the same diversity factors as people, or a little less. On the other hand, no such diversity factors can be applied to places like concert halls, for example, where the occupancy and the lighting is obviously predictable. Table 1.4 gives diversity factors for application to sensible and latent heat gains; they are only approximate and must be used at the designer's discretion.

In the case of business machines, the CIBSE (1992) offers guidance on the use of processors and the like, in offices. A possible conclusion is that a diversity factor is 0.65–0.70, referred to business machines, would be suitable for assessing the refrigeration load for the whole building.

Table 1.4 Diversity factors for application to sensible and latent heat gains

Application	Diversity factor	
	Lights	People
Peripheral areas of offices to a maximum depth of 6 m, 20–50% glazed	0.7–0.85	0.7–0.8
Core areas of offices and peripheral areas with less than 20% glazing	0.9–1.0	0.7–0.8
Apartments and hotel bedrooms	0.3–0.5	0.4–0.6
Public rooms in hotels	0.9–1.0	0.4–0.6
Department stores and supermarkets	0.9–1.0	0.8–1.0

Where machines or appliances are used in an industrial application the only sure way to establish the maximum combination at any one time, and also the combination at 15.00 h sun time in July, is by personal observation over a suitable period of time and by questioning the operating staff. Assessing the maximum combination is also usually necessary to calculate the required supply air quantity, since this may be the largest item in the heat gains.

Where appliances are used in a random fashion the only approach may sometimes be to determine the mathematical probability that the appliances can be simultaneously in use, as one does with the case of flow from a number of HWS draw-off points. When adopting this technique the disparity in the size of the machines must be considered: one large heat-producing item will dominate the heat gain calculations.

Table A.10 in the Appendix provides an indication of the times of the day and months of the year that the maximum refrigeration load is likely to occur for an office block like the hypothetical one considered earlier. Climatic loads in W m^{-2} of treated floor area, including the corridor, are suggested and these include transmission and solar gains through the building envelope plus a fresh air allowance of 1.5 l s^{-1} m^{-2}. There is a spread of times from 11.00 h sun time in August to 18.00 h sun time in July. No maxima occur earlier or later than this and none other than in the months of July and August. If an allowance is made for people, lights and business machines, with proper attention to diversity, plus latent gain, fan power and duct gain, the building cooling load can be assessed.

The total refrigeration load for the whole building is summarised and expressed by the following equation

$$Q_{ref} = [Q_s + Q_l + Q_{fa} + Q_{sf} + Q_{tfp} + Q_{sd} + Q_{ra} + Q_{rh}]f_p \tag{1.13}$$

where Q_{ref} is the total refrigeration load at a particular time and month, Q_s is the sensible heat gain (with the components for people, lights and business machines subject to diversity factors), Q_l is the latent heat gain (with the component for people subject to a diversity factor), Q_{fa} is the fresh air load, Q_{sf} is the supply fan power, Q_{tfp} is the fan power of any terminal units, Q_{sd} is the heat gain to the supply ducting, Q_{ra} is the return air load (comprising the extract fan power and the heat gain from any extract ventilated luminaires), and Q_{rh} is the reheat load (which should be zero under design conditions for a properly designed system). With water chillers, the factor f_p is to cover the contribution of the chilled water pump power and the heat gain to the chilled water piping. These allowances are usually very small for commercial installations and a typical value for f_p would be 1.01 or 1.02. The refrigeration load is not known to this sort of accuracy and the value of f_p is often taken as unity.

The fresh air load is generally large enough (about 15% of the total) to indicate to the designer the time of the day and the month of the year when the total cooling load will be greatest. In the absence of any other information it is usual to calculate the maximum refrigeration load for 15.00 h sun time in July, in the UK, because this is the time when the enthalpy of the outside air is highest and when the fresh air load will also be greatest.

EXAMPLE 1.11

Determine the refrigeration load for the hypothetical office block, at 15.00 h sun time in July, assuming that it is air conditioned by a fan coil system comprising one unit in each module to effect sensible cooling only, supported by a low velocity, ducted, supply and extract system to deal with the latent gains (see Chapter 2). Make use of following design data and the relevant results from earlier examples.

Outside design state:	28°C dry-bulb, 19.5°C wet-bulb (sling), 10.65 g kg^{-1}, 55.36 kJ kg^{-1}
Room design state:	22°C dry-bulb, 50% saturation, 8.366 g kg^{-1}, 43.39 kJ kg^{-1}
Fresh air allowance:	1.4 l s^{-1} m^{-2}, referred to the total floor area of 13 997 m^2
Population density:	9 m^2 per person, referred to the floor area, diversity factor 0.75
Illumination:	500 lux, dissipating 17 W m^{-2}, referred to the floor area, diversity factor 0.8. There are no extract-ventilated luminaires
Business machines:	20 W m^{-2}, referred to the floor area, diversity factor 0.7

Temperature rise due to supply fan power:	0.65 K
Temperature rise due to supply duct heat gain:	1.35 K
Temperature rise due to extract fan power:	0.2 K
Total power liberation by the fans in the terminal units:	70 kW.

Answer

(1) Sensible heat gains, Q_s. For the given orientation of the building (major axis aligned north–south) the sol-air temperatures for the east and west walls have already been determined for 15.00 h sun time in July in Example 1.7. It is now necessary to establish them also for the north and south walls and the roof, at the same time and month.

Orientation	North	South	Roof
Time of heat gain (h)	15.00	15.00	15.00
Time lag (h)	9	9	5
Time of relevant sol-air temperature (h)	06.00	06.00	10.00
Tabulated sol-air temperature (Table A.9)(°C)	16.5	14.5	44.5
Correction (K)	+3.5	+3.5	+3.5
Actual sol-air temperature (°C)	20.0	18.0	48.0
Tabulated 24 h mean sol-air temperature (°C)	20.0	22.5	27.5
Correction (K)	+3.5	+3.5	+3.5
Actual 24 h mean sol-air temperature (°C)	23.5	26.0	31.0

When using Table A.5 to determine the solar gains through glass note that at 15.00 h sun time the windows facing east will have their Venetian blinds open but those on the west windows will be closed. This will affect the value of the shading factor, f_s, used in Equation (1.1).

Noting also that there are 36 modules per floor on each of the east and west faces and that there are 12 storeys, the sensible heat gains for the whole building at 15.00 h sun time in July are calculated.

East glass:	$36 \times 12 \times 3.168 \times 3.3 \times (28 - 22)$	27 098
West glass:	$36 \times 12 \times 3.168 \times 3.3 \times (28 - 22)$	27 098
North wall:	$(13.5 \times 3.3 \times 12) \times 0.45 \times [(23.5 - 22) + 0.2 \times (20.0 - 23.5)]$	192
South wall:	$(13.5 \times 3.3 \times 12) \times 0.45 \times [(26.0 - 22) + 0.2 \times (18.0 - 26.0)]$	577
East wall:	$36 \times 12 \times 4.752 \times 0.45 \times [(26.0 - 22) + 0.2 \times (28.0 - 26.0)]$	4 065
West wall:	$36 \times 12 \times 4.752 \times 0.45 \times [(26.0 - 22) + 0.2 \times (18.0 - 26.0)]$	2 217
Roof:	$(86.4 \times 13.5) \times 0.45 \times [(31.0 - 22) + 0.77 \times (48.0 - 31.0)]$	11 595
Infiltration:	$0.5 \times (86.4 \times 13.5 \times 2.6 \times 12) \times (28 - 22)/3$	36 392
East glass (solar)	$0.95 \times 0.91 \times 154 \times (36 \times 12 \times 3.168)$	182 203
West glass (solar)	$0.74 \times 0.91 \times 270 \times (36 \times 12 \times 3.168)$	248 832
Total sensible gain through the envelope:		540 269
People:	$[(86.4 \times 13.5 \times 12)/9] \times 90 \times 0.75$	104 976
Lights:	$(86.4 \times 13.5 \times 12) \times 17 \times 0.8$	190 356
Business machines:	$(86.4 \times 13.5 \times 12) \times 20 \times 0.7$	195 955
Total sensible gain at 15.00 h sun time in July, Q_s		1 031 556 W

(2) Latent heat gains, Q_l. These arise from the natural infiltration of outside air and the latent heat emitted from the people in the building. They are expressed by Equations (1.10)–(1.12).

Infiltration. The volume of the building is $86.4 \times 13.5 \times 2.6 \times 12 = 36\ 392\ \text{m}^3$. Hence the latent heat gain by natural infiltration is

$$Q_{li} = 0.8 \times 0.5 \times 36\ 392 \times (10.65 - 8.366) = 33\ 248\ \text{W}$$

The number of people in the building is $86.4 \times 13.5 \times 12/9 = 1555$. Each emits 50 W of latent heat and a diversity factor of 0.75 is applied. Hence the latent heat gain from people is

$$Q_{lp} = 0.75 \times 1555 \times 50 = 58\ 312\ \text{W}$$

The total latent heat gain is

$$Q_l - 33\ 248 + 58\ 312 = 91\ 560\ \text{W}$$

Since the fan coil units only do sensible cooling the auxiliary ducted supply air must deal with all the latent heat gains. Assuming that a practical supply air state is 13°C dry bulb and 7.702 g kg^{-1} (see Figure 1.4) and using Equation (2.4) (derived in the Appendix) the necessary supply air flow rate, $\dot{v}_{13}$ at 13°C, can be calculated

$$\dot{v}_{13} = [91.560/(8.366 - 7.702)] \times [(273 + 13)/856] = 46.071\ \text{m}^3\ \text{s}^{-1}$$

This supply rate must include the minimum fresh air quantity, under summer design conditions. The treated floor area is $86.4 \times 13.5 \times 12 = 13\ 997\ \text{m}^2$ and hence the minimum fresh air delivery is 1.4 × 13 997/1000 = 19.596 $m^3 s^{-1}$. This is 42.5% of the supply air quantity. The position of *M*, the mixture state in Figure 1.4, formed by outside air at 28°C and 10.654 g kg^{-1} mixing with recirculated air, is now established. The recirculated air is at the room temperature plus 0.2 K, for the rise through the extract fan, and at the room moisture content of 8.366 g kg^{-1}

$$t_g = 0.425 \times 28 + 0.575 \times 22.2 = 24.66°C$$

$$g_m = 0.425 \times 10.65 + 0.575 \times 8.366 = 9.337 \text{ g kg}^{-1}$$

The enthalpy of the air at the mixture state, h_m, is conveniently determined now for later use when calculating the cooling load. This may be done by interpolation in the CIBSE (1986f) psychrometric tables, or from a psychrometric chart, or by use of the following equation, according to Jones (1994)

$$h = (1.007t - 0.026) + g(2501 + 1.84t) \tag{1.14}$$

where t is the dry-bulb temperature (°C) and g is the moisture content (in kg kg^{-1} of dry air). Hence, using the equation

$$h_m = (1.007 \times 24.66 - 0.026) + 0.009337(2501 + 1.84 \times 24.66) = 48.58 \text{ kJ kg}^{-1}$$

(3) Fresh air load, Q_{fa}. If no fresh air were introduced and all the extracted air were recirculated, the cooling load would be from the enthalpy of the air returned to the air handling plant (state *R′*) down to the enthalpy of the air leaving the cooler coil (state *W*). The introduction of air from outside means that an additional cooling load is imposed, the enthalpy of the fresh air provided (state *O*) having to be reduced to the enthalpy of the air returned to the air handling unit (state *R′*). This is clarified in Figure 1.4 where the continuous line from the mixture state *M*, to the off-coil state *W*, represents a cooling load including the fresh air load. On the other hand, the broken line from the recirculated air state, *R′*, to the off-coil state *W*, shows what the cooling load would be if no fresh air were introduced and all the air were recirculated.

The fresh air load, Q_{fa}, is given by

$$Q_{fa} = (\dot{v}_{fa}/v_s)(h_0 - h_{r'}) \tag{1.15}$$

where $\dot{v}_{fa}$ is the volumetric flow rate of fresh air handled (in $m^3 s^{-1}$), v_s is the specific volume at the supply state (in $m^3 kg^{-1}$), h_o is the enthalpy of the fresh air and $h_{r'}$ is the enthalpy of the recirculated air (both in kJ kg^{-1}). It is convenient to use the specific volume of the air at the supply state, *S*, rather than that of the outside air, *O*, because the total quantity of air handled is usually expressed at the supply state, *S*.

The temperature rise through the extract fan is 0.2 K. Since the extract state has the same moisture content as the room state, namely 8.366 g kg^{-1}, the enthalpy of the recirculated air state is determined by interpolation in CIBSE (1986f) psychometric tables as 43.59 kJ kg^{-1}. Alternatively, a CIBSE psychrometric chart would probably give a value of about 43.6 kJ kg^{-1}. At a practical supply air state of 13°C dry-bulb and 7.702 g kg^{-1}, the specific volume of the air is 0.820 $m^3 kg^{-1}$. Hence the fresh air load is calculated by Equation (1.15)

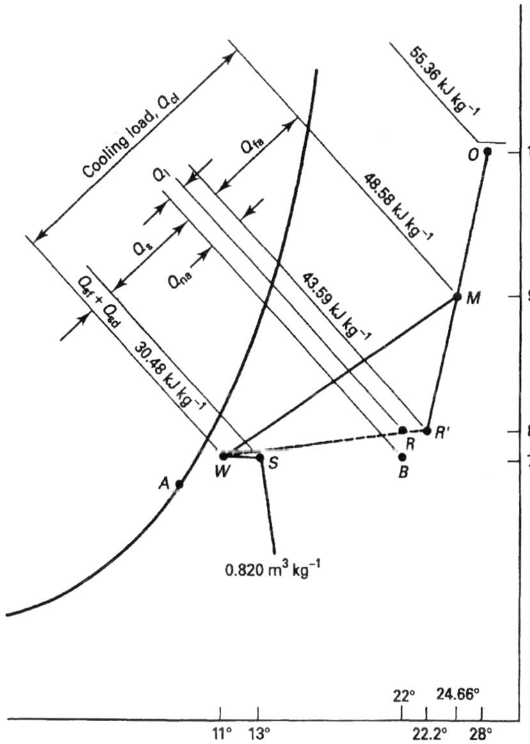

Figure 1.4 The psychrometry of a cooling load: Q_{sf} is the supply fan gain, Q_{sd} is the supply duct gain, Q_s is the sensible heat gain, Q_l is the latent heat gain, Q_{ra} is the return air load and Q_{fa} is the fresh air load. *B* is a notional state that allows the sensible and latent cooling to be shown separately for the conditioned space. *A* is the apparatus dew-point and is in a straight line with *M* and *W*. The contribution from the terminal units is not shown.

$$Q_{fa} = (19.596/0.82)(55.36 - 43.59) = 281.3 \text{ kW}$$

(4) Supply fan power, Q_{sf}. In Jones (1994) it is shown that the temperature rise through a fan is about 1 K/kPa of fan total pressure if the fan motor is outside the airstream and 1.2 K/kPa if the fan motor is in the airstream. Hence Equation (2.3) (derived in the Appendix) can be used to determine the corresponding heat gain to the airstream

$$Q_{sf} = \dot{v}_s(\Delta t_{sf})\, 358/(273 + t_s) \tag{1.16}$$

where Q_{sf} is the heat input to the airstream (in kW or W), $\dot{v}_s$ is the volumetric flow rate of air supplied (in m^3 s^{-1} or l s^{-1}) at a temperature t_s and Δt_{sf} is the temperature rise through the supply fan. For the example considered

$$Q_{sf} = 46.071 \times 0.65 \times 358/(273 + 13) = 37.5 \text{ kW}$$

(5) Terminal unit fan power, Q_{tfp}. The ducted air supply of 46.071 $m^3 s^{-1}$ at 13°C and 7.702 g kg^{-1} deals with all the latent heat gain and is also able to offset some of the sensible heat gain. Using Equation (2.3) this is calculated as: 46.071 × ((22 – 13) × 358)/(273 + 13) = 519.0 kW.

The remainder of the sensible heat gain (1031.6 – 519.0 = 512.6 kW), is dealt with by the fan coil units, located one to each building module, which provide the individual thermostatic control necessary. The cooling output of the fan coil units is distributed to the rooms by means of fans within the units. These fans cause the temperature of the air they handle to rise and this is an additional contribution to the cooling load, which must be dealt with. The whole of the electrical power absorbed from the mains is liberated and its value can only be determined from catalogue data. In the design data for the example, the total value of all the power liberated by the fans in the terminal units is given as 70 kW.

(6) Supply duct heat gain, Q_{sd}. The temperature rise that occurs by heat gain to the supply air duct is not known at an early stage of the design and a reasonable allowance has to be made. The designer should always check this later, after the duct system has been designed, and rectify any discrepancy. In the design data for this example the temperature rise in the supply air duct is given as 1.35 K. The contribution this makes to the total cooling load is then calculated by Equation (2.3)

$$Q_{sd} = 46.071 \times 1.35 \times 358/(273 + 13) = 77.85 \text{ kW}$$

(7) Return air load, Q_{ra}. The recirculation air flow rate is 46.071 – 19.596 = 26.475 $m^3 s^{-1}$, expressed at the supply air temperature of 13°C. Hence, since the temperature rise through the extract fan is 0.2 K and there are no extract-ventilated luminaires, the return air load is given by

$$Q_{ra} = 26.475 \times 0.2 \times 358/(273 + 13) = 6.63 \text{ kW.}$$

(8) Re-heat load, Q_{rh}. This is zero.

(9) Total cooling load, Q_{ref}. Referring to the psychrometry in Figure 1.4, an air cooling load, Q_{cl}, may be determined by

$$Q_{cl} = (\dot{v}_s/v_s)(h_m - h_w) \tag{1.17}$$

The supply state, *S*, is 13°C dry-bulb and 7.702 g kg^{-1} and there is a 2 K rise from the off-coil state *W* to the supply state. Hence the state *W* is 11°C dry-bulb and 7.702 g kg^{-1}. From psychrometric tables the enthalpy is read as 30.48 kJ kg^{-1} or as 30.5 kJ kg^{-1} from a psychrometric chart. The enthalpy of mixture state, *M*, was determined earlier, in item (2), as 48.58 kJ kg^{-1}. A cooling load is then calculated from Equation (1.17)

$$Q_{cl} = (46.071/0.82)(48.58 - 30.48) = 1016.9 \text{ kW}$$

As a check, this is to be compared with the refrigeration load determined by summing the components of the load, according to Equation (1.13)

$$\begin{aligned} Q_{ref} &= [Q_s + Q_l + Q_{fa} + Q_{sf} + Q_{tfp} + Q_{sd} + Q_{ra} + Q_{rh}]f_p \\ &= [1031.556 + 91.560 + 281.300 + 37.500 + 70.000 + 77.850 + 6.630 + 0]1.01 \\ &= 1596.40 \times 1.01 = 1612.36 \text{ kW} \end{aligned}$$

It is clear that this is not the same as the cooling load determined by the psychrometry, using Equation (1.17). This is because the psychrometry only dealt with the sensible heat gains offset by the auxiliary air supply and ignored the power liberated by the fans in the terminal units. In item (5), above, it was determined that the part of the sensible heat gains dealt with the fan coil units was 512.6 kW. The power emitted by the unit fans was quoted in the design data as 70 kW. Hence

Psychrometric cooling load:	1016.9
Sensible cooling by the fan coil units:	512.6
Power liberated by the terminal unit fans:	70.0
Subtotal:	1599.5
1% allowance for chilled water pumps and pipes:	16.0
Total refrigeration load:	1615.5 kW

The agreement is good, the two answers being within much less than 1% of each other. This does not mean that the cooling load is known to this accuracy. It is not. The purpose of the check is to verify that no significant error has been made. If a discrepancy of more than about 1%, or perhaps 2%, emerges, the reason for this should be established and the error corrected.

The specific refrigeration load, referred to the floor area, is 1 612 360 W/13 997 m^2 = 115 W m^{-2}

In the case of all-air systems, such as variable air volume, the cooling load determined from the psychrometry is the refrigeration load, to which must be added the small addition for chilled water pumps and pipes. In the case of all-air, constant volume, re-heat systems, no allowance can be made for diversity in any of the loads, including the solar gain through glass. This is because such systems cancel the unwanted cooling, at partial load conditions, by re-heating. The refrigeration machine stays at virtually full duty all the time it is running.

Exercises

1 Determine the mean surface density per unit of floor area for a double module (4.8 m wide × 6.0 m deep) in the hypothetical office block, making the same assumptions as in Example 1.1.
(*Answer* 248 kg m^{-2})

2 Assuming the major axis of the hypothetical office block shown in Figure 1.1 is aligned east west, calculate the maximum sensible heat gain through a south-facing module, by CIBSE methods. Assume the maximum gain occurs at 13.00 h sun time in August (see Table A.10). Use the relevant design data from Example 1.11 and take the outside design state to be 28°C dry-bulb, 19.5°C wet-bulb (sling) at 15.00 h sun time in August. Use Equation (1.2), with a diurnal range of 9.5 K, to determine the outside dry-bulb at 13.00 h sun time. Refer to tables in the Appendix as necessary.
(*Answer* 1371 W)

3 Repeat Example 1.11 to show that with 2.6 l s^{-1} m^{-2} of floor area as the fresh air supply rate, the maximum total refrigeration load occurs at 15.00 h sun time in July and is 132 W m^{-2}.

Notation

Symbols	Description	Unit
A	Area of a building element	m^2
A_f	Area of a floor	m^2
A_g	Area of glass (if a wooden frame) or area of the hole in the wall (if a metal frame)	m^2
A_f	Area of a roof	m^2
A_w	Area of a wall	m^2
D	Diurnal range	K
I_s	Intensity of scattered radiation normally incident on a surface	$W\ m^{-2}$
I_δ	Intensity of direct solar radiation normally incident on a surface	$W\ m^{-2}$
Q	Heat gain through an opaque building element	W
Q_{cl}	Air cooling load	kW
Q_f	Heat gain through a floor, exposed on its underside to the outside air	W
Q_{fa}	Total fresh air load for a building	kW
Q_g	Sensible heat transmission through glass	W
Q_l	Total latent heat gain for a building	kW
Q_{li}	Latent heat gain due to the natural infiltration of air from outside	W
Q_{lp}	Latent heat gain due to people	W
Q_r	Heat gain through a roof	W
Q_{ra}	Total return air load for a building	kW
Q_{ref}	Total refrigeration load for a building	kW
Q_{rh}	Total re-heat load for a building	kW
Q_s	Total sensible heat gain for a building	kW
Q_{sd}	Total sensible heat gain to the supply air ducting for a building	kW
Q_{sf}	Total supply fan power load for a building	kW
Q_{sg}	Cooling load due to solar heat gain through glass	W
Q_{si}	Sensible heat gain due to the natural infiltration of air from outside	W
Q_{tfp}	Total fan power of terminal units in a building	kW
Q_w	Sensible heat gain through a wall	W
U	Thermal transmittance coefficient of a building element	$W\ m^{-2}\ K^{-1}$
U_f	Thermal transmittance coefficient of a floor exposed on its underside to the outside air	$W\ m^{-2}\ K^{-1}$
U_g	Thermal transmittance coefficient of glass	$W\ m^{-2}\ K^{-1}$
U_r	Thermal transmittance coefficient of a roof	$W\ m^{-2}\ K^{-1}$
U_w	Thermal transmittance coefficient of a wall	$W\ m^{-2}\ K^{-1}$
V	Volume of a room, module or building	m^3
f	Decrement factor	–
f_c	Air point control factor for cooling load due to solar gain through glass	–
f_p	Factor for chilled water pump power and heat gain to chilled water piping	–

f_s	Shading factor for cooling load due to solar gain through glass	–
g	Moisture content of an air water vapour mixture	kg kg^{-1} dry air or g kg^{-1} dry air
g_0	Moisture content of the outside air	g kg^{-1} dry air or kg kg^{-1} dry air
g_r	Moisture content of the room air	g kg^{-1} dry air or kg kg^{-1} dry air
g_s	Moisture content of the supply air	g kg^{-1} dry air or kg kg^{-1} dry air
h	Enthalpy of an air–water vapour mixture	kJ kg^{-1} dry air
h_m	Enthalpy of mixed air at state *M*	kJ kg^{-1} dry air
h_o	Enthalpy of outside air at state *O*	kJ kg^{-1} dry air
h_w	Enthalpy of air off a cooler coil at state *W*	kJ kg^{-1} dry air
h_{so}	Outside surface film coefficient	W m^{-2} K^{-1}
$h_{r'}$	Enthalpy of air at the return air state, *R′*	kJ kg^{-1} dry air
n	Air change rate due to the natural infiltration of air from outside	h^{-1}
n_p	Number of people	–
q_l	Total latent heat gain	W
q_{lp}	Latent heat emission per person	W
q_{sg}	Specific cooling load due to solar gain through glass	W m^{-2}
t	Air temperature	°C
t_{15}	Outside air temperature at 15.00 h sun time	°C
t_{eo}	Sol-air temperature at the time heat enters the outside surface of a wall or roof	°C
t_θ	Outside air temperature at time θ	°C
t_{em}	24-h mean sol-air temperature	°C
t_o	Outside air temperature	°C
t_r	Room temperature	°C
t_s	Temperature of the air supplied to a building	°C
$\dot{v}_{fa}$	Volumetric flow rate of fresh air expressed at the supply air state	m^3 s^{-1}
v_s	Specific volume of air at the supply state, *S*	m^3 kg^{-1} dry air
$\dot{v}_s$	Volumetric flow rate of air supplied to the building at state *S*	m^3 s^{-1}
Λ_c	Equivalent temperature difference	K
α	Absorption coefficient for solar radiation	–
δ	Angle between the horizontal and a window	degrees
θ	Sun time (0–24)	h

References

ASHRAE, *Handbook of Fundamentals*, SI edition, Chapter 26, Non-residential cooling and heating loads, 1993.

Carrier Air Conditioning Company, *Air Conditioning System Design Manual*, McGraw-Hill, 1965.

CIBSE Guide, A9, Estimation of plant capacity, Cooling Load Tables, 1986a.

CIBSE Guide, A9, Estimation of plant capacity, pp. A9–12, 1986b.

CIBSE Guide, A3, Thermal properties of building structures, Equation (A3.36), 1986c.

CIBSE Guide, A3, Thermal properties of building structures, Tables A3.16–A3.21, 1986d.

CIBSE Guide, A2, Weather and solar data, Table A2.33(d), 1986e.

CIBSE Guide, Guide C1, Properties of humid air, 1986f.

CIBSE, Information technology in buildings, *CIBSE Applications Manual AM7*, pp. 24–25, 1992.

IES Code, 1973.

Jones, W. P., *Air Conditioning Engineering*, 4th edn, Equation (2.22), pp. 24, 130, 153, 165, Edward Arnold, London, 1994.

Stewart, J. P., 1948: Solar heat gains through walls and roofs for cooling load calculations, *Trans ASHVE*, **54,** 361–388, 1948.

2

System Characteristics

2.1 System classification

Most systems in use are for commercial applications, principally office blocks, and this has driven the design ethos. Where systems have been required for commercial applications other than office blocks, a system originally developed for treating offices has often been adapted to suit the different use, for example, providing fan coil units for hotel bedrooms. However, some commercial systems have been specifically developed for a particular application, such as the use of roof-top units for treating the malls in shopping centres. For industrial applications the approach has been quite different, the needs of the industrial process generally dictating the form of the system design.

There is no generally agreed classification of systems but the following arrangement is not uncommon.

1. Unitary systems
   (a) self-contained, air-cooled, room air conditioners
   (b) split systems
   (c) cassette units and variable refrigerant volume systems
   (d) water-cooled air conditioning units
   (e) water loop, air conditioning/heat pump units.
2. All-air systems
   (a) constant volume re-heat and sequence heat systems
   (b) roof top units
   (c) variable air volume systems
   (d) dual duct systems
   (e) multizone units
   (f) air curtains.
3. Air–water systems
   (a) fan coil systems
   (b) chilled ceilings
   (c) chilled beams
   (d) perimeter inductions systems.

Because of the very substantial amount of air conditioning done outside the United Kingdom, particularly for places situated at considerable heights above sea level, the first step to be considered is the influence of altitude upon system performance, from a general viewpoint.

2.2 Altitude effects

For any specific place an increase in altitude is accompanied by a drop in both pressure and temperature. However, it is not possible to establish a simple equation that will accurately predict the influences of both height and temperature on barometric pressure and, since the air temperature in the atmosphere near the surface of the earth is much affected by seasonal changes and the local topography, it is unwise to attempt to forecast barometric pressure for an unfamiliar place. Reference should always be made to local meteorological data, where this is available. When such information is not to hand, useful reference can be made to the CIBSE guide (1986a) which tabulates approximate barometric pressures against altitudes. There can be discrepancies but in some instances the agreement is good. For example, the altitude of Tehran is 1220 m above sea level and the mean barometric pressure is about 875 mbar. Interpolation in the table yields 873 mbar, which is well within the probable variation of atmospheric pressure ($\pm 5\%$) arising from changes in the weather.

A fall in barometric pressure has a principal influence on the following: psychrometric properties; air mass flow rate; heat transfer coefficients for air; evaporation rates; air pressure loss in ducts and plant; pressure gauge indications; available net positive suction head (NPSH); electric motor cooling.

PSYCHROMETRIC PROPERTIES

For atmospheric pressures of 95 kPa or less, psychrometric charts and tabulated data for the accepted standard pressure of 101.325 kPa should not be used if progressively serious error is to be avoided. Psychrometric charts by Martin (1972) are published at intervals of 2.5 kPa down to 72.5 kPa and these are very convenient to use. Alternative approaches are to use the corrections for changes in barometric pressures published in the CIBSE guide (1986b), or to employ the ideal gas laws described by Jones (1994) to calculate the desired properties, or even, if the situation demands, to construct a special psychrometric chart.

EXAMPLE 2.1

Air enters a cooler coil at 39°C dry-bulb, 24°C wet-bulb (sling) and 87.5 kPa. If 2 $m^3 s^{-1}$ of air at 16°C dry-bulb, 15°C wet-bulb (sling) leaves the coil, calculate the cooling load and compare the answer with that obtained when the barometric pressure is 101.325 kPa.

Answer
The CIBSE guide (1986b) quotes tabulated additive corrections to be applied to enthalpy values read for standard barometric pressure. The corrections are expressed in terms of adiabatic saturation temperatures but can be interpreted as the same as sling wet-bulb values without significant loss of accuracy. Thus at an atmospheric pressure (p_{at}) of 87.5 kPa and a wet-bulb of 24°C sling, the correction is 7.90 kJ kg^{-1} and at 15°C wet-bulb it is 4.34 kJ kg^{-1}. Taking the enthalpies at 39°C dry-bulb, 24°C wet-bulb and 16°C dry-bulb, 15°C wet-bulb from the CIBSE guide (1986c) as 71.03 kJ kg^{-1} and 42.08 kJ kg^{-1}, respectively, we can deduce that the enthalpies at 87.5 kPa are

$$h_{\text{on coil}} = 71.03 + 7.90 = 78.93 \text{ kJ kg}^{-1}$$

$$h_{\text{off coil}} = 42.08 + 4.34 = 46.42 \text{ kJ kg}^{-1}$$

To establish the specific volume $v_{\text{off coil}}$ of the air leaving the coil we can use an approximate formula quoted in the CIBSE Guide (1986c) as

$$v_{\text{off coil}} = (T/p_{\text{at}})(0.287 + 0.461 \text{ g}) \text{ m}^3 \text{ kg}^{-1} \tag{2.1}$$

where T is the absolute dry-bulb temperature and g is the moisture content, also given in the CIBSE Guide (1986c) as

$$g = [(\mu/100)(0.624\, p_{\text{ss}})]/(p_{\text{at}} - 1.004\, p_{\text{ss}}) \text{ kg kg}^{-1} \tag{2.2}$$

in which μ is the percentage saturation p_{at} is the barometric pressure and p_{ss} the saturation vapour pressure at the dry-bulb temperature, all pressures being in kPa.

The value of μ is not necessarily known, since conditions on and off the coil are usually quoted as dry- and wet-bulb temperatures. However, for a given pair of such values the percentage saturation is approximately independent of atmospheric pressure, so μ can be read from psychrometric tables at 101.325 kPa, without much error.

At 16°C dry-bulb, 15°C wet-bulb (sling) and 101.325 kPa, μ is 90% and we can use this value also for 87.5 kPa. The value of p_{ss} depends solely on temperature and has nothing to do with barometric pressure. Therefore, this can be read directly from psychrometric tables for 101.325 kPa at 100% of saturation and any desired temperature. Thus at the off-coil state, $p_{\text{ss}} = 1.817$ kPa (read from CIBSE tables at 16°C saturated) and by Equation (2.2) the moisture content of the air leaving the cooler coil at 87.5 kPa is

$$g_{\text{off coil}} = [(90/100)(0.624 \times 1.817)]/(87.5 - 1.004 \times 1.817) = 0.01191 \text{ kg kg}^{-1}$$

Hence, by Equation (2.1) the specific volume off the coil at 87.5 kPa is

$$v_{\text{off coil}} = [(273 + 16)/87.5][0.287 + 0.461 \times 0.01191] = 0.9661 \text{ m}^3 \text{ kg}^{-1}$$

The cooling load can now be calculated as

$$(2/0.9661)(78.93 - 46.42) = 67.30 \text{ kW}$$

Referring to psychrometric tables at 101.325 kPa, for comparison, it is seen that the cooling load would be

$$(2/0.8322)(71.03 - 42.08) = 69.57 \text{ kW}$$

As a matter of interest, if the psychrometric charts by Martin (1972), which are based on the ideal gas laws, had been referred to, values of enthalpy and specific volume yielding a cooling load of (2/0.965)(79 – 46.5) = 67.36 kW would have been obtained, the discrepancy being largely attributable to error in reading values from the chart. If the values had actually been calculated by means of the ideal gas laws, as described by Jones (1994), a load of (2/0.9661)(78.76 – 46/27) = 67.26 kW would have been obtained.

The superficial deduction from this example is that the cooling load is not much influenced by altitude, which may be true. However, the performance of the cooling coil is considerably influenced, and the cooling capacity of the air supplied to the conditioned room is reduced, by altitude effects because air density reduces. Plotting the on and off coil states on a psychrometric chart for 87.5 kPa shows that the sensible/total heat removal ratio across the coil is about 0.72 whereas the same states on a chart for 101.325 kPa give a ratio of 0.81. This will affect the air-side film resistance and the overall U-value for the cooler coil. It can also be seen that the mean coil surface temperatures (apparatus dew points) are 14.2°C at 87.5 kPa and 14.1°C at 101.325 kPa. This is not of much significance in this example but might acquire greater importance for other entering and leaving psychrometric states for the cooler coil. Therefore, reliance should not be placed on calculated coil duties without plotting the performance on a psychrometric chart for the correct atmospheric pressure.

AIR MASS FLOW RATE

Air mass flow rate is probably the most important effect of barometric pressure changes upon system performance. It is the air mass flow rate that transfers heat between cooler coils or condensers and airstreams and removes the sensible and latent heat gains from the conditioned space. Therefore, it is of vital importance that the correct air density or specific volume be used in calculations.

EXAMPLE 2.2

If the air leaving the cooler coil in Example 2.1 suffers a rise in temperature of 2 K because fan power and duct heat gain on the way to the conditioned space, calculate the room temperature and humidity maintained in the face of sensible and latent heat gains of 19.82 and 2.67 kW, respectively, (a) when the barometric pressure is 87.5 kPa and (b) when it is 101.325 kPa.

Answer

(a) The specific heats of dry air and water vapour change very little with a fall in pressure and so, using the off-coil moisture content of 0.01191 kg kg^{-1} previously calculated, the specific heat of the supply air is $1.012 + 1.89 \times 0.01191 = 1.035$ kJ kg^{-1} K^{-1}. The temperature maintained in the room, t_r, is then obtained from

$$19.82 = (2/0.9661)(1.035)(t_r - 18) \text{ from which } t_r = 27.3°\text{C}$$

Taking the latent heat of evaporation of water as 2454 kJ kg^{-1} the moisture content in the room, g_r, is obtained from

$$2.67 = (2/0.9661)(g_r - 0.0119) \times 2454 \text{ from which } g_r = 0.01244 \text{ kg kg}^{-1}$$

giving a humidity of about 46%.

(b) The answer can be obtained by similar calculations to (a) or by using the following two formulas (derived in the Appendix)

$$\text{Supply air quantity (m}^3\text{ s}^{-1}\text{) at temperature } t = [\text{Sensible heat gain (kW)}/(t_r - t_s)] \times [(273 + t)/358] \tag{2.3}$$

Supply air quantity ($m^3 s^{-1}$) at temperature t =
[Latent heat gain (kW)/($g_r - g_s$)] × [(273 + t)/856] (2.4)

where t_s is the supply air temperature in °C, g_r and g_s the room and supply air moisture contents in g kg^{-1} and t is the temperature in °C at which the airflow is expressed. These will yield answers of 26°C dry-bulb and 50%, at 101.325 kPa. Equations (2.3) and (2.4) are only valid at the standard, sea level barometric pressure of 101.325 kPa.

HEAT TRANSFER COEFFICIENTS FOR AIR

In order to express both the latent and sensible heat transfer to a cooler coil in terms of a U-value and a logarithmic mean temperature difference (LMTD), air-to-air, an enhanced value of the air film coefficient, h_a, is sometimes obtained by dividing it by the value of the sensible total ratio, S for the cooling process. If r_w and r_m are the water film resistance and metal resistance, respectively, both referred to the external surface area of the coil, then

$$U = 1/(r_w + r_m + S/h_a) \quad (2.5)$$

Thus the value of 0.72, obtained for S in Example 2.1 at 87.5 kPa, will give a slightly larger U-value for the coil than at sea level when S = 0.81, in the same example.

Heat transfer by forced convection is expressed in terms of the Nusselt number (Nu), equal to $h_a d/k\theta$; the Reynolds number (Re), equal to $ud\rho/\mu$; and the Prandtl number (Pr), equal to $c_p\mu/k$; in which d is a relevant linear dimension, θ is a temperature difference, k is the thermal conductivity of air, u is its velocity, p its density, μ its absolute viscosity and c_p its specific heat capacity. It transpires that for forced convection over parallel plates, corresponding to the fins on a cooler coil,

$$(\text{Nu}) = 0.36\,(\text{Re})^{0.8}\,(\text{Pr})^{0.33} \quad (2.6)$$

Little work has been done to establish the values of c_p, μ and k, at sub-atmospheric pressures but it is generally thought that they do not alter very much from their values at sea level. Therefore, it can be concluded that the influence of (Pr) is negligible and that heat transfer depends principally on (Re). It appears that for air, the dependence of heat transfer is upon $(\text{Re})^{0.8}$, and this implies that it is, in turn, proportional to the mass velocity, $u\rho$, to the 0.8 power. The change in the value of S alone, or the mass velocity alone, can influence the U-value of a coil significantly, but their combined effects can sometimes almost cancel each other.

EXAMPLE 2.3

Given that, for the cooler coil in Example 2.1, working at sea level, r_w = 0.00425 m^2 K W^{-1}, r_m = 0.0035 m^2 K W^{-1} and h_a (dry) = 64 W m^{-2} K^{-1}, all referred to the external surface area, calculate its U-value at (a) sea level and (b) a barometric pressure of 87.5 kPa.

Answer

(a) From Example 2.1, S = 0.81 at sea level and so h_a (wet) = 64/0.81 = 79 W m^{-2} K^{-1}, and r_a (wet) = 0.01266 m^2 K W^{-1}.

$$U = 1/[0.01266 + 0.0035 + 0.00425] = 49.0 \text{ W m}^{-2} \text{ K}^{-1}$$

(b) From Example 2.1, $S = 0.72$ at 87.5 kPa and whereas $v = 0.9661 \text{ m}^3 \text{ kg}^{-1}$ at the same pressure, it equals $0.8322 \text{ m}^3 \text{ kg}^{-1}$ at sea level. Thus the factors affecting the value of h_a are: $0.81/0.72 = 1.125$ for the change in sensible heat ratio and $(0.8322/0.9661)^{0.8} = 0.887$ for the mass flow effect. Therefore, h_a (wet) at 87.5 kPa $= 79 \times 1.125 \times 0.887 = 78.8 \text{ W m}^{-2} \text{ K}^{-1}$ and r_a (wet) $= 0.01269 \text{ m}^2 \text{ K W}^{-1}$.

$$U = 1/[0.01269 + 0.0035 + 0.00425] = 48.9 \text{ W m}^{-2} \text{ K}^{-1}$$

EVAPORATION RATE

The water cooling effect in a cooling tower or an evaporative condenser depends on the evaporation rate of the water circulated. At a given dry-bulb temperature saturated air has a larger moisture content at a smaller barometric pressure. An airstream passing through a cooling tower can therefore take up more moisture, as it approaches saturation, at a higher altitude. The effect is fairly small, amounting to only about 3% at a height of 3000 m above sea level.

AIR PRESSURE LOSS IN DUCTS AND PLANT

For all practical purposes, the pressure loss through both ducting and plant is proportional to the density of the airstream. Thus the loss calculated at sea level conditions should be multiplied by the ratio of the barometric pressures and by the inverse ratio of the absolute temperatures, to determine the loss at altitude.

EXAMPLE 2.4

A system of plant and ducting has a total pressure loss of 2 kPa, calculated for sea level conditions of 20°C and 101.325 kPa. Calculate the total loss at a barometric pressure of 87.5 kPa and a temperature of 39°C.

Answer

$$\text{Total pressure loss} = 2 \times [87.5/101.325] \times [(273 + 20)/(273 + 39)] = 1.62 \text{ kPa}$$

PRESSURE GAUGE INDICATIONS

Since a pressure gauge compares the measured pressure with the ambient pressure, higher pressure indications will be given at higher latitudes. The correction is simple.

AVAILABLE NET POSITIVE SUCTION HEAD (NPSH)

This decreases with altitude, as Equation (4.20) shows, giving an increased risk of cavitation within a pump and a subsequent loss of performance. It is countered by reducing the resistance on the suction side of the pump and by increasing the position head, that is, the distance between the surface of the water in the open tank, for example a feed and expansion tank or cooling tower pond, and the centre-line of the pump suction branch.

ELECTRIC MOTOR COOLING

The mass flow rate of air for cooling a motor is proportional to the air density and so decreases with altitude according to reduction in pressure and changes in absolute temperature. These effects may be dealt with by using insulation able to withstand higher operating temperatures, or by using an oversized motor or one of special design.

2.3 Unitary systems

The distinction drawn between air handling units and air conditioning units is that the latter may contain a refrigeration compressor, with or without its condenser, and the former do not. This section is primarily concerned with room air conditioning units, sometimes termed window units, and with the larger, air conditioning packages, free-standing in the conditioned space or suspended at high level in it or just outside it, that are used to treat large rooms, small commercial premises, or even whole individual floors of office blocks.

SELF-CONTAINED, AIR-COOLED ROOM AIR CONDITIONERS

Room air conditioners have had a very wide usage all over the world for many years, their chief advantages being relatively low capital cost for a small cooling load and the possibility of installing them room-by-room, as required, without the need to air condition the rest of the building and without taking up a lot of space for ducts and central plant. These advantages disappear as the cooling load increases and the scope of the air conditioning gets larger. Their chief defects are relatively short life of 3–10 years; noise, which increases with age as bearings wear, fixings are loosened by vibration and corrosion sets in; comparatively poor air distribution; poorer automatic control and poorer air filtration than is provided by central systems; low rate of fresh air ventilation; and, in the case of air-cooled units, a sometimes unsightly interference with the building facade caused by the need to cut openings in it to accommodate the condensers. Heating is commonly electrical, proving costly to run, but better quality units frequently offer an LTHW option. Heat pump units that reverse the roles of the evaporator and condenser in cold weather are also available. Units of this type are very popular, and one reason at least is their relatively easy installation, provided a hole can be cut in an outside wall or the unit allowed to sit on the sill, the windows being raised sufficiently and any gaps around the unit being blocked.

Room air conditioners are available in the range from 1.75 kW of refrigeration to 10.5 kW. The standard of test adopted in the United States to establish unit ratings is often 26.7°C dry-bulb, 19.4°C wet-bulb (sling) in the room and 35°C dry-bulb, 23.9°C wet-bulb (sling) outside, but not every manufacturer states the basis upon which the outputs offered have been established. The cooling capacities listed in manufacturers' published leaflets and catalogues are seldom sufficient for a designer to assess the conditions of temperature and humidity likely to prevail in the conditioned space. At best, little more than an estimate of temperature is possible and that only with doubtful accuracy.

Air-cooled, room air conditioning units invariably carry out latent cooling as well as sensible cooling for much of the time they are in operation. It is usual, as part of a self-contained unit, to collect the condensate in a tray beneath the cooler coil and to pipe this to a ring on the periphery of the condenser fan. The ring has a serrated edge and slings the condensate over the condenser fins, improving the heat rejection by the condenser at the expense of some increased risk of corrosion. The corrosion risk may be worse in coastal regions because of the salt content of the air used to cool the condenser. A more expensive alternative is to provide a proper condensate drainage system for all the units used.

EXAMPLE 2.5

An air-cooled room conditioner has the following characteristic performance when the outside air temperature is 28°C dry-bulb and the volumetric flow rate of air supplied is 150 l s^{-1}, expressed at the room temperature

Room temperature, t_f (°C)	18	21	24
Sensible cooling capacity (W)	1324	1355	1379
Latent cooling capacity (W)	558	646	741
Total cooling capacity (W)	1882	2001	2120
Absorbed compressor power (W)	942	974	1003
Coefficient of performance	2.00	2.05	2.11

If one of these units is installed in a west-facing module on an intermediate floor of the hypothetical office block described in Section 1.2, make an approximate assessment of the room dry-bulb temperature that should prevail under conditions of design sensible heat gain.

Answer

From Example 1.9, the calculated sensible heat gains are 1394 W when the room temperature is 22°C. The gains through the glass and wall, and by infiltration, depend on the inside-to-outside air temperature difference, whereas the solar gains and those from people, lights and machines do not. Hence the following table can be established

Room temperature, t_r (°C)	18	21	22	24
Sensible heat gains (W)				
Glass + wall + infiltration	175	122	105	70
Solar + people + lights + machines	1289	1289	1289	1289
Total	1464	1411	1394	1359
Unit sensible cooling capacity (W)	1324	1355	–	1379

Sensible cooling capacity and sensible heat gain are plotted, in Figure 2.1. The intersection is at 23.3°C and this is the room temperature maintained under outside design conditions.

The above calculation is only an approximation. Under conditions of reduced sensible heat gain the room temperature will vary between the limits of the differential gap related to the two-position control exercised over the compressor in the unit. The thermal inertia of the building structure will play a part in reducing such swings in temperature. The usual way of controlling the unit is to allow the fan to run continuously when the unit is in use but to vary cooling capacity by two-position control over the compressor, in sequence with any heating capacity provided. Sometimes the fan is also switched on–off, in sequence with the compressor, but this is not recommended because it gives variable air movement and variable noise, in addition to some loss of control over temperature.

The volume of the treated module is 37.44 m^3 (Figure 1.1) and it is worth noting that the supply airflow rate of 150 l s^{-1} represents an air change rate of 12.5 h^{-1}. This is reasonable, in terms of air movement for comfort and for the removal of the sensible heat gains. The use of Equation (2.3) shows that the supply air temperature will be 15.7°C, which is reasonable for this sort of unit.

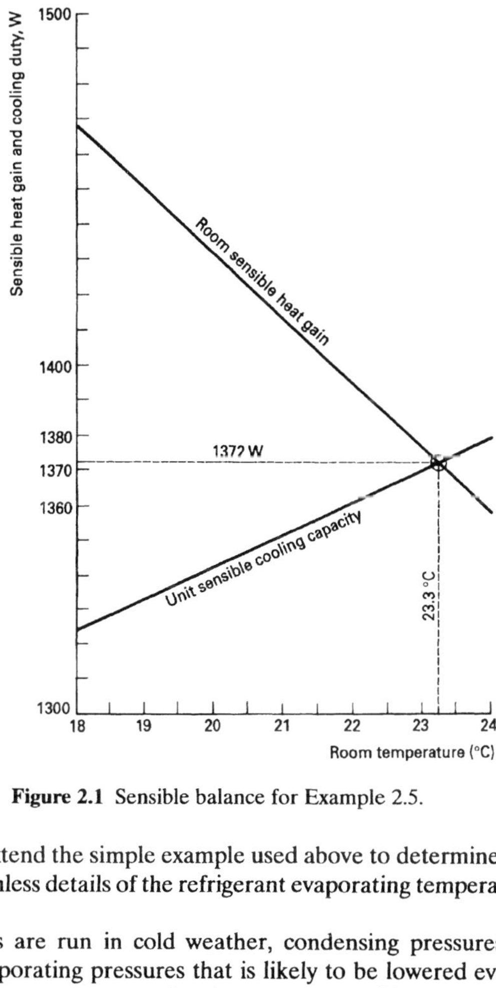

Figure 2.1 Sensible balance for Example 2.5.

It is not possible to extend the simple example used above to determine the approximate humidity in the room, unless details of the refrigerant evaporating temperature are provided by the manufacturer.

If these sort of units are run in cold weather, condensing pressures will fall with a consequent drop in evaporating pressures that is likely to be lowered even more because the dry- and wet-bulb temperatures entering the evaporator will be less, by design, in winter than in summer. The windings of the motor driving the hermetic compressor are cooled by the suction gas and the mass flow of this, and the cooling capacity, reduces in proportion to the fall in evaporating pressure. This reduction in cooling effect far outweighs any possible improvement resulting from the drop in refrigerant suction temperature. If frost forms on the evaporator fins the effect is exacerbated since an extra resistance to airflow is provided, inducing another drop in the load and so in the evaporating pressure. Operating a unit with a very dirty air filter produces a similar effect.

The outcome of inadequate motor cooling is burnt-out windings, and the presence of a safety cut-out thermostat, buried in the windings, is not always sufficient protection. Repeated motor starts at short intervals, or continued running at low evaporating temperatures, tends to bake the winding insulation without necessarily reaching the set point of the cut-out at its location. Eventually the insulation fails and the motor burns out. The chemical products of a burn-out poison the refrigeration system and it follows that it must be very carefully cleaned out before a new hermetic compressor is fitted and the system recharged. Failure to clean the system properly will result in further, early burn-outs.

One method of guarding against the risk of motor burn-out is to use a hot-gas valve to maintain the evaporating pressure in the face of a falling duty accompanied by cold ambient air temperatures, by imposing a false load on the evaporator. The best way of doing this is to inject the hot gas into a hot-gas header, specially made on the side of the evaporator receiving the liquid refrigerant. However, packaged units seldom have this facility. The best alternative is to inject the gas between the expansion valve and the distributor, although even this is not always possible with small commercial units as there is no room between the two. On larger sizes of unit, self-acting hot-gas valves to inject in this manner are sometimes available.

Before selecting a packaged air-conditioning unit, other than the small room conditioner discussed earlier, with a direct-expansion cooler coil, it is generally advisable to calculate the minimum expected conditions onto the cooler coil and the corresponding cooling load, as well as the design maximum duty. If this information is supplied to the manufacturers, they should be able to assess the risk of frost formation and the need for a hot-gas valve.

A cheaper alternative to hot-gas valves is cylinder unloading, actuated by changes in suction pressure. However, this method is not always as effective and is not available for the small sizes of unit considered so far.

SPLIT SYSTEMS

Room air conditioners and the larger, direct-expansion units are also available either with separate, air-cooled condensers or with remote, air-cooled, condensing units. Such installations are often a convenient alternative to single, self-contained packages. Hermetic condensing sets are made in a range of sizes able to deal with as much as 60 kW of refrigeration. Air-cooled condensers, however, have a much wider application for duties that can exceed 500 kW of refrigeration.

Packaged units generally are designed to supply about 50 l s^{-1} of air to a conditioned room for each kW of refrigeration. This is a restriction that limits the choice of the designer and tends to reverse the usual order of the design process, i.e. a packaged unit is selected and the designer estimates what it can do, rather than the selection being aimed at achieving a particular performance. There is often some difficulty in doing this since catalogue data are seldom adequate for a full technical exploration and sometimes the best that can be hoped for is merely to get an idea of the room temperature that will be maintained as in Example 2.5. Conditions of humidity can rarely be forecast with assurance but fortunately this is of secondary importance in comfort conditioning as long as the temperature is satisfactory.

Table 2.1 shows the typical performance of two similar commercial air-handling units, of different sizes, which would be piped up to air-cooled condensing sets having the sort of performances listed in Table 2.2. The duty of the direct-expansion cooler coil in an air-handling unit is given in terms of evaporating temperature, and that of a condensing set in terms of saturated suction temperature. If the actual pressure prevailing at the compressor suction connexion were saturated, which it is not, the corresponding temperature would be called the saturated suction temperature. In fact, it is several degrees warmer because it is superheated by the action of the thermostatic expansion valve. It is customary to size the suction line for a pressure drop relating to a fall of 1 K in saturated temperature and it can, therefore, be inferred that the saturated temperature in the evaporator is 1 K higher than the saturated suction temperature.

Table 2.1 Typical performance of a direct-expansion air-handling unit

Unit size	Entry wet-bulb (°C)	Evaporating temperature (°C) −1		+1		+3		+5		+7		+9	
		Cooling capacities (kW)											
		T	*S*	*T*	*S*	*T*	*S*	*T*	*S*	*T*	*S*	*T*	*S*
1	22.0					22.6	11.4	20.7	10.53	18.8	9.77	16.9	9.02
	19.5			20.0	12.86	18.5	12.2	16.5	11.28	14.5	10.4	12.5	9.61
	17.0	18.5	14.8	16.5	13.78	14.6	12.8						
2	22.0							24.3	12.47	21.8	11.6	19.4	10.7
	19.5					21.8	14.6	19.4	13.49	16.9	12.5	14.6	11.6
	17.0	21.9	17.9	19.5	16.55	17.1	15.4	14.7	14.4				

T = total cooling duty; *S* = sensible cooling duty.
No. 1: fan power = 600 W; by-pass factor = 0.18; air quantity = 700 $l\ s^{-1}$.
No. 2: fan power = 900 W; by-pass factor = 0.25; air quantity = 950 $l\ s^{-1}$.

In Table 2.1 it is assumed that the entering air dry-bulb temperature in all cases is 27°C. Interpolation is allowable but extrapolation is not. Actual performances can differ considerably from these figures.

If the performance of a condensing set is plotted in terms of kW of refrigeration against saturated suction temperature plus one degree, its intersection with the evaporator curve, plotted for kW of cooling against evaporating temperature, gives the duty achieved when the pair of units is piped together.

For a given evaporating temperature the mean coil surface temperature (apparatus dewpoint) of the cooler coil in an air-handling unit is constant as long as the cooling load is also constant, regardless of variations in the entering dry-bulb. If a constant entering wet-bulb is regarded as synonymous with a constant entering enthalpy, an approximate assessment can be made of coil performance at various entering dry-bulbs, other than that on which the tabulated data is based.

Although the intersection of the characteristic performance curves for the evaporator and the condensing set will give the total cooling load for a particular entering wet-bulb temperature, it will not yield the sensible and latent proportions, the slope of the process line on the psychrometric chart and hence the state of the air leaving the cooler coil. However, this can be calculated if the tabulated data includes the by-pass factor or the sensible component (see Table 2.1).

EXAMPLE 2.6

Nine-hundred-and-fifty $l\ s^{-1}$ of air at 27°C dry-bulb, 17°C wet-bulb (sling) enters the cooler coil of a size 2 air-handling unit having a performance as given in Table 2.1. If the unit is piped up to a size 2 condensing set (see Table 2.2) and the air temperature onto the condenser is 29.4°C, determine the following, assuming that the thermostatic expansion valve is large enough to pass the correct flow rate of refrigerant:

(a) The state of the air supplied to the room by the air handling unit.
(b) The state of the air supplied if the entering air state is 23°C dry-bulb, 17°C wet-bulb, 9.618 g kg^{-1}.

Table 2.2 Typical performance of an air-cooled condensing set

		Air temperature entering the condenser (°C) 29.4			35.0		
Unit size	Saturated suction temperature (°C)	Cooling capacity (kW)	Saturated condensing temperature (°C)	Compressor motor power (kW)	Cooling capacity (kW)	Saturated condensing temperature (°C)	Compressor motor power (kW)
1	−1.1	10.1	50.0	4.0	9.5	53.3	4.1
	+1.7	11.2	51.7	4.3	10.5	55.0	4.4
	+4.4	12.2	53.3	4.6	11.5	56.7	4.7
	+7.2	13.4	55.0	4.9	12.5	58.3	5.0
	+10.0	14.5	57.2	5.2	13.6	60.6	5.3
2	−1.1	13.8	45.6	5.1	12.7	50.0	5.3
	+1.7	15.3	47.8	5.5	14.1	52.2	5.7
	+4.4	16.9	50.0	5.9	15.4	54.4	6.1
	+7.2	18.4	52.2	6.2	16.9	57.2	6.6
	+10.0	20.2	53.9	6.6	18.4	58.9	7.0

Note. Actual performances can differ considerable from the above figures.

Answer

(a) Adding 1 K to the saturated suction temperatures given in Table 2.2 enables the cooling capacity of the condensing set to be plotted against evaporating temperature and, using the data from Table 2.1, the characteristic curve of the cooler coil in the air-handling unit can be plotted on the same coordinates (Figure 2.2). At the point of intersection, *P*, the duty is 16 k W and the evaporating temperature is 3.9°C.

At the entering air state of 27°C dry-bulb, 17°C wet-bulb, the specific volume is 0.8607 m^3 kg^{-1} and the enthalpy is 47.21 kJ kg^{-1}. The air mass flow rate is thus 0.95/0.8607

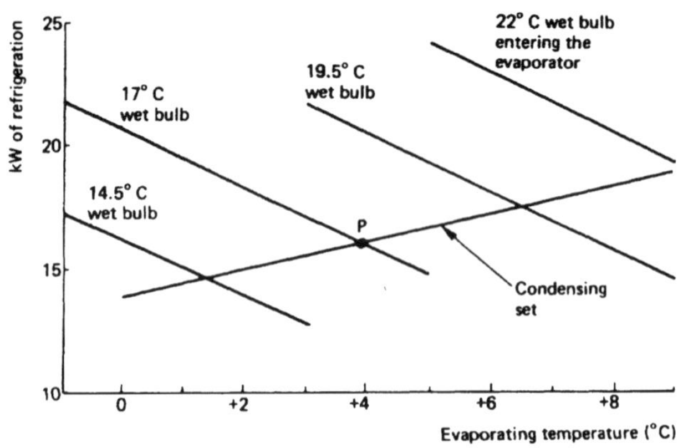

Figure 2.2 Typical performance characteristics for an air-cooled condensing set, with air onto the condenser at 29.4°C dry-bulb and an air cooler coil at various wet-bulbs (see Tables 2.1 and 2.2).

= 1.1038 kg s^{-1} and the enthalpy drop across the cooler coil is 16/1.1038 = 14.5 kJ kg^{-1}. From Table 2.1, the by-pass factor is 0.25 hence the enthalpy at the apparatus dew-point, A, is 47.21 – 14.5/0.75 = 27.88 kJ kg^{-1}. Figure 2.3 shows the psychrometry of this. The apparatus dew-point is 9.4°C and since the by-pass factor is 0.25, the temperature at state W, leaving the coil, can also be calculated as 27 – 0.75(27 – 9.4) = 13.8°C. The fan power, from Table 2.1, is 900 W and so the temperature of the air supplied to the room is 13.8 + (0.9/0.95) × (273 + 27)/358 = 14.6°C at state S, from Equation (2.3).

The sensible component of the cooling load is [0.95 × (27 – 13.8) × 358]/(273 + 27) = 14.96 kW and the latent component is 16 – 14.96 = 1.04 kW. Since the moisture content at state R, onto the coil, is 7.86 g kg^{-1} the moisture content off the coil is 7.86 – (1.04/0.95) × (273 + 27)/856 = 7.48 g kg^{-1} by Equation (2.4). The state of the air supplied to the room is thus defined as 14.6°C dry-bulb and 7.48 g kg^{-1}.

(b) If the state of the air entering the cooler coil, M, is 23°C dry-bulb, 17°C wet-bulb (sling),47.6 kJ kg^{-1}, 0.8515 m^3 kg^{-1} and the total cooling load is unchanged at 16 kW (because the entering wet-bulb is unchanged at 17°C), then by the earlier reasoning the mean coil surface temperature stays at 9.4°C and the evaporating temperature remains at 3.9°C.

Figure 2.3 shows this. The coil leaving state, W', can then either be established geometrically on the psychrometric chart because WW' is parallel to RM or can be calculated as 12.8°C dry-bulb and 8.066 g kg^{-1} by the methods used in answering part (a). The supply temperature is then 12.8 + 0.8 = 13.6°C.

Most air-handling units deal with a mixture of fresh and recirculated air. This poses a problem in determining the performance of the sort of unit that has just been considered. Although a match between the sensible cooling capacity of the air supplied to the conditioned room and the sensible heat gains suffered in it can be achieved thermostatically, by proportional reheating or cycling under two-position controls, it is possible that the latent capacity will exceed the latent gains in many cases. As a result the room wet-bulb temperature will fall. Figure 2.2 shows that this will reduce both the total cooling capacity and the evaporating temperature, with a risk of frosting and its disastrous consequences.

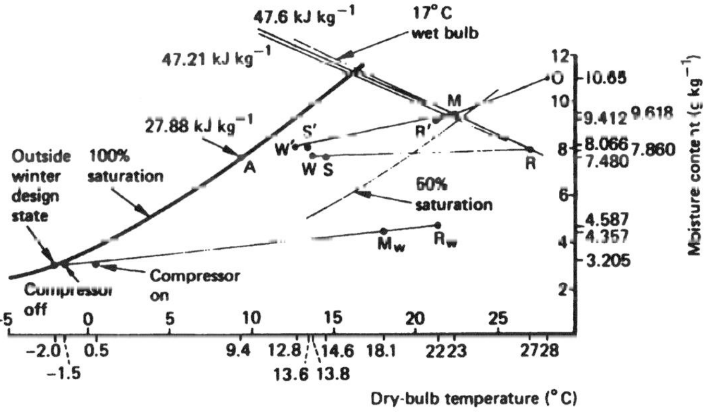

Figure 2.3 Psychrometry for Examples 2.6–2.8.

EXAMPLE 2.7

For the plant used in Example 2.6, determine the sensible and latent cooling capacities of the air supplied to a room conditioned at 22°C dry-bulb, assuming that the outside state is 28°C dry-bulb, 19.5°C wet-bulb and ignoring any effect that the reduction in outside dry-bulb has on the capacity of the condensing set. Of the air supplied, five-sixths is recirculated and one-sixth is fresh and hence 950 l s^{-1} of air at 23°C enters the cooler coil.

Answer

Plotting O and R' on Figure 2.3, the line OR' passing through M (which has a moisture content of 9.618 g kg^{-1}), the moisture content in the room can be determined as follows:

$$10.65 - (10.65 - 9.618)6/5 = 9.412 \text{ g kg}^{-1}$$

The sensible gains to the room that can be dealt with are

$$[0.95 \times (22.0 - 13.6) \times 358]/(273 + 23) = 9.65 \text{ kW}$$

and the latent gains that may be offset are

$$[0.95 \times (9.412 - 8.060) \times 856]/(273 + 23) = 3.714 \text{ kW}$$

These are substantial latent gains and the slope of the room ratio line would be 0.72, whereas, in Examples 1.9 and 1.10, the sensible and latent heat gains to a single, west-facing module in the hypothetical office block were 1.934 and 0.134 kW, respectively, yielding a slope of 0.91 for the room ratio line. The result of using the split system will be to cause the moisture content of the room state to fall, reducing the mixture wet-bulb onto the cooler coil, until some sort of balance is ultimately achieved, with the room temperature controlled at 22°C but its humidity a good deal lower than the customary 50%. When the wet-bulb onto the cooler coil falls, the latent component in its total cooling capacity reduces, as we may verify by Table 2.1, but it is difficult to predict the exact performance without access to full technical information, seldom, if ever, provided in a catalogue. The risks of low entering wet-bulbs can also arise, quite apart from the mismatch of latent capacity and load, because of the presence of the fresh air itself in the mixture and the advent of colder, non-design weather.

Larger air handling units would have a mixing chamber with motorised dampers that are automatically controlled to optimise the use of the refrigeration plant. Figure 2.4 shows a simplified arrangement of this. When the outside enthalpy, h_o, exceeds the enthalpy of the air in the conditioned room, h_r, the cooling load is minimised by using as little fresh air as possible, whereas if h_o is less than h_r, but greater than the enthalpy of the air leaving the cooler coil, h_w, the cooling load is reduced by using 100% fresh air. Finally, if the outside air has an enthalpy less than that normally leaving the cooler coil then the dampers can be controlled to mix fresh and recirculated air to give a temperature of t_w, off the cooler coil, without using the refrigeration plant, which can be switched off.

For the smaller sizes of air handling units, it is not uncommon to operate with fixed proportions of fresh and recirculated air throughout the year. This saves capital cost but involves increased running costs, as well as increasing the chance of frosting on the cooler

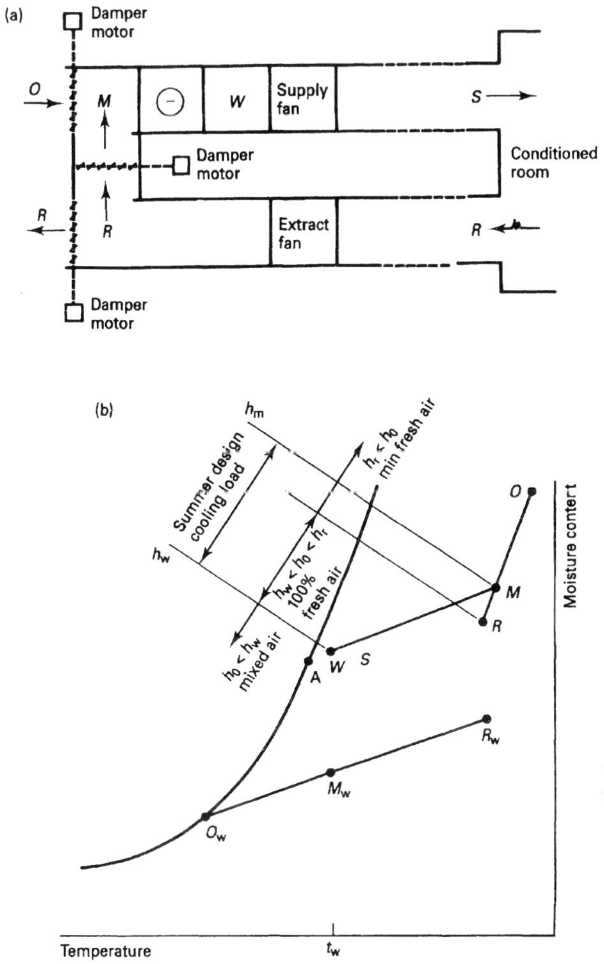

Figure 2.4 Using motorised dampers allows the refrigeration plant to be switched off in winter when $h_o < h_w$. (a) Plant arrangement. (b) Psychrometry, ignoring any temperature rise across the extract fan.

coil and burning out the compressor motor. Figure 2.2 shows that the evaporating temperature (and hence the coil surface temperature) falls as the wet-bulb onto the coil reduces. Figure 2.3 shows that if the proportions of fresh and recirculated air are kept fixed throughout the year the wet-bulb entering the coil becomes progressively lower as the outside state gets colder: in winter the wet-bulb of M_w is obviously less than that of M, in summer. When air handling plants operate in winter with a fixed minimum proportion of fresh air it becomes necessary to establish the minimum on-coil condition, in order to take the necessary steps to ensure that the refrigeration plant can operate safely.

EXAMPLE 2.8

Five west-facing modules on an intermediate floor of the hypothetical office block (see Section 1.2) are to be grouped to form a dining room for thirteen people. If the split system considered in Examples 2.6 and 2.7 is to be used for air conditioning the dining room, with fixed proportions of minimum fresh and recirculated air, determine the lowest on-coil state for the cooler coil. The temperature rise across the supply fan is 0.8 K and any temperature rise across the extract fan is ignored in this case, for simplicity.

Answer

First, the summer design heat gains determined in Example 1.9 must be revised, to take account of the changed occupancy, assuming an internal temperature of 22°C with a humidity of about 50%: it can be assumed that there are no business machines in the dining room but that the lights are unchanged.

Gains for glass, wall and infiltration: 5 × 105 =	525 W
Solar gain: 5 × 576 =	2880 W
Lights: 5 × 245 =	1225
Revised population: 13 × 100 (see Table 1.3) =	1330 W
Total revised sensible gain:	5930 W

Infiltration: 5 × 0.8 × 0.5 × 37.44 (10.65 – 8.366):	170 W
People: 13 × 60 (see Table 1.3):	780 W
Total revised latent gains:	950 W

The minimum condition will occur when the cooling capacity of the supply air, after allowing for the temperature rise from fan power, in winter or autumn, equals the reduced sensible heat gain at that time, without refrigeration. Multiple calculations may be needed but, anticipating that the heat balance will occur in an afternoon in, say, February, the reduced outside air temperature (denoted by t_o) and the changed solar gain through glass can be considered and the corresponding heat gains established.

The transmission gain through a single module of the building fabric was calculated as 105 W for an outside-to-inside temperature difference of 6 K. Assuming a room temperature of 22°C the revised transmission gain through five modules of the building fabric is $5 \times (105/6) \times (t_o - 22) = 87.5 \times (t_o - 22) = 87.5t_o - 1925$ W.

Referring to Table A5, it is seen that in February the peak solar gains through glass are 188 W m^{-2}, at 14.00 h sun time. A generalised sensible heat gain can now be established, in terms of t_o, for an afternoon in February.

Transmission:	$87.5t_o - 1925$
Solar: 0.74 × 0.91 × 188 × 3.168 × 5 =	2005
Lights: 5 × 245 =	1225
People:	1300
Total sensible gain in a February afternoon:	$2605 + 87.5t_o$ W

Since the temperature rise across the supply fan is 0.8 K the temperature of the air supplied by the air handling unit will be $t_s = t_m + 0.8$, just after the refrigeration plant is switched off.

Without refrigeration, the sensible cooling capacity in kW of the air supplied is $[0.95 \times (22 - t_m - 0.8) \times 358]/(273 + t)$, according to Equation (2.3). The value of t in the Charles' law correction bracket does not materially affect the answer but complicates the arithmetic. Hence a value of, say, 17°C can be assumed for t, without significant loss of accuracy. The natural cooling capacity of the air handling unit, with the refrigeration plant off, can then be expressed by

$$[0.95 \times (22 - t_m - 0.8) \times 358]/(273 + 17) = 24.86 - 1.173t_m$$

Noting that five-sixths of the air handled is recirculated and only one-sixth is from outside, $t_m = (5/6) \times 22 + t_o/6$. Substituting this for t_m in the above equation the natural cooling capacity of the air handling unit, in kW, is expressed by $3.36 - 0.2t_o$.

A heat balance is now struck between the revised sensible heat gain and the natural cooling capacity of the air handling unit: $2.605 + 0.875\,t_o = 3.36 - 0.2t_o$, whence $t_o = 0.7$°C. If the fixed proportion of fresh air had been greater the balance temperature would have been higher.

The refrigeration plant might be controlled by an outside air thermostat with a differential of ±1 K. The refrigeration plant would then run until the outside temperature fell to −1.3°C, when it would be switched off. The dry-bulb temperature onto the cooler coil, immediately prior to this, would be $(22 \times 5/6) + (-1.3/6) = 18.1$°C. When the outside temperature rose to +0.7°C the plant would be switched on again. The temperature of the air onto the cooler coil would then be $(22 \times 5/6) + (0.7/6) = 18.4$°C. There is a risk here that the refrigeration compressor could be switched on and off too frequently and its motor windings burnt out. This risk must be dealt with and a timer delay introduced if necessary. The maximum number of starts allowable for hermetic and semi-hermetic reciprocating compressors is 4–6 per hour.

To continue the analysis, the dry-bulb temperature of −1.3°C must be coupled with a moisture content in order to determine the natural latent cooling capacity of the air handling unit, with the refrigeration plant off. One could argue that an outside air temperature of −1.3°C would occur during stable, cold weather when a winter design state of, say, −2°C saturated had prevailed for some time. CIBSE psychrometric tables show that the outside moisture content is 3.205 g kg^{-1}.

An accurate estimate of the latent heat loss due to the infiltration of drier air from outside into the building is not possible, so it is ignored in this case. Consequently, the latent gain is due to people only, namely, 0.780 kW.

Since five-sixths of the air is recirculated and one-sixth is from outside, the moisture content of the air supplied is $(5g_r/6 + 3.205/6) = 0.833g_r + 0.534$ g kg^{-1}. This can be used with Equation (2.4) to give the natural latent cooling capacity of the air handling unit and then equated to the latent gain, to yield the balance value of the moisture content in the room:

$$[0.95 \times (g_r - 0.833g - 0.534) \times 856]/(273 + 17) = 0.780$$

whence $g_r = 4.862$ g kg^{-1}. Reference to CIBSE psychrometric tables (or a chart) shows that with a room dry-bulb of 22°C and this moisture content, the humidity is about 29% and the wet-bulb is 12.4°C (sling).

Just before the refrigeration plant is switched off the moisture content onto the air cooler coil is $(5 \times 4.862/6 + 3.205/6) = 4.586$ g kg^{-1}. With 18.1°C dry-bulb onto the coil and this moisture content the wet-bulb is about 10.3°C. This is the minimum state onto the cooler coil, when the refrigeration plant is running. Reference to Figure 2.2 shows that the evaporating temperature will be less than −1°C.

The 14.5°C wet-bulb line in Figure 2.2 gives an indication of the theoretical evaporator performance for the much lower entering wet-bulb temperature of 10.3°C. It implies that the evaporating temperature is less than −1°C and that there will, therefore, be a distinct risk of frosting and its consequences. It is strongly recommended that evaporator pressure control be provided and the best way of doing this is by hot-gas injection into the evaporator.

A plant of this type and size would probably cycle under two-position control from a thermostat in the recirculated airstream, the single compressor in the condensing set being switched on and off. Larger units might have several compressors, offering step control over room temperature, but eventually the last compressor running will face low load problems.

The balance between an evaporator and a condensing unit will not be achieved at the steady-state level desired unless the thermostatic expansion valve passes the correct mass flow rate of refrigerant. The flow rate through the valve is a function of the pressure drop across it and, when calculating this, allowances must also be made for pipe friction loss, the change of position head in the liquid line before the valve, and the substantial pressure loss through the distributor and the tubes from it that feed the evaporator circuits. Since expansion valves are made in commercial size increments, an exact balance at the desired duty will not, in general, be obtained. If the valve is slightly oversized it will partly close, reducing the flow rate of refrigerant and raising the liquid level in the condenser. Since only the part of the condenser surface above the liquid line is effective in the condensation process, the condensing pressure will rise until it reaches a value that is consistent with the flow rate through the valve, the load on the evaporator and the capacity of the compressor. If the valve is undersized, stability would be reached at a higher condensing pressure than anticipated and less than the design duty would be achieved.

The fluid that is thermodynamically desirable in the system is the refrigerant but oil must also be present for the proper mechanical working of the compressor. The need to minimise the compression ratio, and the power absorbed, plus the necessity of ensuring oil return to the compressor, are the two restraints that limit the lengths of refrigerant pipe runs and the vertical distances between the items of plant. It follows that the location of the condensing set in relation to the air-handling unit is critical, and condensers, evaporators and compressors should normally be as close to each other as possible for the proper running of the system. However, the use of electronic expansion values and microprocessors has introduced some revision to this. See the section entitled Cassette Units and VRV Systems.

When a dry expansion refrigeration plant under thermostatic control is switched off, the thermostatic expansion valve is left open and the evaporator is filled with liquid, some of which may flow along the suction line. Slugs of liquid can then enter the compressor and damage it when it next starts. To prevent this happening, a pump-down control system must be adopted. Under this mode of operation a solenoid valve, located in the liquid line immediately before the expansion valve, is closed when the controlling thermostat is satisfied but the compressor is not switched off. Instead, it continues to run, pumping refrigerant vapour from the evaporator and so boiling off all the liquid within it, until the suction pressure has fallen to a pre-set, low value at which a pressure sensor switches off the compressor. During the period that follows, leakage past the expansion valve and back-leakage through the compressor can cause the suction pressure to rise. If the compressor were started again when the upper end of the differential of the low pressure cut-out was reached, there would be a risk of short-cycling, which means the compressor pumping-down repeatedly at intervals of time dependent on the leakage rate. A non-cycling relay is therefore included in the control circuit to prevent this. As a consequence, the compressor cannot restart after pumping-down unless both the low pressure cut-out and the controlling thermostat require it and, of course, the necessary high pressure cut-out and oil pressure safety switch permit it.

The refrigerant pumped out of the evaporator is delivered to the condenser or the liquid receiver and stored as a liquid until needed. Although a shell-and-tube, water-cooled condenser usually has enough space in its shell to store the quantity of liquid involved, many others, for example air-cooled, have not. A liquid receiver is then essential and it must be big enough to contain all the refrigerant from the rest of the system. It is strongly recommended that direct-expansion air-conditioning systems include pump-down control, particularly if the compressor is likely not to be running for any length of time. These considerations apply equally to air-conditioning units of size larger than exemplified here. Pump down must never be used with water chillers.

With many compressors the crankcase contains a mixture of oil and refrigerant. In such instances, since the crankcase is under suction pressure, foaming may occur when first starting up after a shut-down or when a large fall in the refrigeration load causes a drop in suction pressure. Refrigerant in solution in the oil during the shut-down period suddenly reverts to the gaseous phase in the form of myriads of small bubbles throughout the body of the oil, forming a substantial quantity of foam that lacks the lubricating properties of oil and may cause damage to the bearings and cylinders. This is highly undesirable and is to be prevented by including an electrical crankcase heater, wired to be energised whenever the compressor is off. The oil in the crankcase is then generally free of dissolved refrigerant, since it has been evaporated away during the shut-down period. Crankcase heaters may be fitted within the crankcases of open or semihermetic machines but in the case of hermetic compressors they are wrapped around the outside of the crankcase.

Condenser pressure should be controlled to stabilise the performance of the system under conditions of low ambient air temperature. The two preferred methods are liquid level back-up in the condenser and, as a second best, variable condenser fan speed under solid-state control.

CASSETTE UNITS AND VRV SYSTEMS

So-called cassette units are ceiling-mounted room units connected to a remote air-cooled condensing set. They can work as air conditioning units or as heat pumps and offer the advantage that no floor space is occupied. The cooler coil cannot lie flat, in a plane parallel to the ceiling, because of the obvious problems this would present for condensate drainage. Instead, the unit fan is an open, uncased, forward-curved centrifugal impeller, with the cooler coil wrapped around its periphery (see Figure 2.5). Air is drawn from the room into the centre of the unit and blown outwards from the impeller, over the cooler coil and thence across the ceiling in four directions. Condensate flows down the fins of the coil and is removed by a small, plastic, condensate drainage pump having a submerged suction branch located in the drain tray under the coil. Up to 6 m of condensate lift can be provided. About 300 mm of ceiling void depth is needed to accommodate the units, with an additional projection beneath the ceiling into the room of as little as 40 mm. Alternative arrangements, with conventional condensate disposal by gravity, are also in use.

Indoor units with total cooling capacities in the approximate range of 4–20 kW are available. As many as 16 room units can be connected to a single, outdoor, air-cooled condensing unit and this is possible by adopting a variable refrigerant volume (VRV) system. The cooling capacity of each indoor unit is regulated by a motorised, electronic expansion valve which exercises proportional plus integral control over room air temperature.

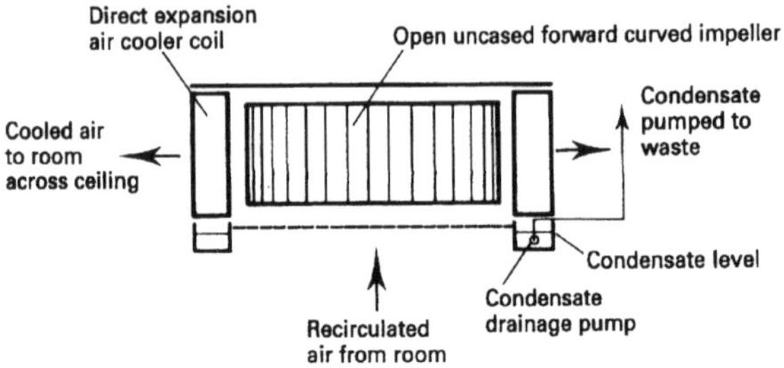

Figure 2.5 Simplified diagram of the cooler-coil arrangement for one type of cassette unit.

Temperatures, pressures and flow rates of refrigerant are measured throughout the system, and at the room units, and the information is transmitted to a microprocessor at the outdoor condensing unit. This then regulates the speed of the compressor, through an inverter, to match the capacity of the condensing set to the total load on the system. With a 50 Hz supply, the frequency fed to the compressor motor can normally be varied from 30 to 90 Hz with the possibility of 115 Hz for boost performance in the heat pumping mode.

It is claimed that total pipe lengths up to 80 m, of which not more than 40 m may be vertical, are possible between the indoor units and outdoor condensing set. The sizes of the copper refrigerant pipes (liquid line and suction line) are not very large and are significantly smaller than corresponding chilled water pipes. Pipe sizing must be carefully done to achieve good performance from the system and care must be taken to ensure that the oil is returned to the compressor, where it is needed. One manufacturer uses variable speed scroll compressors but reciprocating machines are also adopted.

WATER-COOLED AIR-CONDITIONING UNITS

When water-cooled units are fed from a cooling tower their capacities are related to outside wet-bulb rather than dry-bulb temperature, so condensing pressures are lower than with air-cooled units. The consequent reduction in compression ratio reduces their power requirements and, since they are working against a lower head pressure, they are likely to be quieter. Water-cooled units are available as single air-conditioning packages up to about 230 kW of refrigeration. Their catalogue ratings take into account the balance between evaporators and condensing sets and capacities are expressed in terms of condensing temperature, cooling water flow rate and entering temperature. Published ratings are sometimes in error and it is not always wise to adopt the lowest cooling water flow rate listed, even if the associated, necessary, low water temperature is available. A major advantage of such units over their air-cooled counterparts is a freedom in the choice of location, since a cooling tower can be on the roof and water pumped from it to units throughout the whole of the building. The condensers in the units are generally coiled-tube-in-shell, rather than shell-and-tube as used in the larger and better quality units, necessitating chemical cleaning when they become fouled with use. It is much better to ensure clean water for the condensers by interposing a plate heat exchanger in the cooling water circuit between the tower and the condensers, if it can be afforded, even at the cost of raising the condensing temperature by a few degrees.

A further significant advantage is that the onset of warmer weather has less impact on the capacity of a water-cooled unit than on an air-cooled one because the wet-bulb

Table 2.3 Typical performance of a commercial, single-package, water-cooled air-conditioning unit, for an airflow rate of 950 l s^{-1} entering the unit and a by-pass factor of 0.22

Entering state		41°C condensing			43°C condensing		
Dry-bulb (°C)	Wet-bulb (°C)	Total cooling (kW)	Sensible cooling (kW)	Compressor motor power (kW)	Total cooling (kW)	Sensible cooling (kW)	Compressor motor power (kW)
28	20	18.0	13.6	5.40	17.7	13.4	5.70
	19	17.4	14.5	5.33	17.0	14.2	5.63
	18	16.8	15.4	5.27	16.3	15.0	5.57
	17	16.2	16.2	5.20	15.7	15.7	5.50
22	18	16.8	8.4	5.27	16.3	9.7	5.57
	17	16.2	9.3	5.20	15.7	10.7	5.50
	16	15.5	11.8	5.13	15.2	11.6	5.43

temperature changes more slowly than the dry-bulb. For example, an increase of one degree in the dry-bulb at the design state to 29°C, at constant moisture content, causes a rise of only 0.4°C in the sling wet-bulb.

Table 2.3 shows abbreviated details of performance for a commercial, water-cooled, air-conditioning unit, which must be interpreted in conjunction with the requirements for cooling water listed in Table 2.4. Most catalogues also quote performances for air quantities that are roughly 25% above and below the nominal air quantity in Table 2.3.

Unfortunately, as with published data for air-cooled units, tabulated errors and inconsistencies are not uncommon. Interpolation within tables is generally permissible but extrapolation is decidedly risky and this is particularly so with low entering wet-bulbs when most of the cooling is sensible. When a cooler coil does only sensible cooling, because the mean coil surface temperature must be above the dew-point of the air flowing over the coil the apparatus dew-point does not lie on the saturation curve and the usual geometrical method of determining the by-pass factor cannot be used. There is often also doubt as to whether the air quantities quoted are referring to the state entering or leaving the unit. Since there is something like 5% difference in the air densities at these two states a corresponding, minimum inaccuracy is immediately set for the tabulated data.

Selecting a unit depends upon the choice of a temperature for the cooling water fed to the condenser in the unit and the source of this water is usually a cooling tower.

Table 2.4 Condenser cooling water requirements

Condensing temperature (°C)	Water flow rates in l s^{-1} for various entering water temperatures in °C						
	15.0	17.5	20.0	22.5	25.0	27.5	30.0
41	0.27	0.30	0.35	0.41	0.49	0.61	0.84
43	0.24	0.26	0.29	0.33	0.39	0.48	0.76

EXAMPLE 2.9

Determine the total and sensible cooling duties and the necessary cooling water flow rates of a water-cooled, air-conditioning unit with a performance as listed in Tables 2.3 and 2.4. Assume an ambient wet-bulb of 19.5°C sling outside and a state of 23°C dry-bulb, 17°C wet-bulb sling entering the unit. Take a condensing temperature of 41°C.

Answer
By interpolation, Table 2.3 shows that when condensing at 41°C the total cooling duty is 16.2 kW for the entry condition given and the sensible duty is 10.4 kW. The compressor motor power is 5.2 kW and so the heat rejected at the condenser will be 21.4 kW. A practical cooling tower can cool water to within 5–8°C of the ambient wet-bulb and it can, therefore, be assumed that a water temperature of 27.5°C is attainable. To condense at 41°C and give the listed duties a water flow rate of 0.61 l s^{-1} is needed (Table 2.4) and it can be calculated that this will rise to 27.5 + [21.4/(0.61 × 4.187)] = 35.9°C. This cooling range of 8.4 K is possible for a cooling tower.

It is generally unwise, because of lack of confidence in the accuracy of all tabulated data, to adopt the extreme values of performances. It is much better to choose plant items to operate in the middle of their listed duties. This is also true of cooling water flow rates where it is sometimes risky to select the lowest quoted flow rate, even if the associated low temperature is available. It is generally a good policy to try and verify the psychrometric performance of the equipment chosen because in doing so any inconsistencies in the tabulated information will become apparent.

EXAMPLE 2.10

Examine the psychrometric performance of the water-cooled unit considered in Example 2.9 and determine the sensible and latent heat gains that could be dealt with if the conditioned space were maintained at 22°C dry-bulb and 50% saturation.

Answer
The state of the air entering the unit is 23°C dry-bulb, 17°C wet-bulb, 9.618 g kg^{-1}, 0.8515 m^3 kg^{-1}, but if it is accepted that the unit handles a constant quantity of 0.95 m^3 s^{-1} of air at this state then the fan blows through the cooler coil. The mass flow rate of air is thus 0.95/ 0.8515 = 1.116 kg s^{-1}. The catalogue will quote fan power, probably with the assumptions of a clean filter and a wet coil. If it is assumed here that the fan power is 600 W, the temperature rise across the fan can now be calculated by Equation (2.3) as [(0.6/0.95) × (273 + 23)]/ 358 = 0.5 K. Thus, the state entering the cooler coil is 23.5°C dry-bulb, 9.618 g kg^{-1} and 48.11 kJ kg^{-1}. The enthalpy leaving the coil is, therefore, 48.11 – (16.2/1.116) = 33.59 kJ kg^{-1} and, as the by-pass factor is 0.22, the enthalpy at the apparatus dew-point is 48.11 – [16.2/(1.116 × 0.78)] = 29.50 kJ kg^{-1}. These states can be identified on a psychrometric chart and it can be seen that the state leaving the coil is 13°C dry-bulb, 8.116 g kg^{-1}. In the absence of supply air ducting outside the conditioned space, this is also the supply state. Consequently

Sensible gain absorbed in the room = [0.95 × (22 – 13) × 358]/(273 + 23) = 10.34 kW

Latent gain absorbed in the room = [0.95 × (8.366 – 8.116) × 856]/(273 + 23)
= 0.687 kW

This unit would be able to cope with the sensible heat gains of 5.93 kW calculated and used in Example 2.8 for an air-cooled unit but, because the latent cooling capacity is less than the figure of 0.95 kW calculated in the same example, the humidity in the room would be higher than the value of 50% assumed. A precise solution is not readily obtained but the indications are that the unit capacity would balance the room load at a comfortable condition of 22°C dry-bulb and less than 60% saturation, in the process of which the state of the air entering the cooler coil would be a little different from that used in Example 2.9.

WATER LOOP AIR CONDITIONING/HEAT PUMP UNITS

These units are basically water-cooled, room air-conditioning units with a manually or thermostatically selected heat pump facility. Water is circulated in a two-pipe system and units either reject heat into the room from their condensers when they are cooling or absorb heat from it when heating. Since units in different parts of a building may, in a temperate climate, be both heating and cooling simultaneously, energy is conserved by transference through the water loop. Most of the time there will be a net imbalance of energy, and surpluses and deficits of heat in the water system must be corrected by a cooling tower and a boiler. Because such units are sensitive to variations in water flow, it is absolutely essential that plate heat exchangers be incorporated to separate the dirty water circuits through the cooling tower, particularly, and the boiler, from the necessary clean water circuit through the units. A closed cooling tower could be used directly, instead of an open tower and a plate heat exchanger, but closed towers are invariably bulky, heavy and expensive. Further, their water circuit usually needs a glycol additive for winter-frost protection and this imposes penalties on heat transfer.

The fans in the units may be run at high, low or medium speed with typical airflow rates of 120–250 $l\ s^{-1}$, depending on the fan speed. For the case of medium fan speed and air entering the units at 22°C, nominal total cooling capacities are in the range 1.4–4.2 kW, sensible heat ratios are 0.94–0.81 and electrical power consumptions are 0.59–1.63 kW, implying cooling coefficients of performance of 2.4–2.6. The units are quite suitable for dealing with the sensible gains in the peripheral areas of buildings to a depth of 6 m inward from the windows.

Heating capacities for an entering air temperature of 19°C and a medium fan speed are 2.33–6.03 kW with respective electrical power consumptions of 0.53–1.68 kW, implying heating coefficients of performance of 4.4–3.6. Thus, when heat pumping from the warm side of the building to the cool side, about one-quarter to one-third of the heating is from an electrical source, by virtue of the power consumed by the unit compressors.

Fresh air should be provided by an auxiliary, central ventilation system handling filtered, tempered air but, since one advantage of using terminal units of this type is that plant space is cut to a minimum, fresh air is commonly admitted through a hole in the wall behind the unit. Although having a central ventilation system is better because air distribution and filtration are superior, as much as 22% of the supply air quantity can be introduced through the unit locally.

Most units are available as floor or wall-mounted options or for installation at high-level above suspended ceilings. Some terminals are designed for use with auxiliary, local duct systems, fitted with acoustic linings or attenuators. Selection procedures are likely to be similar to those for water-cooled air-conditioning units for summer operation. In winter, on the other hand, allowance must be made for the work done by the refrigeration compressor when assessing the boiler power. The critical feature of their performance is the water flow rate: for some units this can be as low as $0.06\ l\ s^{-1}$ at 27°C but for others twice this quantity is needed. During commissioning it is essential that the correct water flow rate be achieved and demonstrating this may be difficult; it is of little use to measure the water temperature rise because this is dependent on the cooling load which in turn depends on many things and may be even more difficult to determine. It is better to use reliable water regulation devices, possibly to aim at having about 10% more than the nominal water quantity at each unit, regulation downward then being practical. Flushing through the piping system, with the terminals out of the circuit, is vital before they are connected.

It is feasible, sometimes, to use much of the distribution piping in a two-pipe LTHW heating system, after checking that its condition and size is adequate in all respects, if the building served is to be upgraded from heated to air-conditioned, by replacement of the radiators with water loop terminal units and by the addition of a cooling tower, plate heat exchangers, a condensate drainage system in plastic or copper from the units, pumps, automatic controls, switchgear and, possibly, an auxiliary ventilation system. It is good practice to introduce random start relays so that large numbers of unit cannot come on together. Controls should include a minimum flow water temperature cut-out, with a probable set-point of about 15°C. Units themselves are generally controlled by cycling the compressor on–off from a return air or room thermostat, the supply fan running continuously.

Figure 2.6 is a schematic diagram of the piping arrangement for a water loop system of air conditioning/heat pump units. If plate heat exchangers were not used, the units would operate in the cooling mode with 27°C flow and 32°C return water temperatures from the

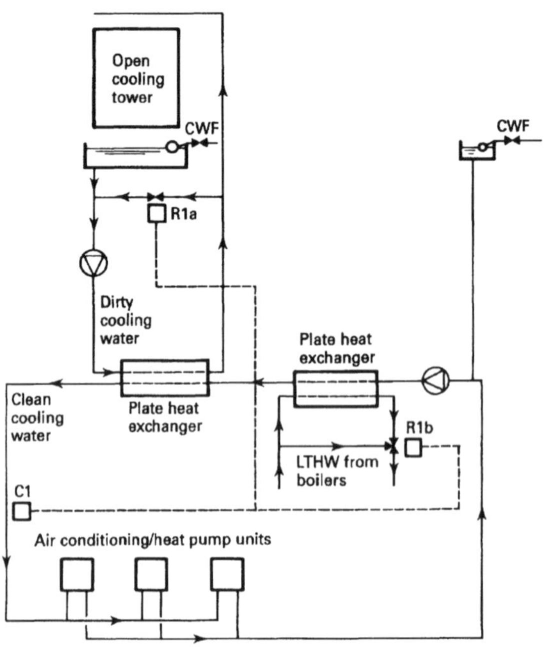

Figure 2.6 Schematic piping diagram for a system of water loop air conditioning/heat pump units.

cooling tower. In the heat pumping mode they would use flow and return temperatures of 24 and 19°C. Since dirty water cannot be allowed for the units, plate heat exchangers must be used and these operating temperatures become 28 and 33°C for cooling with 23 and 18°C for heat pumping. R1a is a motorised, modulating, butterfly valve with a pressure drop when fully open equal to the static lift for the cooling tower and R1b is a three-port mixing valve with an authority of between 0.3 and 0.5 when fully open. C1 is an immersed temperature sensor that regulates R1a and R1b in sequence.

2.4 Constant volume re-heat and sequence heat systems

The constant volume re-heat system is the basic system of air conditioning and nearly all other systems are a variation. Figure 2.7 illustrates the plant and psychrometry. Air at state *R* is extracted from the conditioned room and rises in temperature to state *R′* because of

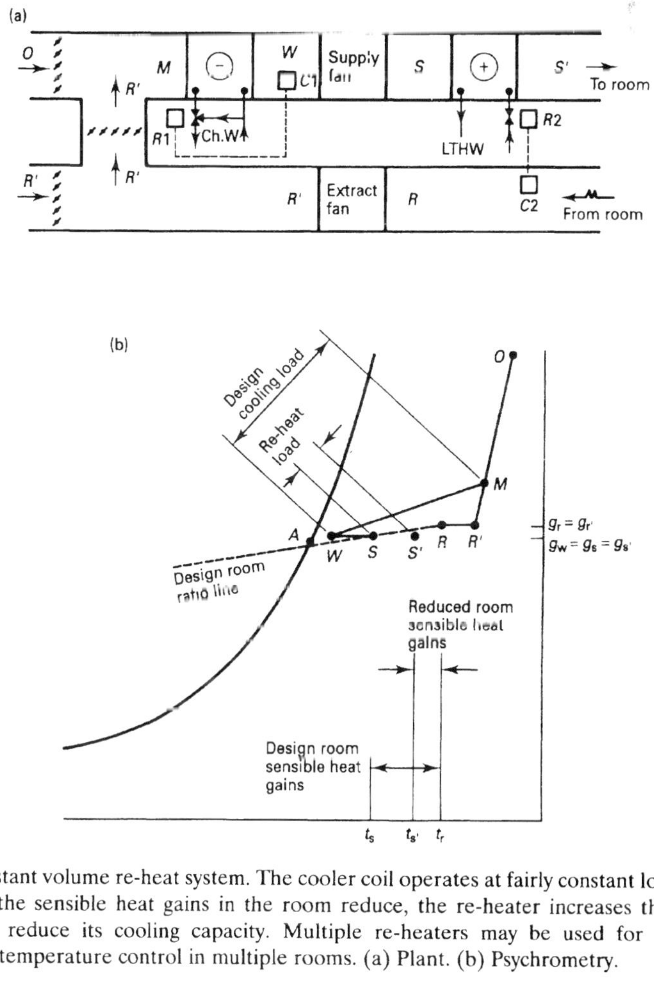

Figure 2.7 Constant volume re-heat system. The cooler coil operates at fairly constant load and cools to state *W*. As the sensible heat gains in the room reduce, the re-heater increases the supply air temperature to reduce its cooling capacity. Multiple re-heaters may be used for independent temperature control in multiple rooms. (a) Plant. (b) Psychrometry.

extract fan power and any other heat gain to the return air system. This air mixes with fresh air at state *O*, forming state *M*, and is cooled and dehumidified to state *W* by the cooler coil. Under design conditions the temperature of the air rises by one or two degrees because of the supply fan power and is delivered to the conditioned room at state *S*. No re-heating occurs. When the sensible heat gain in the room reduces, the same airflow rate is delivered but the supply temperature is increased by re-heating, in order to balance the cooling capacity of the air with the reduced sensible heat gain. The system is simple but wasteful in both capital and running cost. No advantage can be taken of diversity in the components of the sensible heat gain: as the sensible gain reduces the cooling capacity of the supply air is reduced by re-heat. The refrigeration plant continues to run at virtually a constant duty, unwanted cooling being cancelled by wasteful re-heating. Since the benefits of diversity cannot be allowed the refrigeration plant is oversized. Similarly, the running costs are high. The advantages of the system are simplicity and a close control over room temperature. The latter is unnecessary for most commercial applications but offers a required solution for temperature control with industrial conditioning. In this case, if dry steam injection is provided, preferably as close as possible to the air supply terminal for the room, good control over room humidity can also be obtained. Multiple re-heaters can be used in order to give individual temperature control in multiple rooms.

In commercial applications, the objection that no advantage is taken of diversification in the sensible heat gains can be dealt with by arranging for cooling and heating to be in sequence. Figure 2.8 illustrates this. When recirculated air is handled, as is usual, illustrating the psychrometry is complicated and hence, to simplify the picture, it is assumed here that 100% fresh air is handled. The cooling capacity of the cooler coil is controlled in sequence with the heating capacity of the heater battery. As the room temperature tends to rise on fall in the sensible heat gains, sensor C1 reduces the cooling capacity by opening the by-pass port of the motorised three-port valve R1a. Eventually no chilled water is flowing through the cooler coil and, subsequently, the re-heater valve, R1b, is progressively opened, as heat gains turn into heat losses. When the cooler coil valve, R1a, is moved towards the open by-pass position the state of the air leaving the cooler coil slides up the dashed line in Figure 2.8(b) from *W* to *W'*. This gives a warmer supply air temperature, $t_{s'}$, but the heat and mass transfer characteristics of a cooler coil are such that the moisture content of *W'* is greater than that of *W*. It is clear that this will result in the room air moisture content also rising, the room state moving from *R* to *R'*. From a comfort point of view this is not very significant and for commercial applications it is customary to tolerate this rise in humidity up to 60% or thereabouts. A high limit humidistat, C2b, then interrupts the control sequence and allows the cooler coil to respond to the humidistat and bring the humidity down to about 50%. Meanwhile, the system operates as a re-heat system until the high humidity is corrected, whence it reverts to sequence control.

Using multiple re-heaters with sequence control poses difficulties. It is certainly possible to have two re-heaters: the heater that requires the more cooling operates in sequence with the cooler coil, responding to the thermostat in the room sensing the higher temperature. The heater for the other room then works as a re-heater, out of the sequence. It is evident that the larger the number of rooms treated in this way the less the advantage of having sequence control. The only reduction in the refrigeration load throughout the year is that resulting from a fall in the outside enthalpy, as the weather changes, which will reduce the fresh air load. Thus the system is wasteful in its use of energy. It is best used for industrial applications or for small commercial duties, where the waste of energy may be insignificant.

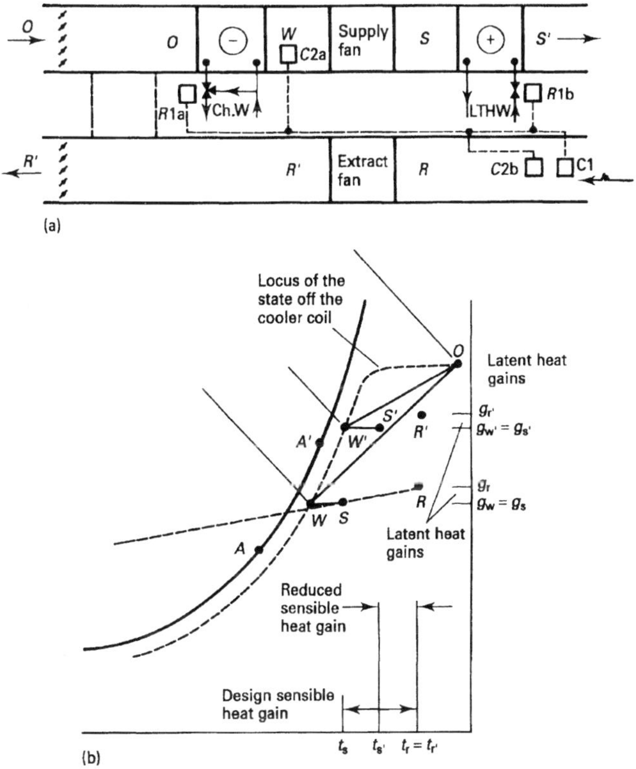

Figure 2.8 Constant volume with heating in sequence with cooling. Temperature sensor C1 controls the heater battery in sequence with the cooler coil, through R1a and R1b. C2b is a high limit humidistat that interrupts the sequence upon rise in room humidity. The plant would normally use recirculated air but the psychrometry is then complicated. For ease of illustration, the psychrometry shows the operation when 100% fresh air is used. (a) Plant. (b) Psychrometry.

EXAMPLE 2.11

Determine the refrigeration load for the hypothetical office block (Figure 1.1), assuming it is air conditioned by a constant volume re-heat system. Make use of the appropriate data for Examples 1.8–1.11, as necessary. Use CIBSE solar gains from the tables in the Appendix. In order to accommodate the ducting of an all-air system in the building it might be necessary to distribute the supply air at medium velocity and a reasonable assumption for the fan total pressure of the supply fan is 1.5 kPa, with the driving motor in the airstream. Hence the temperature rise for supply fan power is $1.2 \times 1.5 = 1.8$ K according to Jones (1994). Allow 1.7 K for the supply duct heat gain.

Extract systems are much simpler than supply systems and less extensive, hence low velocity air distribution is the norm and the fan total pressure is likely to be no more than 200 Pa, with the fan outside the airstream, giving a temperature rise of 0.2 K. Take the fresh air allowance as $1.4\ \mathrm{l\ s^{-1}}$, referred to the treated floor area of $13996.8\ \mathrm{m^2}$.

Answer

Referring to Example 1.11 and using the same design data the refrigeration load for the building when conditioned by a constant volume re-heat system can be determined, remembering that the diversity factor for people, lights and business machines is unity.

Sensible heat gains:	watts
Total sensible gain through the envelope:	535 183
People: $[(86.4 \times 13.5 \times 12)/9] \times 90$:	139 968
Lights: $(86.4 \times 13.5 \times 12) \times 17$:	237 946
Machines: $(86.4 \times 13.5 \times 12) \times 20$:	279 936
Total sensible heat gain:	1 193 033

Latent heat gains:	watts
Infiltration: $(0.8 \times 0.5 \times 36392)(10.85 - 8.366)$:	33 248
People: 1555×50:	77 750
Total latent gain:	110 998

A reasonable off-coil temperature is probably 12°C dry-bulb. We must add an allowance of 1.8 K for supply fallen power and a further 1.7 K for duct heat gain to this in order to give a practical supply air temperature. Hence the supply air temperature is 15.5°C, dry-bulb.

Using Equation (2.3) the necessary supply air quantity is calculated

$$\dot{v}_{15.5} = [1193.033/(22 - 15.1)] \times [(273 + 15.5)/358] = 147.9 \text{ m}^3 \text{ s}^{-1}$$

Using Equation (2.4) the supply air moisture content is established

$$g_s = 8.366 - (110.988/147.9) \times [(273 + 15.5)/856] = 8.113 \text{ g kg}^{-1}$$

The practicality of this must be checked on a psychrometric chart.

Noting that the minimum fresh air allowance is $1.4 \times 13\,996.8/1000 = 19.6$ m^3 s^{-1}, which is 13.25% of the supply air quantity the mixture state, *M*, onto the cooler coil is determined

$$t_m = (0.1325 \times 28) + (0.8675 \times 22) = 22.80°\text{C}$$

$$g_m = (0.1325 \times 10.65) + (0.8675 \times 8.366) = 8.669 \text{ g kg}^{-1}$$

The return air state, *R′*, has a temperature of 22.2°C and a moisture content of 8.366 g kg^{-1} and its enthalpy is then conveniently calculated by means of Equation (1.14):

$$h_{r'} = [(1.007 \times 22.2) - 0.026] + 0.008366[(2501 + 1.84) \times 22.2] = 43.59 \text{ kJ kg}^{-1}$$

Hence

$$h_m = 0.1325 \times 55.36 + 0.8675 \times 43.59 = 45.15 \text{ kJ kg}^{-1}$$

The off-coil state is 12°C dry-bulb, 8.113 g kg^{-1}.

A psychrometric chart, tables, or Equation (1.14) give the enthalpy as 32.53 kJ kg^{-1}.

Reference to tables or a chart shows that the specific volume is 0.8279 $m^3\ kg^{-1}$ at the supply state.

Figure 2.9 illustrates the psychrometry and shows that the contact factor for the cooler coil is 0.9 which is a practical value and implies six rows of tubes. The design cooling load is then calculated:

$$\text{Design cooling load} = (147.9/0.8279)(45.15 - 32.53) = 2254\ \text{kW}$$

A check on this load is as follows:

	kW
Sensible heat gain:	1193
Latent heat gain:	111
Supply fan power and duct gain $147.9 \times 3.5 \times 358/(273 + 15.5)$:	642
Extract fan power $0.8675 \times 147.9 \times 0.2 \times 358/(273 + 15.5)$:	32
Fresh air load $(0.1325 \times 147.9/0.8279)(55.36 - 43.59)$:	279
	2257

This is to be compared with the total refrigeration load calculated in Example 1.11, which was 1607 kW when a fan coil system was used and only 71% of the design cooling load of the above constant volume re-heat system.

Constant volume re-heat systems should never be used for applications of this size.

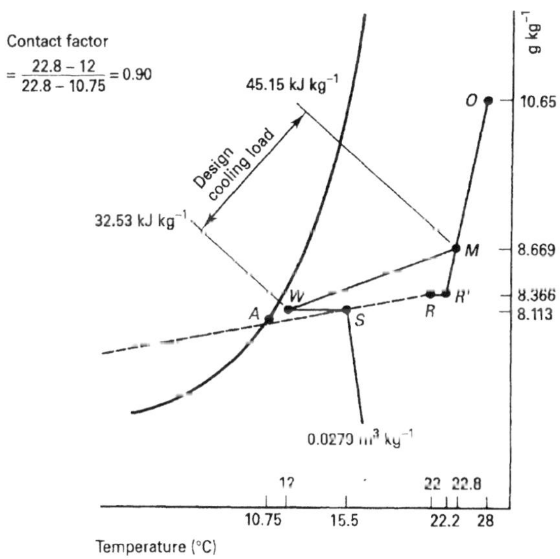

Figure 2.9 Psychrometry for Example 2.11.

2.5 Roof-top units

Although almost any weatherproofed, air-handling or air-conditioning unit can be located on a roof, the term 'roof-top unit' tends, by common usage, to have a special meaning. It is generally understood to denote a self-contained, air-conditioning unit that comprises mixing box, filter, direct-expansion cooler coil, air-cooled condensing unit, direct-fired air heater battery (gas or oil) and supply fan with motor, drive and switchgear. The unit is weather-proofed, prewired and factory tested. It only requires electrical connexions and a gas or oil feed, after erection on the roof, to operate. Condensate from the cooler coil runs to waste on the roof, preferably in a gully.

Such units are available in a range of capacities from 5 to 100 kW of refrigeration and from 17 to 150 kW of heating. The filters provided as standard are generally washable polyurethane foam or oil-impregnated glass fibre, but electrical filters can also be obtained. Although end connexions for ductwork are possible the standard and most convenient arrangement is to have supply and extract spigots on the underside of the unit to connect with ductwork running beneath the roof.

Anti-vibration mountings for the moving parts are generally built into the unit, as is also thermal/acoustic lining, but before locating a unit it is always wise to establish that the mass, span and stiffness of the roof is adequate for the prevention of a noise and vibration problem in the occupied space below (see Sections 7.12 and 7.25).

The advantages of roof-top units are that they are simple and cheap, no boiler plant is needed, duct distribution systems can be reduced to a minimum, multiple units can be used to provide multiple zone control and they are ideal for single-storey commercial buildings in a highly competitive application, such as a shopping centre development. One of their limitations is that the degree of control offered is simple, just on/off cooling, in sequence with one- or two-step heating, but this is not necessarily a disadvantage for the correct application. Two-speed fans are sometimes available, permitting operational economy in winter when the refrigeration plant is off. It is also possible to include an assembly of variable-position, automatic, motorised dampers that offer varying proportions of fresh and recirculated air and allow the refrigeration plant to be switched off when the outside air is cool enough. Some manufacturers will not let their units be operated with 100% fresh air in winter because, in cold weather, flue gases condense on the primary surfaces of the direct-fired heater battery, with a consequent risk of corrosion. Other units, designed specifically for the American market, have burners that are not easily set up to meet the UK need to establish a flame within three seconds of ignition.

2.6 Variable air volume systems

HISTORY AND DEVELOPMENT

Since the aim is to match sensible cooling capacity to varying sensible heat gain or loss, most commercial air conditioning systems run at partial load for the majority of their operational lives. The requirement in modern office blocks is for an air conditioning system that provides thermostatic control on a modular basis and in a way that gives flexibility in partition re-arrangement: future tenants must be able to move partitions to suit their needs, without compromising control over temperature in an office, no matter how many building modules it occupies. Hence system development has been to develop terminal units that can offer the individual thermostatic control required. Constant volume re-heat systems provide terminal units that vary the supply air temperature in response to changes in the sensible heat gain or loss but the system is intrinsically wasteful in both refrigeration power and thermal energy (see Section 2.4). An alternative is to keep the supply air temperature constant and to reduce the mass flow rate of air supplied as the sensible heat gains diminish, in accordance with the thermal balance implied by Equation (2.3). In this way it would appear that energy need not be wasted by cancelling unwanted cooling with re-heat.

There is a further advantage implied by the definition of fan power:

$$W_f = P_{tF}\,\dot{v}_t/\eta \tag{2.7}$$

where W_f is the fan power (W or kW), p_{tF} is the fan total pressure (Pa or kPa), $\dot{v}_t$ is the volumetric airflow rate ($1\ s^{-1}$ or $m^3\ s^{-1}$) and η is the fan total efficiency, as a fraction.

Equation (2.7) suggests that if the average supply airflow rate is, say, half the summer design rate, and if total pressure loss in the system is proportional to the airflow rate squared, then the annual averaged fan power will be one-eighth of the design fan power, with obvious savings in electrical energy consumption. The reality is not as good as this because the fan total pressure cannot be reduced as much as suggested and there is a limitation on the minimum airflow achievable, set by the air distribution system, but savings in fan power are possible in properly designed and operated systems.

Early designs for office blocks used two-pipe, perimeter-induction systems for the peripheral strip of floor area, with a width of 4.5–6.0 m inward from the window (see Section 2.10). Deeper areas, and any core, were treated by a separate, constant volume, re-heat system, usually with two re-heaters per floor, to suit the letting arrangements. With the development of the variable air volume system (VAV) the re-heat system for the core was replaced by the VAV system This was soon modified: the perimeter-induction system was dropped and the VAV installation extended to deal with the whole floor. The VAV system has no intrinsic ability to provide heating and this introduced certain winter heating problems but these have since been largely solved.

ROOM AIR DISTRIBUTION

Conventional systems for the supply of constant volumes of air at about 13°C in rooms with ceiling heights that range from 2.6 m up to about 3.5 m are well established and are dealt with in detail in Chapter 5. The supply of variable volumes of air to a room is also discussed in detail in Section 5.16. The essence of the difference between constant and variable air supply is that the Coanda effect (that counters the anti-buoyancy of the cold jet and keeps it on the ceiling) diminishes as the supply airflow rate is turned down, in response to thermostatic control when the sensible heat gains reduce. Figure 5.13 illustrates this: the risk is that the cold supply air might be dumped into the room and cause complaints of draughts when the flow rate is reduced to its minimum value, say 40% of its design value, and not re-attach to the ceiling until it had risen to, say, 60% of this. It follows that the design of VAV supply air terminals has been greatly influenced by these considerations.

In this respect, considerable difficulty has been experienced with variable air distribution because the sensible heat gains from business machines have been over-estimated in the past. Using the data from Example 1.9 suppose that, mistakenly, an allowance of 50 W m^{-2}, referred to the floor area, had been made for the heat gain from business machines the design sensible gain would have risen from 1394 to 1826 W. Even under design conditions the terminals would have been handling only 76% of the rate for which they had been selected.

Forty per cent of 1826 W is 730 W and this represents 52% of 1394 W. The terminals would cause draughts at low duties.

Catalogue data for VAV units should always quote maximum and minimum throws. It is essential that units are not selected to have throws more than the minimum listed, otherwise draughts are inevitable. Where a problem like this arises during the layout planning stage of a design, it is often possible to solve it by rearranging the positions of the air distribution units. For example, if a linear slot VAV diffuser with a minimum listed throw of 1.95 m when handling 80 $l\,s^{-1}$ is arranged with its axis along the major centre-line of the 6 × 2.4 m module in the hypothetical office block, draughts will occur because the distance to the walls of the module, or to the opposing airstream from the VAV unit in the adjoining module if there is no partition separating them, is only 1.2 m. On the other hand, if the linear diffuser is turned through 90°, the distance to the window or the corridor wall is 3 m, which exceeds the minimum listed throw and no draughts will occur (see Figure 2.10).

PRESSURE DEPENDENCE

If one air terminal throttles its supply air flow rate, when several VAV terminals are fed from a common duct, less air flows in the main duct and the static pressure therein rises, causing other terminals to deliver more air, even though there is no thermostatic demand at the terminals for such an increase. If nothing is done to deal with this, the system is pressure dependent and is unstable. Such systems should never be used with medium or high velocity

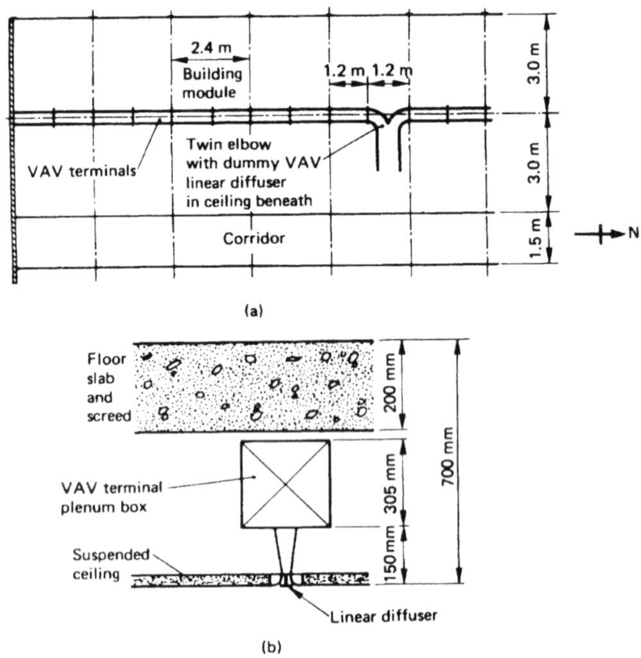

Figure 2.10 (a) VAV terminal and duct layout. (b) Cross-section of a VAV terminal in a ceiling void.

ducted air distribution because static pressure in the ducts will vary considerably with a consequent unpredictability in the performance of the VAV terminals. However they may be used in small, low velocity systems, because static pressure variations will be comparatively small. There will still be some fluctuation in the performance of individual terminals but this must be recognised and accepted.

If a VAV terminal has an in-built pressure regulator that maintains a constant static pressure on the upstream side of the throttling section, the performance of individual VAV units will be stable, even though the duty of other terminals is varied under thermostatic control. Such terminals are independent of pressure variations and are used in high, medium and low velocity systems of ducted air distribution. The principle is illustrated in Figure 5.11(b). Many different types of pressure-independent terminal are available, using self-acting techniques, counter weights and springs, electrical actuation, pneumatic actuation, and so on. Variable geometry, pressure independent, VAV terminals are best.

HEATING

A variable air volume system has no inherent heating ability: a fall in air temperature causes the terminal unit to throttle the supply airflow rate and this continues, as the sensible heat gains diminish, until the unit is eventually delivering its minimum stable airflow rate. It would certainly be possible to warm the air in the duct to a temperature higher than the room temperature (in order to offset heat loss) and to reverse the action of the thermostat, so that the terminal increased the air flow rate on fall in temperature. However, this would not be satisfactory if the duct system was feeding air to some units which were dealing with heat gains and some with heat losses. For example, a shadow falling across the face of a building, on a cold but sunny spring day, could give heat gains to some offices and heat losses to others.

Three heating methods that do not sacrifice thermostatic control have been in common use.

(1) *VAV with compensated perimeter heating*. Radiators, finned tube, convectors or the like, are located around the perimeter and fed with LTHW, the flow temperature of which is compensated against outside air temperature to deal with the heat loss through the building envelope and by infiltration. Meanwhile the VAV system copes with the heat gains from solar radiation, people, business machines and lights. This method is the cheapest of the three and is fairly effective. Since the heating output of the perimeter system is related to outside air temperature no advantage is taken of heat gains that could offset heat loss. Hence the method wastes some thermal energy. On the other hand, heat is supplied where it is wanted, at the perimeter, and downdraughts are dealt with. The system is also very good at providing boosted heating, before occupancy commences, when systems are operated intermittently.

Attempts have been made to take advantage of the benefit of casual gains by fitting the perimeter heating units with thermostatic radiator valves, on a modular basis, while retaining the overall compensated control. Results have been mixed. In principle it is not a good solution because thermostatic radiator valves are self acting with a slow response and are not in the same control loop as the VAV terminals, which have a comparatively fast response. Nevertheless, some good results have been reported.

(2) *VAV units with terminal re-heaters*. Although more expensive in capital cost then the method described above, this has tended to be the favoured solution. On fall in room

temperature the VAV terminal throttles to its minimum airflow rate. Upon further fall in temperature the airflow rate stays constant and the re-heater warms the air. Control over the airflow rate and the re-heater output is in sequence and may be simple proportional or proportional plus integral. Sometimes, when electrical re-heating is used, control over the heater is two-position. Since only the minimum airflow rate is re-heated, the system does not waste thermal energy in the way that a constant volume re-heat system does.

A difficulty is that when, as is usual, the VAV terminal is above the suspended ceiling, heat is provided in the wrong place: the best place is beneath the windows where downdraughts, cold infiltrating air from outside and so-called cold radiation, can be dealt with. One way of dealing with this is to locate the VAV terminal so that the slot of its linear ceiling diffuser is close to, and parallel with, the window line. Cool air is blown inwards across the ceiling as it is throttled to its minimum value. After this, while the airflow rate is held constant, warm air is blown down the window, mixing with any cold downdraught and infiltrating air and tempering the effect of cold radiation. An alternative is to locate the VAV terminal, with its re-heater, in the floor and to blow warm air at its constant minimum flow rate, up the perimeter wall and window. This may be less attractive because of the floor area occupied. However, with the increased use of double glazing in commercial offices conforming with the Building Regulations, problems with downdraughts at windows will be reduced.

(3) *Double duct VAV*. As the sensible heat gains in the room reduce and the temperature falls, the VAV terminal receives air from the cold duct (see Section 2.7) and throttles the flow rate to its minimum value. No air is drawn from the hot duct. On further fall in room temperature air is drawn from the hot duct and mixed, under thermostatic control, with air from the cold duct, the total delivery of air to the room staying constant at its minimum value.

The system is expensive in both capital and running cost. It uses a lot of building space because of the need to accommodate the extra duct and any problems that a double duct system may have are added to any that a VAV system may have. It has not been popular.

SYSTEM TYPE

There are many different forms of VAV terminal unit available on the market, but the following two typify a basic difference of approach.

(1) A self-contained unit, often actually forming part of the air distribution ducting, in which air is handled directly at medium velocity and high pressure in the process of being thermostatically controlled.
(2) A split system that uses two devices, the first a pressure-reducing valve and the second a thermostatically controlled, motorised, variable volume diffuser.

An example of the first type comprises a linear diffuser, 75 mm in width, flush-mounted in the suspended ceiling and fed from an acoustically lined plenum box above, several such boxes usually being connected in series to form part of the distribution ducting (see Figure 2.10). Proportional thermostatic control is self-acting, the static pressure of the air in the duct itself being used as a power source, providing it is above a minimum value of about 100 Pa and less than a maximum of about 1250 Pa. There are thus no electrical or pneumatic connexions required and the thermostatic element, a bimetallic strip, and its set point adjustment are unobtrusively located in the end part of the unit, where entrained room air permits a representative assessment of room temperature. Constant volume as well as variable volume units are available and it is also possible to have as many as five slave units linked with one master control unit, where the thermostat is fitted, to provide automatic control of larger rooms at a reduced cost.

There are two bellows in such units and each may be thermostatically controlled from its own side of the diffuser slot, independently of the other. The resulting important advantage of this type of unit is that there is complete freedom for partition rearrangement, even after the system has been commissioned and the building occupied. Also, all the control modifications may be done from below and ceiling panels need not be removed, it being unnecessary to change duct connexions to the units since they will have been chosen to provide controlled, flexible air conditioning on a modular basis; partitions can be erected along the centre line of the linear diffuser and independent control achieved in the offices on each side.

The other principal form of VAV system is available in many versions, including motorised, circular, square or rectangular ceiling diffusers and pneumatically-actuated side-wall grilles. There are electric reheat options and slots or grilles that permit distribution upwards at the window sills. The pressure reducing valves can cope with upstream pressures as high as 1 kPa when they are of the inflatable, neoprene bellows type that are operated by compressed air at a pressure of 100 kPa. The diffusers are able to deal with upstream duct pressures from about 12 to 60 Pa and the side-wall grilles can manage with 25–125 Pa. A modification to units of this type (2) introduces an induction principle to the diffusers. The unit comprises a plenum chamber fitted with high velocity nozzles inducing air from the ceiling void, which is warmed by the recessed light fittings, at a rate that is varied under thermostatic control. Obviously, no heating is possible until the lights are switched on.

Self-acting, square and linear VAV ceiling diffusers are also a possibility for small, low velocity installations, where pressure dependence is tolerated. These terminals use the thermal expansion of a solid or liquid within the temperature sensor to achieve thermostatic control and also to provide the power source for throttling the diffuser. Volumetric airflow rates, per diffuser, are from 50 to 445 l s^{-1} for square diffusers and from 35 to 115 l s^{-1} for linear slot diffusers.

VAV units of type (1) have a typical performance as shown in Table 2.5. The minimum pressures quoted are the static pressures in the duct immediately upstream of the unit. The maximum allowable static pressure in the duct is about 1250 Pa. Each unit has a nominal length of 1200 or 1500 mm and a width of 290 mm, but the depth of the plenum box varies according to the air quantity handled (Table 2.6) To obtain the overall height of the unit, the height of the approach to the linear diffuser, i.e. 150 mm, must be added to the plenum depth (see Figure 2.10).

Table 2.5 Typical performance of a self contained VAV unit

	Air quantity (l s^{-1})										
	20	30	40	50	60	70	80	90	100	110	120
Minimum throw (m)	0.6	0.6	1.0	1.3	1.6	2.0	2.3	2.6	2.9	3.2	3.6
Maximum throw (m)	1.5	2.4	3.2	4.0	4.8	5.6	6.5	7.3	8.1	8.9	9.7
Minimum pressure (Pa)	370	330	290	255	225	190	165	135	150	100	100

Table 2.6 Variation in plenum box depth in a self-contained VAV unit

	Unit size			
	1	2	3	4
Maximum inlet air quantity ($l\ s^{-1}$)	330	470	710	940
Plenum box depth (mm)	168	216	305	400

EXAMPLE 2.12

Referring to Example 1.9 and the design brief in Section 1.2, as necessary, determine the maximum supply air quantities needed for the modules on intermediate floors of the hypothetical office block. Assume that the dry-bulb temperature leaving the cooler coil is 10°C and allow 3 K rise for duct heat gain and fan power. Assume also that a drawthrough cooler coil is used. The system is VAV.

Answer
From Equation (2.3), the maximum required supply air quantities are:

West-facing modules: $[1394/(22 - 13)] \times [(273 + 13)/358] = 123.7\ l\ s^{-1}$

East-facing modules: $[1330/(22 - 13)] \times [(273 + 13)/358] = 118.1\ l\ s^{-1}$

Table 2.5 shows that two units will be needed, each handling 61.8 and 59.0 $l\ s^{-1}$ with minimum throws of 1.7 and 1.6 m from the west and east modules, respectively. A satisfactory arrangement is for the units to be positioned on the smaller centre-line of each module with their own centre-lines parallel to the windows. The minimum throws are then less than the distance to the corridor wall or the window and no draughts will be felt (see Figure 2.10). Each unit has a nominal length of 1.2 m and so a pair connected in series can be accommodated in the 2.4 modular width.

The maximum air quantity required for the whole building, i.e. the supply fan duty, will not be the sum of the maximum individual values. Instead, it is necessary to evaluate the maximum simultaneous sensible gain to all the modules, taking proper account of diversification for the loads from lights, people and business machines. Example 1.11 showed that the maximum sensible gain for the whole building, including corridors, was 1026 kW. In fact, it can be shown that the biggest simultaneous gain is 1031 kW and occurs at 16.00 h sun time in July, but the slight error is insignificant in comparison with the doubt that exists in the diversity factors chosen.

The supply fan should therefore deliver a maximum quantity of

$$[1026/(22 - 13)] \times [(273 + 13)/358] = 91.1\ m^3\ s^{-1} \text{ at } 13°C$$

It is most important to appreciate that the benefits of diversity in the solar heat gain through windows can only be achieved if the air-handling plant and main duct system serve opposing faces of a building, for example, east and west.

The difference between floor-to-floor and floor-to-ceiling heights is 700 mm, which must include the structural slab and screed, a void for ducts and extract-ventilated light fittings, the suspended ceiling and its supports. Allowing 200 mm for the slab and screed, plus 75 mm for the ceiling structure, leaves a clear space of 425 mm beneath the soffit of the slab, although it must be remembered that, in places, downstand beams may encroach upon this and due provision must be made when planning the duct layout. Table 2.6 shows that a size 3 plenum box is the largest that will fit (Figure 2.10b). Such a unit can handle a maximum of 710 $l\,s^{-1}$ at its inlet and this sets a limit on the number of VAV terminals that may be fed in series, i.e. 11 units for the west face and 12 for the east. Ducts can be arranged to feed the end of a series group of nine units, for the west face, or to feed the middle of a series group of 18 units at a twin-elbow. Figure 2.7(a) shows this and also that the half-module containing the twin-elbow will receive no air, which is unavoidable. If the minimum size of an office is two modules wide, say, then where a twin-elbow was present the three live linear diffusers would deliver a total of $4 \times 62 = 248\ l\,s^{-1}$, or 83 $l\,s^{-1}$ each, when under thermostatic control during periods of peak heat gain for the west face.

FAN-ASSISTED VAV TERMINALS

The problem of draughty conditions at reduced airflow, experienced with some forms of VAV terminals, has led to the introduction of fans in terminal units, aiming to prevent the supply airflow from leaving the ceiling and dumping. There are two versions available:

(1) *Series flow*. This is illustrated in Figure 2.11(a). The fan handles a constant airflow rate and consequently the terminal is not a variable volume unit. As the room temperature falls the VAV regulator throttles the flow of ducted, cool air and the balance of airflow is drawn from the ceiling void to preserve constant flow through the fan. After the VAV regulator reaches its minimum rate any further fall in room temperature actuates the re-heater.

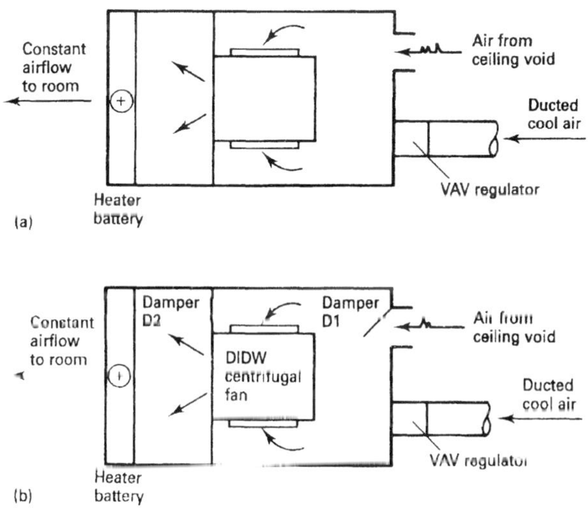

Figure 2.11 Fan-assisted air supply terminals – diagrammatic illustration. (a) Series arrangement. (b) Parallel arrangement.

(2) *Parallel flow*. This is illustrated in Figure 2.11(b). As room temperature falls, the VAV regulator throttles the flow of ducted, cool air. Damper D1 is kept closed by the pressure of the ducted air supply into the unit and damper D2 is kept open by the same pressure. The fan is off and a reducing airflow rate is fed to the room. Eventually the VAV regulator is delivering its minimum rate to the terminal. The fan is switched on and delivers enough airflow to the room to give satisfactory air distribution without dumping. Damper D1 is opened by the negative pressure developed at the fan suction and handles the difference between the fan duty and ducted minimum airflow rate. Damper D2 is closed by the positive pressure at fan discharge. On further fall in room temperature the heater battery is actuated and warms the minimum constant airflow supplied to the room.

An advantage claimed for the series arrangement is that air drawn from the ceiling void is warmer than room air because of the heat gain from the electric lights and that this represents some saving in thermal energy when there is a demand for heating. Also, the fan in the main air handling unit does not have to develop such a high fan total pressure because the terminal fans are providing the pressure required at the terminal units. Against this is the fact that the terminal fans are single phase, permanent split capacitor motors with operating efficiencies that will be less than that of the three phase motor driving the fan in the main air handling plant. The main objection to the series arrangement is that it is not a variable volume system and hence the overall, annual energy required for the fan power will be greater.

FRESH AIR SUPPLY

Referring to Example 2.12 it is seen that the total supply airflow rate for the hypothetical office block is 91.1 $m^3 s^{-1}$, over a total treated floor area of $86.4 \times 13.5 \times 12 = 13\,997$ m^2. This represents a specific supply rate of 6.5 $l\,s^{-1}\,m^{-2}$. In Example 1.11 a specific fresh air supply rate of 1.4 $l\,s^{-1}\,m^{-2}$ referred to the treated floor area, was assumed. This represents 21.5% of the supply rate as minimum fresh air. When the system throttles to, say, 30% as a typical minimum supply rate, the fresh air supply falls to $0.3 \times 1.4 = 0.42$ $l\,s^{-1}\,m^{-2}$, which for a population density of 9 m^2 per person is only 3.78 $l\,s^{-1}$ for each person and is clearly inadequate. (The CIBSE recommendations for each person are 8 $l\,s^{-1}$ when there is no smoking and 16 $l\,s^{-1}$ when there is some smoking.) The refrigeration load for the whole building determined in Example 1.11 was 1610 kW and was for a fan coil system. If a VAV system is used the refrigeration load will increase, to about 1763 kW, principally because the supply air distribution system is at medium velocity and the supply fan total pressure will be about 1 kPa, instead of about 500 or 600 Pa for the low velocity system used with a fan coil system.

There are several ways of dealing with the difficulty posed by the reduction in the amount of fresh air supplied as the system throttles and the following are worth considering, in terms of the impact on the refrigeration load, and hence on the cost:

(1) Use 100% fresh air. This will increase the refrigeration load of 1763 kW by about 60% with large increases in the capital and running costs.
(2) Increase the minimum fresh air proportion at design load so that, when the system is throttled to, for example, 30%, the necessary minimum fresh air will still be provided. This increases the fresh air rate under design conditions to $1.4/0.3 = 4.67$ $l\,s^{-1}\,m^{-2}$, and the refrigeration load goes up by nearly 40%.

(3) Increase the fresh air allowance as in (2), above, but use a pair of run-around coils (see Chapter 9) having an effectiveness of 50%. This increases the refrigeration load by about 20%. The extra capital cost of the run-around coils must be borne. The supply and extract fan powers will increase because the fans must develop larger fan total pressures in order to cope with the extra resistance to airflow posed by the run-around coils.
(4) Increase the fresh air allowance as in (2) but use a thermal wheel with an effectiveness of 85% (see Chapter 9). The load then rises by about 12%. The extra costs mentioned in (3) above must be borne.
(5) Continuously monitor the fresh air rate flowing in the fresh air duct and increase this as the total system airflow rate falls, so as to provide never less than the minimum fresh air necessary. The difficulty here is the poor quality of performance of multi-leaf motorised dampers in the mixing boxes of air handling plants.

In terms of the total amount of fresh air supplied to the building in a year the picture is not quite so bad. Reference to Figure 2.4 shows that if variable proportions of fresh and recirculated air are used the system operates with a good deal more than the minimum proportion for much of the time.

DUCT SYSTEM DESIGN

Generally, to avoid noise and to minimise energy consumption by fan power, it is best to design VAV systems as medium velocity, in the range of 10–12.5 m s^{-1}, rather than as high velocity, in the range of 15–20 m s^{-1}. Exceeding a velocity of 15 m s^{-1} should be avoided and 20 m s^{-1} should never be exceeded. Run-outs feeding VAV terminals, of whatever type, should be sized for 10 m s^{-1}, or less.

To assist in balanced air distribution, main ducts should be sized by static regain where building space permits. It is wise to keep a distance of at least six duct diameters between successive fittings and to adopt conical tees. An air-tight supply system is essential and this means ducting must be pressure-tested after installation. It is possible that, with the conservative approach to sizing suggested above, the fan total pressure may not be high enough to bring the system within the range of pressures that codes of practice regard as necessary for pressure tests. Pressure tests should be carried out nevertheless. Well-designed systems are likely to have fan total pressures of the order of 1.0–1.25 kPa, rather than 1.5 kPa upwards. Duct layouts should be simple and a circular, spirally-wound duct is preferable to rectangular ducting, the cost of the latter, if air-tight, being some four times that of the former. A spirally-wound flat oval duct is acceptable, to a limited extent, and is better than rectangular. When using a rectangular duct, or flat-oval duct, never exceed an aspect ratio of 3:1. Extract systems are generally much less extensive than supply duct schemes, since the location of the supply terminals dictates the quality of the air distribution. Extract duct systems should never be sized by high velocity methods. Steel ducts under large suction pressures are structurally unstable and, in any event, suction pressure reducing valves, where used, will tend to get fouled with dirt and become increasingly unreliable in performance. Supply air terminals will not give good air distribution and will perform badly if erected above egg crate ceilings and, even when located flush with the lower surface of the egg-crate, air distribution is likely to be poor. When such a form of installation is being considered it is wise to test the proposed arrangement in a full-scale mock-up.

The designer must calculate three sets of sensible heat gains:

(1) The maximum for each individual module so that the VAV terminal for the module can be sized.
(2) The sensible gains at the time of maximum refrigeration load for the whole building.
(3) The maximum simultaneous sensible heat gains for the part of the building dealt with by the air handling unit in question.

In doing this it is essential that, to take advantage of the diversity in solar heat gain through glass resulting from the movement of the sun in the sky, each air handling plant should serve opposite faces of the building, such as east and west. The appropriate diversity factors must be applied to heat gains from people, lights and business machines.

The problem with duct sizing is that ducts at the ends of branches feeding VAV terminals must be large enough to handle 100% of the design airflow, diversity factors of 1.0 being applied to gains from people, lights and business machines and full account being taken of solar gain through glass. On the other hand, having calculated the maximum simultaneous sensible heat gain for the part of the building being dealt with, the total airflow rate handled by the air handling unit and the duct system is known, and this is less than the sum of the airflow rates to the individual VAV terminals.

Engineering judgement is needed to size the ducts between the last (index) VAV terminal and the supply fan and the following facts are known:

(1) The airflow rate in the main duct immediately after the fan discharge.
(2) The design airflow rate in each duct feeding a VAV terminal.

Figure 2.12 illustrates one possible approach to sizing. The main duct from A to B can be sized to handle the calculated duty of the air handling plant, say 2000 $l\ s^{-1}$ for the purpose of illustration. Each end duct, B4–B5, C4–C5, and so on, must be sized to handle the design duty of the VAV terminal it is feeding. It would then be prudent to examine the nature of the sensible heat gains dealt with by each group of units such as B1–B5, C1–C5, etc. This might reveal the extent to which it is possible to apply diversity factors to the airflow rates for the relevant VAV terminals. The longer a branch and the larger the number of VAV terminals fed from it, the greater the possibility that appropriate diversity factors can be applied. Failing this, particularly when branches are short and feed only a few terminals, it would be wise to assume that all the units in a branch are likely to handle their design airflow rates simultaneously and to size the branch accordingly. The main duct, from B–E in the figure, might then be sized to handle proportions of the total airflow rate so as to arrive at the air handling plant with the correct total duty. Thus the increment in the figure is 375 $l\ s^{-1}$ for each main section from B–F. This is not strictly correct since the probability that a main duct section may have to handle the full design rate increases as the design airflow reduces. Probability factors might be determined and used to increase the proportions handled by the main duct shown in the figure.

Some general principles to follow are:

(a) Do not exceed a mean, main duct velocity of 15 $m\ s^{-1}$.
(b) Do not exceed a mean duct velocity of 10 $m\ s^{-1}$ in branch ducts.
(c) Velocities should get progressively smaller as the volumetric airflow rate diminishes.
(d) Do not exceed the maximum velocity recommended by the manufacturers for entry to any VAV terminal.
(e) For preference, use an entry velocity that is in the middle of the manufacturer's range for a VAV terminal, provided this is acoustically acceptable.
(f) Use static regain sizing to equalise the static pressure at branches.
(g) Use low velocity to size the extract system of ducting.

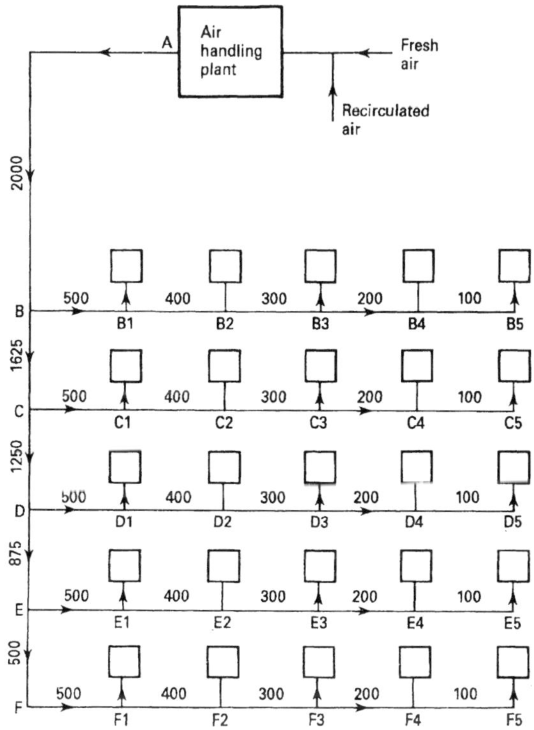

Figure 2.12 The problem of duct sizing for a VAV system. Each VAV terminal is shown handling a notional 100 l s^{-1}, for the purpose of illustration.

COOLER COILS AND AIR FILTERS

(a) *Cooler coils*. If blow-through coils are used the temperature rise through the supply fan is before the coil and although there is still a duct heat gain, the supply air temperature is less than if a conventional draw-through coil is used. This is attractive, implying that the design supply airflow can be reduced. However it may not always be wise to use blow-through coils. The air distribution at the outlet of a fan is very disturbed and must be smoothed out before it can enter the cooler coil. This requires at least 2.5 equivalent duct diameters of straight duct, followed by a smoothing screen, before the air flows into the cooler coil. If this is not done, the uneven airflow over the coil face will cause a reduction in cooling capacity and there will be the risk of condensate carryover. A draw-through coil is likely to have a smoother and more uniform distribution of airflow over its face and a better chance of achieving design cooling performance, without condensate being blown off the fins.

A direct expansion coil can be used in a small VAV system but it is essential that evaporating pressure is stabilised, preferably by means of a properly sized hot gas valve located in the correct position in the refrigeration piping system. Hot gas may be injected into a purpose-made distributor, or into a hot gas header provided by the manufacturer as part of the cooler coil.

(b) *Air filters*. The best filters affordable should be used. An exception is the electrostatic filter which should not be used in VAV systems.

The reason is that a reduced airflow through the ionising section of the filter generates an excessive amount of ozone, which is objectionable and a poison.

VAV IN TROPICAL CLIMATES

It does not follow that VAV is an acceptable solution for all climates. For example in the hot, humid weather of part of the West African coast, with imperfect building construction, natural infiltration can impose a very large latent load which can stay constant throughout 24 h, while sensible heat gains vary. Thus at start-up in the early morning a VAV system can be delivering only a small amount of air to meet the much reduced sensible heat gain but can be facing a very high latent load, simultaneously. Humidity will rise as a result. For instance, outside design conditions in Lagos in March could be taken as 33.5°C dry-bulb, 28°C wet-bulb (sling), 21.74 g kg^{-1}. An inside condition might be 25 ± 1°C dry-bulb, with a maximum of 60% saturation. Under design conditions inside (26°C, 60%) the moisture content is 12.86 g kg^{-1} and, for a module of our hypothetical office block, one air change of infiltration causes a latent heat gain of 263 W (see Example 1.10). At 26°C each person will liberate 70 W of latent heat and so the total latent heat gain becomes 403 W.

EXAMPLE 2.13

A variable air volume system delivers 251 l s^{-1} of air at 20°C dry-bulb, 23.37 g kg^{-1} to an office module, maintaining it at 26°C dry-bulb 60% saturation in the presence of sensible gains of 1837 W and latent gains of 403 W, when the outside state is 33.5°C dry-bulb, 28°C wet-bulb (sling). Determine the humidity in the room if the sensible gains reduce by 80% to 367 W, the latent gains and the supply air state remaining constant.

Answer

For a proportional band of 2 K quoted, the room temperature will be 24.4°C when the sensible gains are only 20% of their design maximum.

$$\text{Reduced supply air quantity} = [0.367/(24.4 - 20)] \times [(273 + 20)/358]$$
$$= 0.068\ m^3\ s^{-1}$$

$$\text{Room moisture content} = [12.37 + (0.403/0.068)] \times [(273 + 20)/856] = 14.4\ g\ kg^{-1}$$

At this moisture content and 24.4°C the humidity is about 75%.

AUTOMATIC CONTROL

The many different types of VAV terminals available offer a wide variety of control methods. Most, however, share the facility of operating several slave units in unison from a master controller or control unit. This is worth doing because of the saving in control costs when dealing with large rooms containing more than one VAV terminal. If partitions are rearranged it is then a simple matter to revise the slave–master unit relationship in order to provide individual thermostatic control for each office. The majority of units require electrical or pneumatic power to actuate their thermostatic throttling mechanism but a few types use duct pressure directly, with a saving in control costs.

FAN PERFORMANCE

No system is ever installed as designed. It follows that the performance of a fan will be different from that quoted in the catalogue, which is based on tests to BS 848 (1980) in the case of fans and on BS 6583 (1985) in the case of air handing units. Great care should be taken with installation: duct connections, on both sides of the fan, should be designed to give the smoothest possible airflow and at least 2.5 equivalent diameters of straight duct should be allowed at fan discharge, according to Keith Blackman (1986). Large margins on fan design duty are bad practice because they prevent the fan from operating at its best efficiency. Small margins, however, are essential to cover the difference between catalogue performance and achievement on site. It is suggested that from 5% to 7.5% should be added to the design volumetric airflow rate and from 10% to 15% to the design fan total pressure, for the purpose of ordering a fan. Similar margins should be considered for the airflow rate and the calculated external resistance, when ordering air handling units. All air handling units used should have been tested to BS6583: 1985.

The performance of a fan in a VAV system is often regulated by varying the fan speed and is then predictable by the fan laws (see Figure 2.13), the relevant equations of which are summarised as follows:

(a) For a given fan, given system and a constant air density.

1. $\dot{v}_{t2} = \dot{v}_{t1}(n_2/n_1)$ (2.8)
2. $P_{tF2} = p_{tF1}(n_2/n_1)^2$ (2.9)

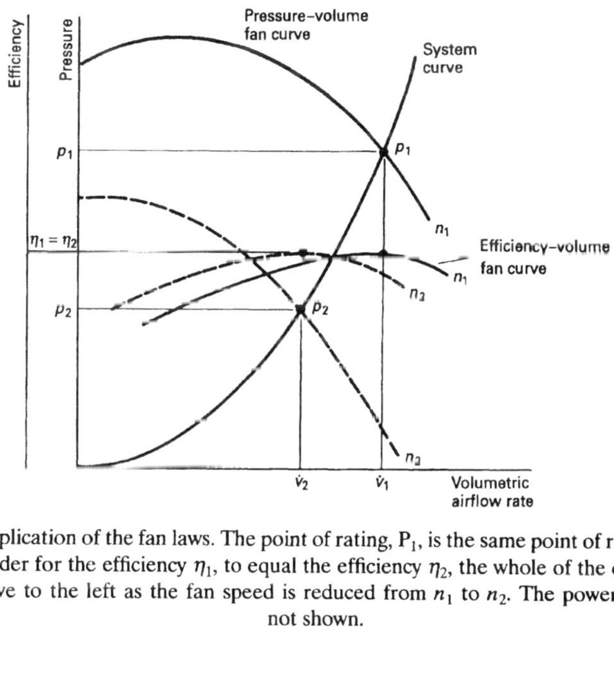

Figure 2.13 Application of the fan laws. The point of rating, P_1, is the same point of rating as P_2, on the fan curve. In order for the efficiency η_1, to equal the efficiency η_2, the whole of the efficiency–volume curve must move to the left as the fan speed is reduced from n_1 to n_2. The power–volume curve is not shown.

3. $W_{f2} = W_{f1}(n_2/n_1)^3$ (2.10)
4. Fan total efficiency is constant.

(b) For a given fan, given system and given fan speed.

1. Volumetric airflow rate is constant.
2. Fan total pressure, fan static pressure and velocity pressure at fan discharge are proportional to air density.
3. Fan power is proportional to air density.
4. Fan total efficiency is constant.

Where $\dot{v}_t$ is a volumetric airflow rate at temperature t, p_{tF} is fan total pressure, W_f is fan power, n is fan speed, and the subscripts 1 and 2 denote the initial and final fan speeds.

The laws are for a given point of rating on the pressure–volume curve for a particular fan. A point of rating is any point on such a curve, but is usually the point on the fan curve where it intersects the pressure–volume curve for the system of plant and duct to which the fan is connected. The performance curve for a system is determined by assuming a square law relating pressure drop and airflow.

METHODS OF VARYING FAN CAPACITY

Methods in use are as follows:

(1) *Variable inlet guide vanes*. See Figure 2.14. Commonly used only for centrifugal fans in the past but now obsolescent with the introduction of comparatively cheap methods of fan speed control. The design and operation of the mechanical linkage for the vanes has often been poor and the performance non-linear: the vanes must close by 45° before the flow is reduced, and hysteresis occurs. In approximate terms, as the vanes close, the pressure–volume curve for the fan rotates clockwise about the origin. The position of

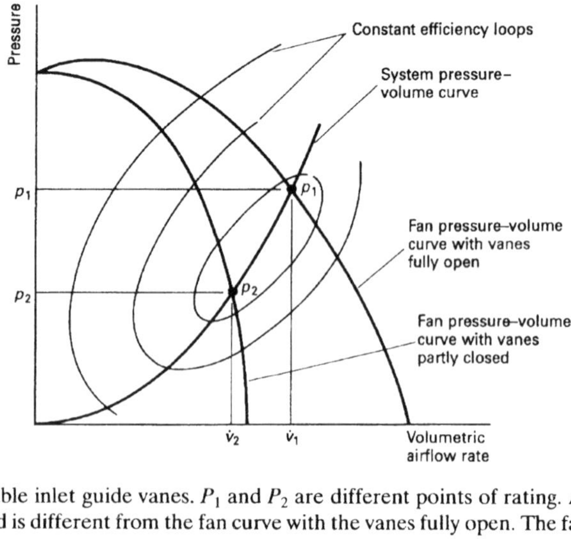

Figure 2.14 Variable inlet guide vanes. P_1 and P_2 are different points of rating. A fan curve with the vanes partly closed is different from the fan curve with the vanes fully open. The fan laws do not apply.

the system curve is unchanged and the intersection with the fan curve occurs at a series of different points of rating as the vanes close: a fan with its vanes partly closed is a different fan from one with its vanes fully open. Hence the fan laws cannot be applied.

(2) *Variable blade pitch angle*. This is used for axial flow fans only. The behaviour of the pressure–volume curve for the fan is similar to that of a centrifugal fan with variable inlet guide vanes but the engineering is much better and the method is most effective.

(3) *Disc throttle*. This is used only with centrifugal fans. In the case of forward-curved impellers a motorised disc moves within the impeller (see Figure 2.15), altering its effective width and shifting the position of the fan pressure–volume curve down the system curve, intersection occurring at different points of rating.

(4) *Speed variation*. This is the best method of controlling the capacity of centrifugal fans used in VAV systems. Several techniques have been used, not all of which have been efficient: with some electromagnetic methods the current handled by the driving motor remains unchanged as the fan speed is reduced and the power factor actually increases. Such forms of fan speed control should not be used for VAV systems because they nullify a major advantage, namely, economy in the energy consumed by the fans. The best method of efficient fan speed control for centrifugal fans is by frequency variation through an inverter.

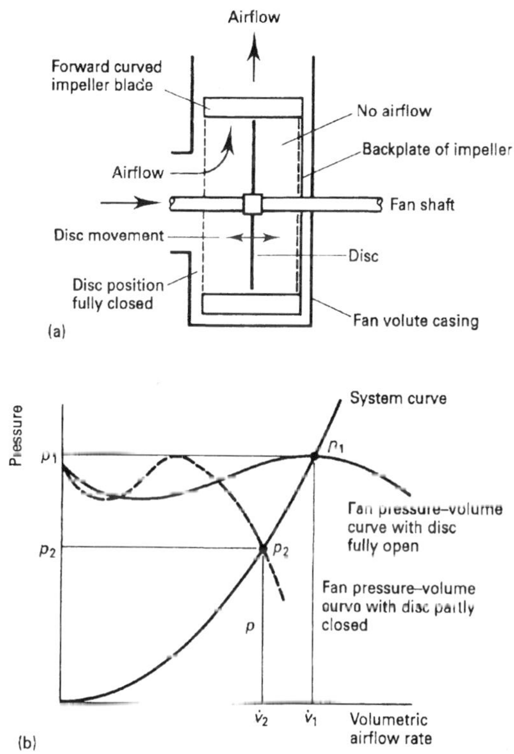

Figure 2.15 Disc throttle control. The fan curve is compressed to the left as the disc throttles the airflow through the impeller. Analogous to the behaviour of fans in parallel.

FAN CAPACITY CONTROL FOR A SIMPLE SYSTEM

To illustrate the principle of fan capacity control in VAV systems Figure 2.16(a) shows a very simple system with only one VAV supply terminal and one extract grille. The supply terminal is under thermostatic control from its room temperature sensor and as the room temperature falls the unit throttles the supply of air to the room and the pressure behind the unit increases. In Figure 2.16(b), showing the characteristic curves, the system curve rotates in an anti-clockwise direction about the origin and the intersection with the fan curve moves from position 1 to position 2, in order to supply a reduced airflow rate, $\dot{v}_2$ instead of the design airflow rate $\dot{v}_1$. If nothing is done to vary fan capacity the pressure drop across the terminal VAV unit rises from $(p_1 - p_3)$ to $(p_2 - p_4)$, increasing the noise produced and making thermostatic control by the unit more difficult. If throttling continues, the system

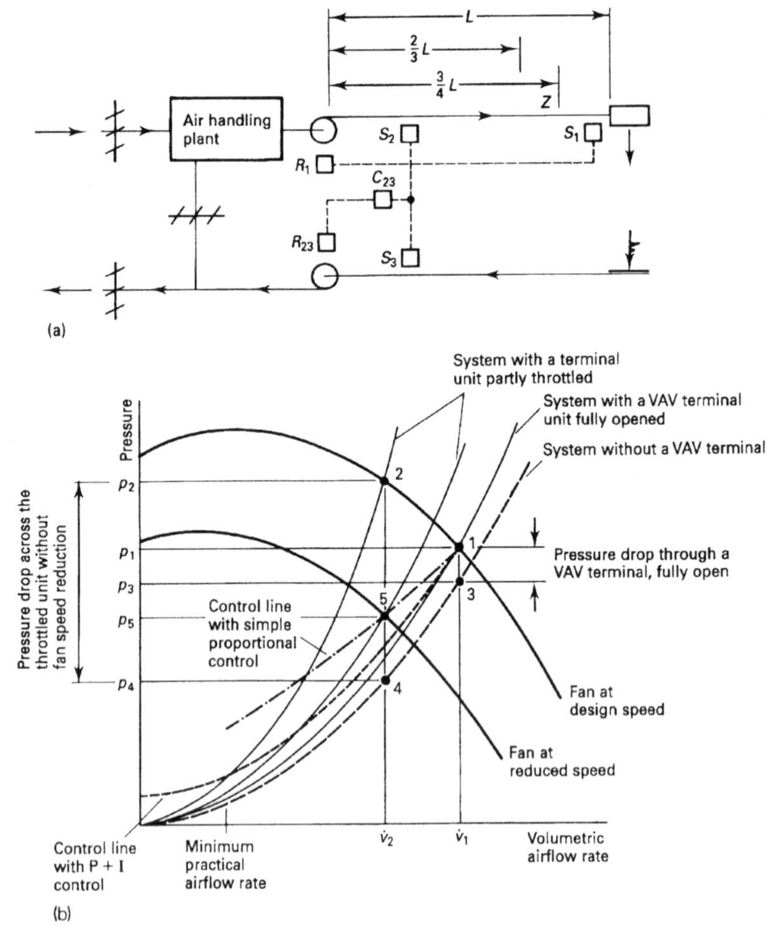

Figure 2.16 Performance of a simplified VAV system with fan speed control. (a) Plant/ducting for a simplified VAV system with only one VAV terminal unit. (b) Pressure–volume characteristics.

curve moves further to the left, the fan total pressure rises and the pressure at fan discharge increases. This will largely nullify the reduced fan power hoped for at partial duty, implied by Equation (2.7).

Figure 2.16 shows how this adverse situation is dealt with by varying the fan speed. S_1 is a pressure sensor, which, for the simple case of one unit, is located immediately before the VAV terminal. If this exercises simple proportional control the pressure sensed rises by the small amount of the proportional band as the unit throttles and this is used to reduce the speed of the supply fan. The point 1 slides down the system curve as the latter rotates, to occupy a new position at 5. The chain-dotted line through such points of intersection could be called a control line. It is seen that the fan total pressure, indicated by p_5, is much less than p_2, for the case when there is no control over fan capacity. The situation may be improved further if the pressure sensor, S_1, can exercise proportional plus integral control over fan speed. This eliminates the offset in the pressure at S_1 and the control line occupies a lower position, as shown. The static pressure behind the terminal unit then remains constant, except for the short periods when the unit is responding to a change of load, as measured by the room temperature sensor.

The capacity of the extract fan must be reduced to match that of the supply fan. Using the pressure signal from the sensor S_1 is not satisfactory: the extract duct system is low velocity whereas the supply system is medium velocity, the extract fan is likely to be a forward-curved centrifugal whereas the supply fan could be a backward-curved aerofoil-bladed centrifugal, or an axial flow fan, and the supply system feeds multiple VAV units whereas the extract system draws air through multiple grilles. Hence the supply and extract fans will not respond in the same way to a pressure signal from S_1. A better solution is to choose two representative sections of ducting, one in the supply system and the other in the extract system, so that their mean velocity pressures correspond to the volumetric airflow rates. Signals from such velocity pressure sensors, S_2 and S_3, can be compared by C_{23} and used to adjust the extract fan speed regulator, R_{23}, to keep the performance of the extract system in harmony with the supply system.

In a real system, with many duct branches and VAV terminals, a position must be chosen for the pressure sensor, S_1. This is not easy. The worst place is at fan discharge: the pressure there must be allowed to fall, not kept constant. A compromise, based on common experience, is to place the sensor at Z, where the design airflow rate is from 50% to 75% of the total design rate. This is usually interpreted as being in a position between two-thirds and three-quarters of the distance from fan discharge to the index terminal. It has sometimes been prudent to fix pressure sensors and stop cocks in three or four such locations, between the two-thirds and three-quarters limits, and to decide on the best choice during commissioning. More complex and expensive arrangements have also been used, involving multiple sensors in many branch ducts, the signal from the lowest static pressure being used, through a selector relay, to regulate the fan speed. It is much better to use VAV terminal units with direct digital control. The static pressure at all VAV units can then be scanned at suitable, short intervals of time and the lowest measured static pressure used to control the speed of the supply fan.

Even with good control over both the supply and extract fans there may be problems with static pressure in various rooms in a building causing difficulties with opening and shutting doors. This is because air is extracted from the building as a whole, without any control over the airflow removed from individual rooms, whereas the supply airflow is specific to particular rooms. A way out of this problem would be to fit VAV extract terminals, one for each VAV supply terminal. The increased capital cost is a clear objection.

NOISE

The noise produced by a VAV terminal is a function of the air quantity and the upstream duct static pressure. If either rises, the other staying constant, the sound power level (see Sections 7.4 and 7.5) also increases. Table 2.7 gives some typical values of sound power level for units tested under ideal conditions of airflow in a laboratory. The units are of the variable geometry type, feeding air through linear diffusers and provided with self-acting bellows for throttling the airflow through the use of duct static pressure as the power source. No account is taken of residual noise in the ducted air supply to the unit. Inadequate plant silencing and regenerated noise arising from air turbulence in the duct system will give greater values than those tabulated. Poorly made units will also generate more sound power.

Table 2.7 Typical values of sound power levels of VAV terminals

		Sound power level (dB: re 10^{-12} W)						
Air quantity	Duct pressure	Mid-octave band frequencies (Hz)						
($l\ s^{-1}$)	(Pa)	125	250	500	1000	2000	4000	8000
20	250	35	28	24	20	16	14	14
	375	36	20	26	22	20	18	19
	500	36	32	28	25	24	24	21
	750	37	36	32	28	27	27	27
	1000	38	38	34	33	32	33	35
	1250	39	39	36	35	34	35	39
40	250	41	34	32	28	26	23	18
	375	41	36	36	30	28	26	23
	500	42	39	37	31	30	28	27
	750	43	42	38	32	32	30	32
	1000	44	43	40	34	34	33	36
	1250	45	44	41	35	35	35	40
60	250	44	43	41	35	33	26	21
	375	45	44	42	36	34	29	26
	500	45	45	43	37	35	31	31
	750	46	48	44	38	36	33	35
	1000	47	49	45	39	37	35	39
	1250	48	50	46	40	39	37	41
80	250	46	46	46	40	37	29	22
	375	47	48	47	41	38	33	26
	500	48	50	47	42	40	36	30
	750	49	53	48	43	41	37	35
	1000	50	54	49	44	42	39	41
	1250	51	55	50	45	43	40	43
100	250	46	49	49	44	41	34	24
	375	49	51	50	45	42	37	31
	500	50	53	51	46	43	39	36
	750	51	55	52	47	44	40	42
	1000	52	56	53	48	45	42	46
	1250	54	57	53	48	46	43	46

Making use of the data in Table 2.7, it is possible to estimate the noise rating (NR value) (see Section 7.9) likely in a typical module of the hypothetical office block (see Section 1.2), assuming that the room effect (see Section 7.8) is zero. Figure 2.17 shows this.

EXAMPLE 2.14

If two VAV terminals installed in a west-facing module of the hypothetical office block each deliver 73 l s^{-1} with a duct pressure of 500 Pa, estimate the NR value likely (a) under design conditions, (b) when the units throttle to 36 l s^{-1} to meet a reduced heat gain but the duct pressure rises to 1250 Pa and (c) when each unit handles 36 l s^{-1} but the duct pressure has been controlled at 250 Pa. In all cases assume a room effect of –5 dB.

Answer

(a) From Figure 2.17 one unit handling 73 l s^{-1} at 500 Pa gives about NR 42. For two units in our softer room the noise level will be 42 + 3 – 5 = NR 40.
(b) Two units each handling 36 l s^{-1} at 1250 Pa will give 38 + 3 – 5 = NR 36.
(c) Two units each handling 36 l s^{-1} at 250 Pa will give 30 + 3 – 5 = NR 28.

Unattenuated residual plant noise, regenerated air noise in the supply duct feeding the units and the fact that the actual installation departs from the ideal test conditions used to establish the table, mean that higher NR values are to be expected.

It is characteristic of VAV systems that they operate at less than their design flow rates for most of the time. Solar gains through glass often represent over 40% of the maximum total gains, for example in Example 1.9 it was 41% for a west module, and vary greatly on both an hourly and a seasonal basis. The peak solar gains can be regarded as prevailing for only about an hour in a day and this can be as little as 10% of the total working hours of the system. It follows that most VAV terminals will be operating at less than their maximum noise rating for the majority of the time. It can be inferred that NR values of 40–45 can be

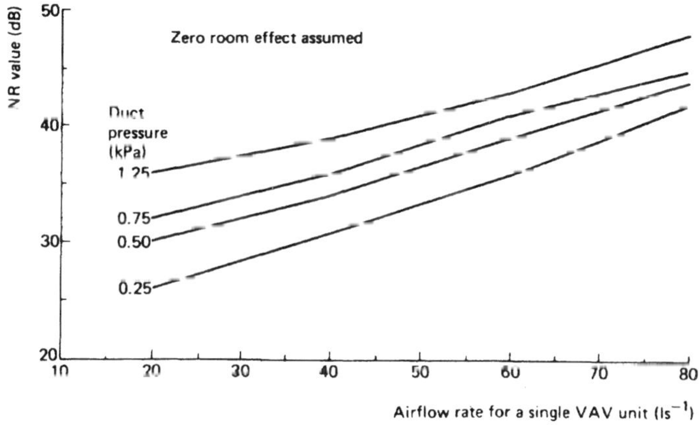

Figure 2.17 Typical NR values produced by a single VAV unit with a performance as given in Table 2.5 and a sound power level as in Table 2.7 (reproduced by kind permission of Haden Young Ltd).

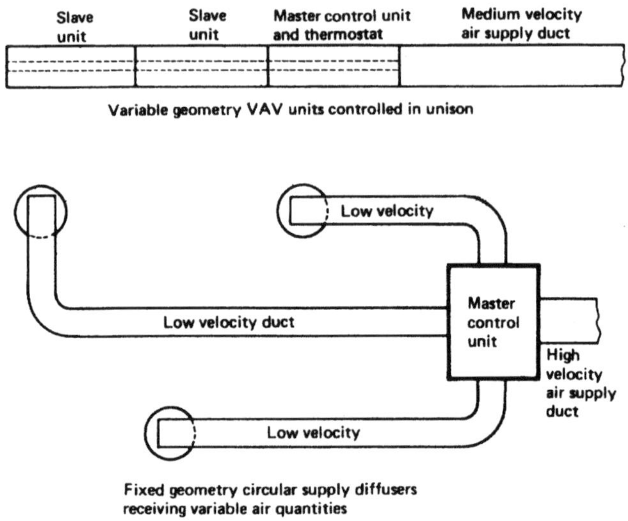

Figure 2.18 Different layouts of VAV systems.

specified for west-facing modules with 40% of their external facades glazed, because the units will give noise levels less than this throughout most of their life. See also Figure 7.18.

PRACTICAL ASPECTS

Since about two-thirds of the capital cost of an all-air system may be in the handling and distribution of air, efforts to minimise design supply air quantities are worthwhile and have the side advantage of assisting in the choice of quiet VAV terminals. When seeking to reduce air quantities it must be remembered that less than about $3\ l\ s^{-1}\ m^{-2}$ supplied to an occupied room sometimes give rise to complaints of inadequate air movement, although as little as $1.4\ l\ s^{-1}\ m^{-2}$ may be enough to keep an unoccupied room fresh. Design supply air quantities may be minimised by selecting 8-row cooler coils with chilled water flow temperatures of 5.5°C to give a leaving air temperature of 9°C. If there is sufficient room (at least 2.5 equivalent diameters) to ensure a smooth expansion onto the upstream face of the cooler coil and a uniform velocity across it then the coil can be installed on the discharge side of the fan and the penalty of temperature rise across it avoided. Thus, assuming a 2 K rise for duct heat gain, air can be delivered to the VAV terminals at 11°C. If a design room temperature of 22.5°C or 23°C is selected each $l\ s^{-1}$ of air supplied at 11°C can then deal with 14.5 to 15 W of heat gain and systems handling $12.5\ l\ s^{-1}\ m^{-2}$, near the limit for VAV, can cope with as much as $187\ W\ m^{-2}$.

Having reduced the supply air quantity to a practical minimum, the secret of an installation low in capital cost is to arrange an economical duct layout to feed the VAV terminals. Furthermore, as few VAV terminals as possible should be used, consistent with avoiding noise problems, providing satisfactory air distribution and offering the facilities for individual, modular, thermostatic control and easy partition re-arrangement that characterises certain types of VAV system. Air zoning is generally unnecessary.

VAV systems probably only run at their design duties for 10% of the time and for the rest of the year they handle much less (see Section 8.3). Fans should therefore be selected to have their peak efficiencies at about 75% of their design flow rates, in order to conserve energy. It is claimed by some manufacturers that their VAV terminals, operating from duct static pressures, control room temperature within a proportional band of as little as 1 K and it is certainly true that properly slaved units, operating in unison from a single control terminal, give better air distribution and control than do multiple ceiling diffusers fed with low velocity air from a single control VAV terminal, for example an octopus box. Figure 2.18 shows the difference in layout. With the second arrangement, individual diffusers are not themselves VAV terminals – they receive variable quantities of air at low velocity from the central control box – and will not necessarily always be in volumetric balance.

Shut-off leakage at VAV units can be as little as 15% of the design duty but this may still be enough to overcool an unoccupied space, although room temperatures should soon reach a comfortable level with the return of the occupants and the switching on of lights. In exceptional circumstances auxiliary heating may be needed. Since VAV systems can inherently deal with reductions in volumetric flow rates, it follows that ducts feeding untenanted floors in an office development can be dampered off, provided this is not carried to excess, and the fan capacity still controlled in a satisfactory way at the reduced duty. It appears that properly designed, variable linear diffusers can give better air distribution for VAV systems than can circular or square diffusers, or side-wall grilles. Also, with properly designed VAV installations the time spent in air balancing during commissioning may be minimised, but not entirely dispensed with. A well thought out balancing procedure must be followed.

2.7 Dual duct systems

The chief advantage of the double duct system is that full cooling and full heating capacity are simultaneously available at all times, under thermostatic control. Further, it has, in common with other all-air systems, the facility of free cooling when the outside air temperature is low enough, which is about 10°C in the UK, the refrigeration plant being off. It is suitable for any application needing an all-air system with thermostatic control in a multiplicity of rooms and it has been used for offices, hotels, hospitals and ships. One advantage over induction or fan coil systems is that since no air is induced or recirculated from the room over a local cooler coil, there is no local contamination at the unit, hence its suitability for hospitals. However, the system is expensive in capital and running cost and needs a large amount of building space to accommodate ducts. These are formidable objections which have restricted its use in comparison with other systems, such as perimeter-induction, fan coil and variable air volume.

The supply air quantity for a dual duct system will always exceed that for a variable volume scheme and because it takes little account of load diversification more plant space is required. Furthermore, since two supply ducts must be accommodated, the designer of a dual duct system will always tend to use higher air velocities than for a single duct system. It follows that noise will often be a problem and, since fan total pressures will be high, the component of the cooling load attributable to fan power will be large, increasing both the size of the installed refrigeration plant and its energy consumption.

Terminal mixing boxes may be used in two forms: either installed beneath window sills, when rather more floor space is used than with an induction or fan coil system; or located above suspended ceilings. Mixing box terminals require minimum operating duct static pressures from about 70–400 Pa.

The only stable dual duct system is one having constant volume regulators on all mixing terminals so that each can deliver a fixed quantity of air, regardless of upstream duct pressure variations caused by changes in the ducted volumes of hot and cold air consequent upon load fluctuations at the other terminals. It should be noted that so-called constant volume regulators are factory-set to deliver a nominally constant amount of air with a tolerance of about ±5%, corresponding to their proportional band. However, it is not uncommon for such factory setpoints to be inaccurate, and a check on site is essential. The design volume at each terminal is the cold air quantity required to meet the local design sensible heat gains and this is the nominal setting of the constant volume regulator. The sum of the terminal cold volumes is the duty of the supply fan. The hot duct quantity is less than the cold duct amount because a bigger temperature difference, supply-to-room, can be adopted to deal with the heat losses, and the room that has the smallest ratio of supply air quantity to design heat loss will need the warmest air. This index room fixes the design hot duct temperature and its design hot quantity will be the same as its design cold quantity. Elsewhere, terminals will require less hot air to satisfy their design heat losses and so, progressing back to the fan, the hot duct will become smaller than the cold duct. ASHRAE (1976) claims that the maximum airflow in the hot duct will seldom occur when meeting the design winter heat loss but will do so during light summer loads or marginal weather, when the hot duct temperature is compensated down from its winter design value. Because the dual duct system, for the building as a whole, takes account of diversity in heat gain, for example that maximum solar gains through east and west modules do not coincide, the maximum total cold duct quantity is less than the fan duty, some small amount of warm air being mixed with cold air to provide a higher supply temperature to those rooms not suffering peak gains. Shataloff (1964) suggested that the ratio of the design cold duct quantity to the fan duty is a basis for deciding on the maximum hot duct quantity (Table 2.8). Exact relationships between hot and cold duct sizes probably depend very much on particular system designs and computer-aided techniques should be able to resolve individual cases.

If excessively high hot duct temperatures are chosen the adverse consequences are stratification in the room, stratification in the hot duct, which often reasserts itself even after a deliberate attempt to remove it by turbulent mixing, excessive duct heat loss and a tendency to underheat some rooms. Air temperatures above about 35°C, with conventional ceiling heights of about 2.7 m, are best avoided. High hot duct temperatures will also aggravate the effects of hot air leakage, as much as 10%, past the nominally closed valve in the mixing box.

Many dual duct installations have a pair of ducts running above the suspended ceiling in the corridor with connexions to terminal units mounted above the ceilings in the rooms on either side. Making connexions that have to cross over the main ducts to feed the terminals

Table 2.8 Suggested relationship between hot and cold duct sizes

Design cold cut quantity / Fan duty	Hot duct area / Cold duct area
1.0–0.9	0.8
0.9–0.85	0.8
0.85–0.80	0.85
0.80–0.75	0.85
<0.75	0.9

poses a problem of accommodation, space being at a premium. Flexible ducts from the mains to the terminals, therefore, tend to be used, very often conveying air at high velocities. Such flexible ducts frequently generate a good deal of noise, particularly if they are sharply bent. In addition, noise break-out through the relatively light-weight material of the flexible ducts is common. The mixing boxes themselves generate noise in the process of reducing the air pressure from the high velocity inlet side to the low velocity, single duct outlet side to the air diffusers. Although the boxes are lined with sound-absorbing material this is often inadequate and additional acoustic lining is recommended for the downstream, low velocity ducting. A further problem is that boxes radiate sound outwards from their casings and this passes unhindered through the light-weight suspended ceiling tiles into the room. The temptation to use high velocities should be resisted and it is recommended that maximum velocities be kept as low as possible, preferably to below 15 m s^{-1} and never over 20 m s^{-1}. High velocity ducts with rectangular sections must be avoided as they regenerate much more noise than do spirally wound ducts at the same velocity. Branch dampers are of little value in a dual duct system where air volumes are constantly changing. Duct layouts should be designed to give smooth airflow around bends and through branch take-offs, where conical fittings are preferred. Square mitred bends are undesirable but, if they must be used, they should be fitted with small section turning vanes. Large section turning vanes should not be used as there is a tendency for vortices to form and act as pure tone sound generators.

Dual duct systems are not recommended.

2.8 Multizone units

A system using a multizone air-handling unit is essentially the same in principle as a dual duct scheme, the difference being that thermostatic mixing of hot and cold air occurs at the plant and not at a terminal unit in the room (see Figure 2.19). Single ducts conveying air at low velocity feed each room or zone, i.e. two or more rooms with similar change patterns for their sensible heat gains. Multizone units are applied to small buildings or parts of buildings where running several low velocity ducts, one to each zone, does not use up too much building space. Multizone systems have been used for treating groups of public rooms in hotels that need an all-air design approach and share similar usage routines. Blow-through air handling units are commonly available with up to 12 zones but best results are achieved with fewer zones that each require approximately equal amounts of supply air. The poorest control performances are obtained when there is a large disparity in zone duties. Damper characteristics of position versus flow rate are by no means linear and it is highly probable that the air quantities delivered to zones will alter as damper positions change under thermostatic control. It is absolutely essential that the flow temperature of the LTHW feeding the heater battery in the hot deck be compensated against outside air temperature. If this is not done control will become difficult when the hot deck dampers are nearly closed and small amounts of very hot air mix with air from the cold deck. It is also desirable to compensate the chilled water flow temperature to the cooler coil in the cold deck. If direct-expansion cooler coils are used (on small jobs) it is vitally important that their evaporating temperatures be stabilised by means of hot gas valves.

The quality of dampers provided with commercial multizone units is usually not good enough to give the best proportional control and the control mode may often degenerate to two-position. Leakage past the dampers in their closed positions is of the order of 15% and introduces additional operating and control problems. Even specially made, tight shut-off dampers can still have 5% leakage.

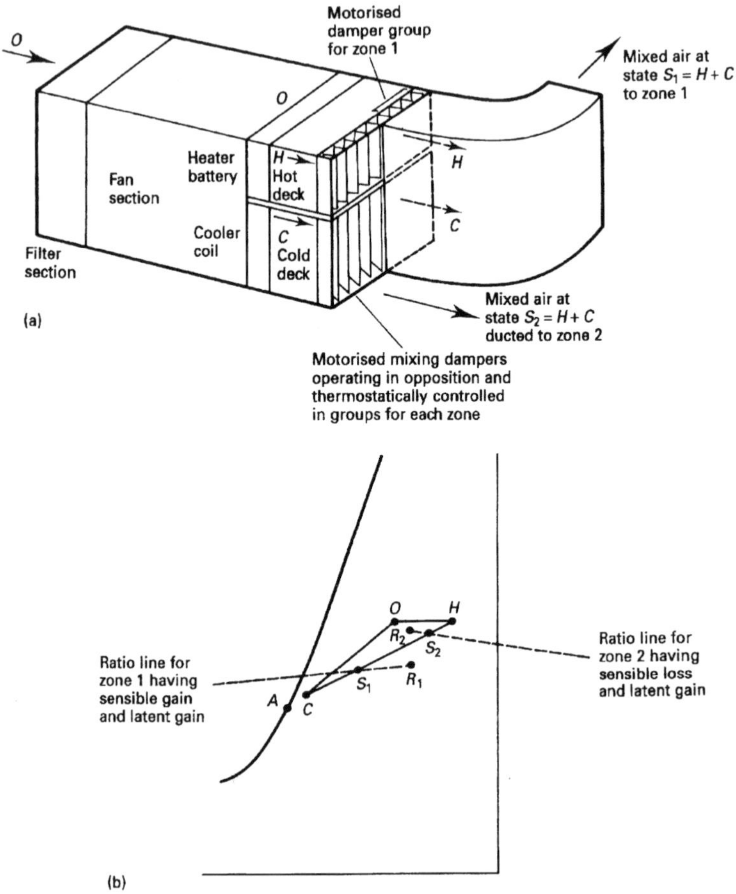

Figure 2.19 Multizone unit. (a) A two-zone unit (b) Psychrometry.

2.9 Air curtains

Air curtains and air doors include a variety of types: complete air curtains (Figure 2.20), industrial warm air curtains (Figure 2.21a), industrial cold air curtains (Figure 2.21b) and simple door heaters (Figure 2.21c). There are other subtypes such as that used above the door of a cold store, blowing air downward and inward to retain the large mass of cold, dense air within the store when its door is opened.

A simplified analysis developing the basic principles of complete air curtains is as follows. Two columns of air, one of density ρ_o and absolute temperature T_o outside the building and the other of density ρ_r and temperature T_r within, share a common stack, height H. The pressure difference across the open door at the base of the column is $\rho_o gH(1 - T_o/T_r)$ and is the stack effect. At the same time the wind speed, u_o, exerts a velocity pressure of $0.5\rho_o u_o^2$ and the combined stack and wind effect on the open doorway is a pressure difference of $0.5\rho_o u_o^2 + \rho_o gH(1 - T_o/T_r)$ which will produce a velocity of airflow, u_t, trying to enter the

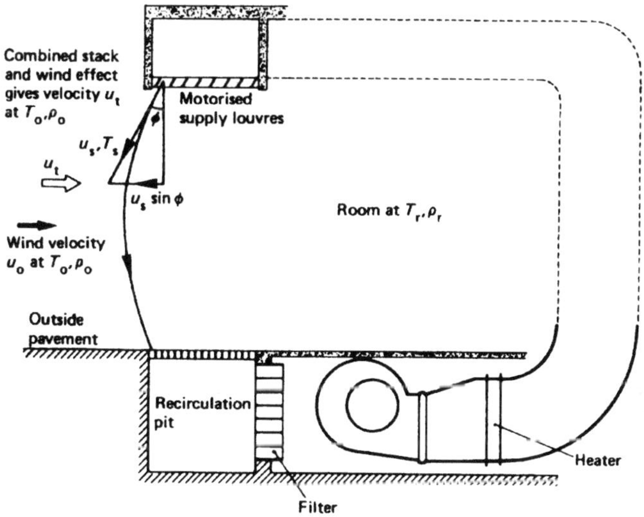

Figure 2.20 Complete air curtain.

doorway and which must be countered by the horizontal, outward component of the airflow supplied by the air curtain. Over the upper half of the doorway the combined effect is held at bay by the outward-curving supply airstream of the curtain but, over the lower half of the doorway, the supply airstream is defeated and is curved inward by the combined effect to be extracted through a floor grille (Figure 2.20). Therefore

$$0.5\rho_o u_o^2 = 0.5\rho_o u_o^2 + \rho_o g H(1 - T_o/T_r)$$

and so

$$u_t = \sqrt{(u_o^2 + 2gH(1 - T_o/T_r))} \tag{2.11}$$

Since force is the rate of change of momentum, a balance exists between the momentum of the airstream trying to enter the doorway and that opposing it. Introducing a coefficient of discharge, c_d, for the quantity of air, Q_t, trying to enter the upper half of the doorway, gives

$$c_d Q_t \rho_o u_t = Q_s \rho_s u_s \sin \phi$$

$$(c_d h w \rho_o u_t^2)/2 = Q_s \rho_s u_s \sin \phi$$

$$\begin{aligned} Q_s &= (c_d h w \rho_o u_t^2)/(2\rho_s u_s \sin \phi) \\ &= (c_d h w T_s u_t^2)/(2T_o u_s \sin \phi) \end{aligned} \tag{2.12}$$

where Q_s, ρ_s, T_s and u_s are the quantity, density, absolute temperature and velocity, through the actual open area of the grille, respectively, of the supply airstream at an angle ϕ to the vertical. The height of the doorway is h and w its width.

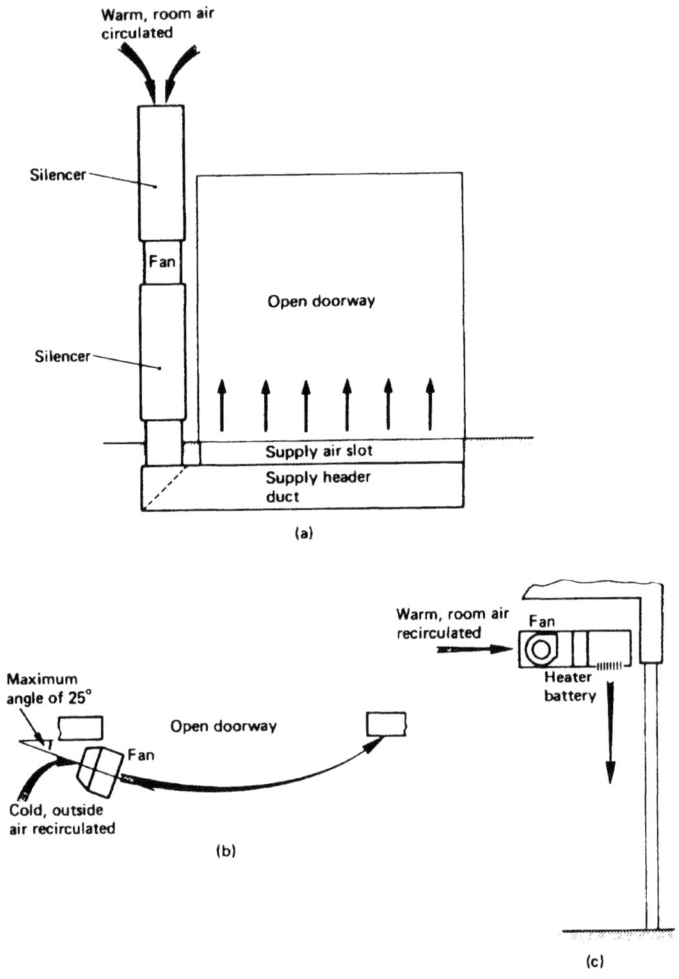

Figure 2.21 (a) Front elevation of an industrial warm air curtain. (b) Plan of an industrial cold air curtain – several fans are stacked on top of each other, the number depending on the door height. (c) Section of a fan coil door heater.

A choice of supply grille is dictated by the equation

$$Q_s = fc_s u_s wd \tag{2.13}$$

in which f is the free area ratio of the grille, c_s its coefficient of discharge and d its depth. A certain amount, depending on the wind velocity and stack effect, of cold outside air is entrained by the air curtain and a corresponding quantity of air from within the building is lost to the outside, some entrainment of inside air also occurring. The net effect is to reduce the supply air temperature from its initial value and so the recirculated air must be warmed.

Deciding on a value to which the air temperature should be raised is difficult but, in the UK, 32°C has been found practical for an outside temperature of –1°C and an inside one of 20°C. Selecting a supply air velocity is also empirical. It must not be so high that serious draughts are felt by people passing through the air curtain in a commercial application, although transient, perceptible warm air movement at head level is generally acceptable. For a door height of 2.5 m a value for u_s of 10 m s^{-1} is often found practicable, but operating experience could dictate other values, hence the desirability of control over the air volume handled by the fan.

EXAMPLE 2.15

A complete air curtain is to be fitted in a doorway of dimensions 2.5 m high × 4.0 m wide to combat a wind speed of 4 m s^{-1} (14.4 km h^{-1} or 8.9 mph) when the outside air temperature is –1°C and the room temperature is 20°C. The stack height is 10 m. Determine the duties of the fan and heater battery and establish the dimensions of the supply grille. Assume the vanes of the supply grille are turned at an angle of 45° to the vertical and a supply air temperature of 32°C.

Answer
From Equation (2.11)

$$u_t = \sqrt{(16 + 2 \times 9.81 \times 10(1 - 272/293))} = 5.48 \text{ m s}^{-1}$$

Taking a coefficient of discharge for the doorway of 0.65

$$Q_s = (0.65 \times 2.5 \times 4.0 \times 305 \times 5.48^2)/(2 \times 272 \times 10 \times \sin 45°) = 15.48 \text{ m}^3 \text{ s}^{-1},$$
from Equation (2.12).

The heater battery duty is to warm this quantity of air from an assumed value of 293 to 305 K and, from Equation (2.3), this equals [15.48 × (305 – 293) × 358]/305 = 218 kW.

Assuming a free area of 60% for the supply grille and a coefficient of discharge of 0.7 Equation (2.13) gives

$$15.48 = 0.6 \times 0.7 \times 10 \times 4.0 \times d, \qquad \text{so } d = 0.921 \text{ m}.$$

The floor grille should be the full width of the doorway, w, as should the supply grille and not less than the depth of the supply grille, d. The floor grille must be specially designed to withstand heavy traffic and must have spacings between its bars that will not catch pedestrians' heels or umbrella tips. The louvres on the supply grille ought to be motorised for easy adjustment to suit the wind velocity and the fan should be variable speed to economise on running costs and to permit the empirical modification to supply velocity, mentioned previously. The pit beneath the floor grille must be accessible for cleaning out and, preferably, should have an arrangement for washing down. Fans may be required to run at fan total pressures of the order of 500 Pa and the noise they produce must not be overlooked.

It is evident from the example that complete air curtains use a lot of air and will be expensive to run. Further, a large amount of space is occupied by plant and ducting. It is also clear that only a modest wind speed can be dealt with. An air curtain must, therefore, be regarded as a device that allows doors to be kept open only when the outside wind velocity and air temperature are within stated design limits. It is not really practicable to consider coping with a wind speed greater than about 14.5 km h^{-1} (9 mph) in a commercial application and it follows that an air curtain must never be used without a door that can be properly shut during weather conditions outside the design scope.

Industrial or air curtains of two types, warm and cold, are also in use. Their purpose is to minimise heat loss and draught through doorways with a high traffic frequency. It is arranged that the air curtain only runs when the door is open, a microswitch turning off the fans when it is closed. Industrial air curtains are claimed to be effective in dealing with wind speeds of up to 32 km h^{-1} (20 mph) but with noise levels in the vicinity of 85 dB(A), in spite of silencing.

Air curtains, or air doors, are also used to keep flying insects out of food processing plants. The average flying speed of an insect is about 1 m s^{-1} and research by the US Department of Agriculture has shown that insects can be effectively screened if the moving airstream of the curtain is at least 125 mm thick; the velocity at a level 1 m from the floor exceeds 6.6 m s^{-1}; and the air curtain covers the full width of the doorway. A typical arrangement is to use recirculated air at high level and to blow it downwards with an outward direction of about 10° to the vertical. The air screen, or curtain, is switched on only when the door is opened and it is, therefore, important that the full airflow is attained as quickly as possible if insects are to be excluded. It is thought that axial flow or propeller type fans are better than centrifugals, because they come up to full duty within 2 or 3 seconds of being switched on, whereas centrifugals take as much as 20 seconds.

Fan heaters installed over the doors in some shops are intended to diminish the worst effects of cold draughts when the doors are temporarily opened to admit customers. They are not installed to permit the doors to remain open for any lengthy period in the winter. Air curtains, air doors and door heaters must never be regarded as allowing the conventional door to be discarded.

2.10 Perimeter induction systems

PRINCIPLE

The induction unit (Figure 2.22a) comprises a plenum box fed with dehumidified primary air through a high velocity duct system, from a central air handling plant. A suitable balancing damper within the unit absorbs any surplus static pressure, residual noise is attenuated as necessary and the primary air is delivered through nozzles at a static pressure of between about 100–700 Pa, depending on the unit type and the form of the nozzles. Each volume of primary air issuing from a nozzle entrains between 3–8 volumes of surrounding air and hence air from the room is induced across a secondary cooler coil which achieves thermostatic control over room temperature by throttling the flow of water. An induction ratio (primary-to-secondary airflow rates) of 4:1 is normal. Figure 2.22(b) shows the air distribution that results in the room.

Primary air is normally a mixture of fresh and recirculated air that is filtered, cooled, dehumidified and re-heated as necessary. Since the air at the units is required at a comparatively high static pressure (to induce the secondary airflow) and the ducted distribution is at high velocity, the fan total pressure tends to be high, of the order of 2 kPa. Consequently, the temperature rise through the fan is also fairly high and, taking into account the duct heat gain, the primary air temperature at the units can be as much as 4 or 5 K higher than the dry-bulb leaving the cooler coil in the air handling plant.

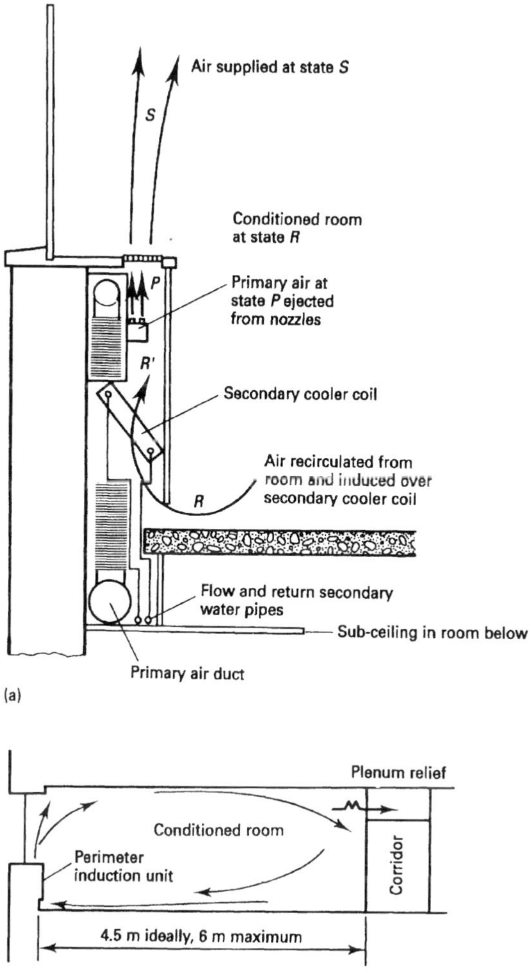

Figure 2.22 The perimeter induction unit. (a) A section of a typical unit. (b) Air circulation pattern with an induction unit.

CHANGEOVER AND NON-CHANGEOVER SYSTEMS

Originally conceived in about 1935 as a changeover system, to deal with the severely cold winters and hot summers of the northern part of the United States, the operational mode is in two forms, one for summer and the other for winter, a changeover between the two modes being necessary. During winter operation, cool unheated primary air is fed to the units and LTHW, compensated against outside air temperature, is supplied to the secondary coils. This gives a large heating capacity to deal with cold winter weather, but a small sensible cooling capacity because the primary air only comprises about 20% of the total amount supplied from the induction unit, the remainder being from the room and drawn over the secondary coil. If a net heat gain occurs when the system is working in its winter mode the flow of LTHW through the secondary coils is throttled and the sensible cooling capacity available is provided by the primary air. Assuming air leaves the primary air cooler coil in the air handling unit at 10°C it will reach the unit at about 14.5°C and will not have much capacity to deal with heat gains when the room temperature is 22°C.

In the summer mode of operation, secondary chilled water is fed to the unit coils at a constant temperature with a value that ensures sensible cooling and avoids the formation of condensate. The primary air is cooled and dehumidified and delivered to the induction units at a temperature which is compensated against outside air temperature by re-heat, thus allowing it to offset any fabric heat loss in marginal weather. During summer design conditions there is no re-heat and the primary air offers its full quota of sensible cooling, the secondary coils providing the remainder.

When the system is changed over, the action of the thermostat controlling room temperature is reversed.

In the UK, with its comparatively mild winter, the large amount of heating capacity available in the winter mode has proved unnecessary. Furthermore, the relatively small amount of sensible cooling capacity offered by the primary air alone, in the winter mode, has been inadequate for dealing with the net heat gains often present, even in cold weather, in office buildings. Changing the system over has introduced additional problems: the thermal inertia of the system of piping and of the building itself means that the changeover cannot be instantaneous. Deciding on the correct time to perform the change is difficult: automatic changeover has been unsuccessful because of the thermal inertia and manual methods have been equally unsuccessful. For these reasons the changeover system has been a failure in the UK and has not been used, except on a very few occasions, immediately after its introduction to the UK, when experience was lacking. A solution to the problem was to operate the system in its summer mode throughout the year – hence the term: two-pipe, non-changeover system. Systems of this type were extensively used in the UK, Europe and elsewhere in the 1960s and 1970s. However, because heat loss is dealt with by primary air at a temperature that is compensated against outside temperature, this has also meant a reduction in the sensible cooling capacity of the primary air, except at summer design conditions. Further, when outside air temperatures are less than the room temperature, the re-heated primary air may actually provide an additional heat gain. Various attempts have been made to remedy this defect by cutting back the compensated temperature of the primary air but with very limited success. As a result the two-pipe, non-changeover induction system has given way to the four-pipe induction system which, in turn, has become obsolescent.

PRIMARY AIR

The primary air in a two-pipe non-changeover system has four functions:

(1) To provide sufficient fresh air for ventilation.
(2) To offset the heat losses.
(3) To deal with the latent heat gain.
(4) To induce enough airflow over the secondary coil to deal with the sensible gains.

The amount of primary air handled affects the capital cost, the running cost and the building space occupied. Hence the designer keeps the primary air as small as possible, provided that the four functions listed above are achieved.

Taking a module of the hypothetical office building as an example the fresh air needs are $1.4\ l\ s^{-1}\ m^{-2} \times 14.4\ m^{-2} = 20.16\ l\ s^{-1}$.

With conventional flow and return temperatures of 85 and 65°C for LTHW, an upper limit for the primary air temperature is 40°C if stratification within the ducts and excessive heat loss from them is to be avoided, but it is wise to use a lower maximum of 35°C, as previously suggested for all-air systems. The heat losses from modules in the hypothetical office building are easily calculated as 549 and 692 W, for intermediate and top floors, respectively. Using Equation (2.3) it is also easily calculated that the primary airflow rates to satisfy these losses are 23.34 and $36.98\ l\ s^{-1}$. End modules, on the north and south faces, would have different losses and would need different airflow rates of primary air.

The latent heat gains were calculated as 134 W (Example 1.10) for a module in the hypothetical building and in Example 1.11 a practical state for air leaving a cooler coil was taken as 11°C dry-bulb, $7.702\ g\ kg^{-1}$. Using $7.702\ g\ kg^{-1}$ as the primary air moisture content and $8.366\ g\ kg^{-1}$ as the room moisture content (for 22°C and 50% saturation) in Equation (2.4) shows that this then requires a primary airflow rate of $67.66\ l\ s^{-1}$. This is likely to be too much for a commercially acceptable choice of induction unit and a lower moisture content should be chosen for the primary air leaving the cooler coil. An off-coil state of 9.8°C dry-bulb, with $7.1\ g\ kg^{-1}$ should be feasible for a six- or eight-row primary coil with a chilled flow temperature at 5.5°C. (Chilled water at this temperature is possible if a water chiller, such as a centrifugal or screw machine, with modulating capacity control is chosen that can give proportional plus integral plus derivative control over the chilled water flow temperature.) Using Equation (2.4) shows that the primary air quantity to deal with the latent gain is then $35.5\ l\ s^{-1}$, which is much more acceptable.

Considering an intermediate floor, the sensible heat gain for a west module was calculated in Example 1.9 as 1394 W. If primary air leaves the cooler coil at 9.8°C dry-bulb and the temperature rise for fan power and duct gain is 4.2 K, it will reach the induction units at 14°C. The use of Equation (2.3) then determines that the sensible cooling capacity of $35.5\ l\ s^{-1}$ air (in design summer conditions when it is not re-heated) is 354 W, leaving 1040 W of sensible cooling for the secondary coil to provide. This is well within the cooling capacity of the secondary coil in a commercially available induction unit.

It is theoretically correct to choose a primary air temperature with the use of reheat if necessary, that allows summer design transmission heat gain, by virtue of air-to-air temperature difference, to be offset by the primary air alone, leaving the secondary coil to cope with the random gains from people, lights and sunshine under local thermostatic control. This permits one primary air reheat zone and schedule to be used for the entire building and allows the problems associated with the movement of shadows cast by adjacent buildings over the sunlit facades to be disregarded. Achieving this invariably means reheating the primary air in summer design weather slightly above the lowest available temperature, with a consequent loss in its capacity to deal with part of the maximum sensible heat gain. This imposes a bigger load on the secondary cooler coil and tends to increase the size of the induction unit needed. The disadvantages of a single primary air zone treatment are, therefore, increased capital cost and increased running cost by wasteful reheat. For these reasons it is common to use the lowest practical temperature at the induction units, for summer design conditions, after allowing for the rise accruing from fan power, about 1 K or 1.2 K per kPa of fan total pressure, and duct heat gains.

INDUCTION UNIT COOLING CAPACITY

The sensible capacity of the secondary coil in the unit depends on the finning, the number of rows (usually one), and the amount of air induced, for a specified chilled water flow rate. Table 2.9 shows the approximate variation in capacity for different flow rates.

It is invariably arranged that the cooler coils in the units only do sensible cooling. It follows that, during normal operation, the outside surface temperature of the coil, its fins, or the piping connected to the coil, must not be equal to or less than the dew-point temperature in the room. Some initial condensation on the secondary coil in the unit is usually tolerable on start-up in the morning, when the room dew-point may be temporarily higher than during the daytime. The condensate drains down the fins into an emergency collection tray beneath the coil. It is generally not necessary to provide a drainage system for this tray because the small amount of condensate re-evaporates once the induction system is working in its normal fashion. The choice of the secondary chilled water flow temperature requires a little thought. It is possible to use a flow temperature that is slightly less than the room dew-point without condensate forming on the external surfaces of the coil and piping. This is because the thermal resistance of the water film within the piping gives an external surface temperature that is slightly higher than the temperature of the water inside the pipe. The manufacturer's advice should be sought regarding the allowable temperature difference, surface-to-water. Condensate collection trays should extend beneath the secondary chilled water control valve and the pipe connections. Pipe lagging and vapour sealing must be provided as necessary.

The induction ratio is a function of the air velocity issuing from the nozzles and their diameter. More air is induced by many small nozzles than by few large ones, at a given velocity. The cooling capacity of the secondary coil is usually regarded as proportional to the difference between the room air temperature (t_r) onto the coil and the entering secondary chilled water temperature (t_w), without appreciable error. Primary air cooling capacity is defined by Equation (2.3) in terms of the primary air temperature (t_p) and t_r. Unit capacities and noise levels are, therefore, expressed in terms of unit size, nozzle arrangement and pressure, for nominal water flow rates and stated values of t_r, t_p and t_w. It must be noted that catalogue ratings are always published for a specified standard form of unit installation and any deviation from this results in a loss of performance.

The costs of distributing and lagging primary airflow are very roughly the same as the costs of the associated induction units, in many cases. So choosing a larger unit, requiring less primary air, could cost about the same as selecting a smaller unit with more primary air. There is some merit in adopting a larger unit, provided it will fit in the building space available, as its cooling capacity can always be increased by feeding more primary air to it, with what may be an acceptable rise in the NR value. Nozzle pressures exceeding about 500 Pa are sometimes regarded as being on the high side because the fan total pressure becomes rather large, with increased fan motor power and running costs. Catalogue acoustic claims are frequently optimistic and it is generally wise to expect a room effect to be less than the catalogue anticipation.

Table 2.9 The variation in capacity of a secondary cooler coil for various chilled water flow rates

Flow rate (%)	10	20	30	40	50	60	70	80	90	100	115	130	145
Capacity (%)	24	48	64	86	88	92	94	96	98	100	102	104	106

NIGHT-TIME HEATING

With a heavy-weight building of slow thermal response there is no case for energy conservation by intermittent operation and so the system should run continuously. However, it is clearly wasteful to run the primary fan all night as there is no need for fresh air. Circulating hot water, with its flow temperature compensated against outside air temperature, through the secondary circuit in winter is worth consideration. Since supply and extract fans are off, heat transfer from the secondary coils is by natural convection, 'gravity heating', but there are several practical problems that must be dealt with if the method is to be successfully adopted:

(1) Automatic control action at the induction units must be reversed, or by-passed, if heating is to be possible at night
(2) The secondary circuit should include a heat exchanger to warm the water when necessary
(3) Tight shut-off valves must be used to prevent warm water entering the primary circuit at night
(4) The heat in the secondary system must be dissipated in the morning, before the refrigeration plant can be allowed to start chilling the primary water.

If the thermal response of the building is quicker, intermittent heating shows energy savings. With a two-pipe, non-changeover system the options are:

(1) Switch everything off at night but start the primary air fan and extract fan next morning at a time, related to outside air temperature, that will enable the system to bring the building up to temperature by the time the people arrive for work. The fresh air dampers are fully closed and the system delivers recirculated air at a boosted temperature to the induction units during the preheating period. The limit of about 40°C on the primary air temperature still applies. When the boost period ceases, the system reverts to normal operation.
(2) Augment the primary air boost by circulating hot water through the secondary coils during the preheating period, taking due regard of the problems. In this way, high primary air temperatures are avoided. Again it is important to ensure that the refrigeration plant does not start until the secondary water circuit has cooled to an acceptably low temperature.

CONTROLS

Water-controlled systems in the past have generally used pneumatically operated two-port modulating valves or three-port mixing valves, with some preference for the former in order to take advantage of the reduction in secondary water flow under partial load. Self-acting valves have often proved a failure in the past for various reasons, one being the wide proportional band needed to work the valves, giving an unsatisfactory variation in room temperature. Modulating solenoid valves have been accepted as a satisfactory alternative. Modulating air dampers, that vary the secondary airflow over the coils and are usually self-actuated from the primary air pressure in the induction units, are also used. Comparative cost studies suggest that an air-damper control system may be up to 5% cheaper than water control, taking account of the costs of induction units, automatic controls, secondary piping, lagging and secondary pumping. No diversity factor may be applied to the secondary water flow if air dampers are used or three-port valves fitted. Secondary piping should be lagged and vapour-sealed on all systems, particularly four-pipe. One valve and thermostat is not always provided for each induction unit, so economy in capital cost is achieved by fitting a valve on each unit but slaving groups of, say, four valves from a common thermostat. If, after installation and commissioning, a rearrangement of partitions is done, the connexions between valves and thermostats can be modified fairly easily, additional thermostats being bought if necessary, at the expense of the tenant. Using one large valve and thermostat to control a group of several units is not as satisfactory since it lacks the flexibility to cope with partition rearrangement. Thermostats are best located behind recirculated air slots or grilles on the units with remote set-point adjustment at the sill. Positioning thermostats on partitions is undesirable because these are often moved at a later date, and locating them on external walls can introduce problems because surface temperatures are less than air temperatures in winter and vice-versa in summer.

PIPING ARRANGEMENT FOR AN INDUCTION SYSTEM

Figure 4.7(a) illustrates a primary–secondary chilled water piping arrangement suitable for an induction system. Water from the chiller at a flow temperature of about 5.5°C flows to the primary air cooler coil and leaves the coil with a return temperature of about 10 or 11°C. The motorised, three-port mixing valve shown, R1, is not always used to control the cooling capacity of the coil – the capital cost of a large valve is saved and it is argued that some increased dehumidification at times other than for summer design conditions is acceptable, as far as comfort is concerned. Against this, the unnecessary extra dehumidification wastes some energy.

Chilled water is required for the secondary circuit to feed the induction units at a flow temperature that will be high enough to ensure the units only effect sensible cooling. With a room state of 22°C dry-bulb and 50% saturation the dew-point is 11.3°C and, taking advantage of the thermal resistance of the water film within the tubes, a secondary chilled water flow temperature of about 11°C might be suitable, as explained earlier. Hence the secondary flow connection is at the point *A*, where the temperature is at about the correct value, under design conditions. A temperature sensor, C2, is immersed in the secondary flow line and exercises control over the motorised three-port valve R2. This valve must not be oversized – the port connecting to the point *A* should be fully open under design conditions and this will only be the case if the return temperature from the primary cooler coil is at the same value as the desired secondary flow temperature to the induction units. The secondary return line must be connected to the primary line at the point *B*, as close as possible to the point *A*, the aim being to keep the pressure drop negligible between *A* and *B*. This will ensure that the primary water flow rate will be virtually constant, regardless of the flow through the secondary circuit.

The primary–secondary chilled water flow relationship is complicated by the fact that the primary water flow rate is based upon the performance of the primary air cooler coil, whereas the secondary flow rate depends on the selection of the induction units. This imbalance must be reconciled. The aim should be to have a primary flow rate that equals or exceeds the secondary flow rate. If the secondary flow rate is greater than the primary rate there will be reverse flow in the primary circuit, between *B* and *A*. In itself this does not matter but it means that the temperature of the water flowing from *A* into the three-port control valve R2 will be higher than the return water temperature from the primary cooler coil. Hence the selection of the induction units will be affected. The problem can be dealt with by re-selecting the induction units, or by fitting a by-pass in the secondary circuit between *Y* and *Z*, or by fitting a by-pass in the primary circuit between *W* and *X*. All these remedies have consequences on: water flow rates, chilled water temperatures, primary coil performance, induction unit performance, chiller performance, and pump duties. These issues must be dealt with.

Each induction unit is controlled from a room thermostat, C3, C4, Cn, etc., which regulates the cooling capacity of the secondary coils. If motorised valves, R3, R4, Rn, etc., are used it is desirable to provide a three-port valve at the end of a branch (as at Rn). This keeps the line alive and improves the response of the system. As the load reduces and two-port valves throttle, the pressure increases near the end of the system and valves find it more difficult to exercise proper control and they become noisy. A pair of pressure sensors, P1a and P1b, is fitted in the system, about three-quarters of the way from the pump discharge to the index unit. (There is an analogy with VAV systems, shown in Figure 2.16.) As the pressure difference sensed by P1a and P1b increases, a motorised valve, R1, located at the pump discharge, could be throttled. An alternative and better arrangement, as shown, is to use an increasing pressure difference between P1a and P1b to reduce the speed of the secondary pump, by means of an inverter.

FOUR-PIPE SYSTEMS

Units have the capacities of their heating and cooling coils controlled in sequence, with a no-capacity gap in between, either by water control valves or by air dampers. There is a tendency for air dampers not to close properly and so some secondary air is induced over both coils simultaneously, actual cooling and heating capacities not reaching rated values. Similarly, four-port valves, a pair of two-port throttling valves in a common valve body, and six-port valves, a pair of three-port valves in a common body, **should be avoided** as heat is conducted through the valve body from the hot to the chilled water, particularly in six-port valves where hot and cold water flow is continuous. A further objection to such multi-port valves is that, since clearances between valve discs and their seatings is small because of the deliberately compact construction, any scale or dirt in the water system prevents valves from closing fully and can cause endless difficulty with loss of performance.

The four-pipe system provides cooled and dehumidified, unheated primary air at all times. There is, therefore, no cancellation of cooling by primary reheat and the problems sometimes experienced with two-pipe systems in selecting units for southern and eastern building faces do not exist for the four-pipe version. Moving shadows across building facades present no difficulties and full heating and full cooling is available at all times. A wider range of set point choice is possible at the thermostats and the response to load changes is quicker than with two-pipe units. Since no heating is used to cancel cooling, the system is more economical to run than a two-pipe system. Four-pipe systems cost between 10–15% more than the equivalent two-pipe systems in terms of induction units, piping, lagging, pumping and controls. A four-pipe unit can be deeper than a two-pipe, occupying about 5% more floor area.

UNIT LOCATION

The induction unit was initially designed for mounting beneath a window, blowing air upwards from the sill and recirculating room air through a grille in the front of the unit. A similar performance is achieved when an open slot, of at least 100 mm height and the full width of the secondary coil, at floor level replaces the front grille. However, other departures from these standard arrangements result in a reduction of performance. When any non-standard installation is proposed, tests on cooling capacity and air distribution should be done before the units are fitted on site in quantity. One unit per module is the usual arrangement but sometimes larger units are selected with sufficient capacity to cope with the gains in two modules, and for installation centrally beneath alternate windows. Partition rearrangement then introduces a problem since, for example, an office could comprise three modules, the two outer ones lacking an induction unit. A partial solution sometimes adopted is to select the units on the same basis but to position them centrally on the mullions so that the module on each side is then dealt with by half an induction unit. The objection to this is that acoustic privacy across the partition dividing a unit is not always certain. Therefore, it may be necessary to try and fit a barrier within the unit itself, between the nozzles and the secondary coil, obviously with great difficulty. Units have been mounted above suspended ceilings, freeing floor space. If the unit is mounted thus, horizontally, an access panel must be devised by the architect. Such a panel must be easily opened for maintenance; safely secured when not open; durable; easily cleaned; aesthetically pleasing. This combination of properties, however, is not always easily achieved.

A horizontal installation at high level could perhaps be accommodated in a clear ceiling space of about 250 mm, depending on the type of unit. Alternative arrangements are possible, with ingenuity, but any will suffer a loss of performance and the air distribution with a high level unit must always be tested to make sure that it will prove satisfactory.

MAINTENANCE

Induction unit coils require lint screens to pick up floor dirt, and fibres from carpets, when they are floor-mounted. Such screens should also be fitted in high-level units because the air circulation pattern in the room carries dirt up to the ceiling. Also, lint screens, secondary coils and nozzles need cleaning every three to six months and if this is not done unit performance gradually declines, with dirt periodically ejected onto the window sill and into the room. Good quality filtration is recommended because, with time, dirt tends to build up within the unit, as it will with any air distribution system. In this respect, units with nozzles smaller than 3 mm seem to retain dirt more readily than do larger nozzles. Routine maintenance is also required for the automatic controls.

PSYCHROMETRY

Figure 2.23 illustrates the psychrometry of two-pipe non-changeover systems. In Figure 2.23(a) the performance shown is for a system operating under design summer conditions. Air extracted from the room at state R is warmed to state R'' by the extract fan power and mixes with minimum fresh air at state O to form state M. The primary air is then cooled and dehumidified to an off-coil state W, which is dry enough to handle the latent heat gains in the treated rooms. A temperature rise of about four or five degrees, caused by the supply fan power and duct heat gain, changes the state from W to P, at which the primary air is delivered to the nozzles of the induction units. About four volumes of secondary air for each volume of primary air are induced over the secondary coils and sensibly cooled from state R to state R'. This mixes with primary air at state P and is supplied to the room at state S. The length of the sensible cooling process line, R–R', alters as the secondary coil cooling capacity is varied in response to signals from the room thermostat.

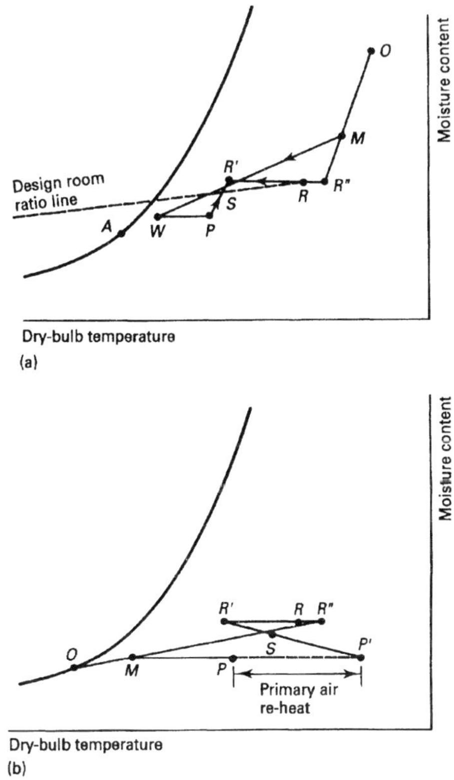

Figure 2.23 The psychrometry of a two-pipe, non-changeover, perimeter induction system. (a) Summer design performance. (b) Winter performance.

In winter (Figure 2.23b), cold fresh air at state *O* is automatically mixed with recirculated air at state *R″* to form state *M*. This has a temperature that is cold enough to chill the water flowing within the tubes of the primary air cooler coil. In this way, referring to Figure 4.7(a), if the primary and secondary pumps are running but the refrigeration plant is off, the water leaving the primary cooler coil is cooled by the airflow over the outside of its tubes to a value suitable for the secondary chilled water flowing to the induction unit coils. The temperature rise from *M* to *P* (Figure 2.23b) is partly due to the heat removed from the chilled water by the primary coil and partly to the supply fan power and the duct heat gain. It is then necessary to heat the primary air according to the compensated schedule, to the right temperature to offset the heat losses. This is shown by the dashed line from *P* to *P′*. Primary air at *P′* issues from the nozzles and induces secondary air over the coils in the units, sensibly cooling it from state *R* to state *R′*.

It is evident that the re-heated primary air can represent a waste of energy, advantage not being taken of casual gains to offset heat losses.

Frost protection is needed to protect the chilled water system in winter and it may be necessary to add glycol to the system. When this is done account must be taken of the consequences: the heat transfer coefficients of all heat exchangers reduces significantly, mechan-

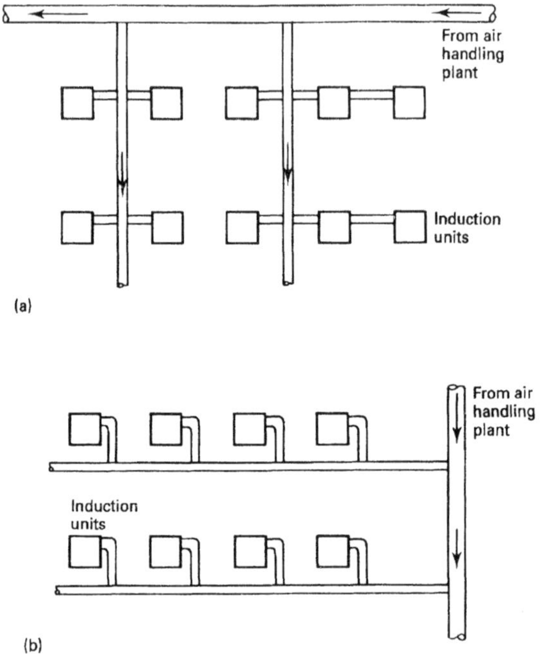

Figure 2.24 Primary ductwork arrangements. (a) Vertical distribution. (b) Horizontal distribution.

ical seals (instead of packed glands) become necessary because the viscosity of a water–glycol solution is different from that of water, zinc in the piping system will be preferentially attacked, in the presence of oxygen.

DUCTING ARRANGEMENTS FOR AN INDUCTION SYSTEM

The extract system is conventional low velocity and is not nearly as extensive or as complicated as the primary system, which is at high velocity and high pressure with fan total pressures of about 2 kPa. The primary air ducting may have a vertical or a horizontal distribution arrangement (Figure 2.24). The vertical arrangement was favoured in the United States but has not always been suitable for use in the UK since the location of the vertical dropping ducts at the mullions between adjoining modules clashed with reinforcing rods or steel joists. However, the vertical arrangement has been used successfully outside the building envelope. When used, it is advisable to size the horizontal header ducts by static regain, in order to obtain the same static pressure at the beginning of each vertical dropping duct and so assist balancing.

The horizontal arrangement, in Figure 2.24(b), has been adopted for many installations because it does not clash with the building structure to the same extent as the vertical form. The dropping ducts should again be sized by static regain, where possible, to simplify balancing by having the same static pressure at the beginning of each horizontal duct.

The vertical arrangement is potentially cheaper than the horizontal, because multiple induction units can be fed from a single dropper, as many as three on each side of the main. A maximum of two units may be blown through.

NOISE FROM UNITS

Selecting a quiet unit is insufficient guarantee of a quiet installation. Regenerated noise in the primary duct system will enter the unit, and hence the room, if the ducting has not been sensibly sized, constructed and installed. Its layout must also be such as to avoid the creation of turbulence. In particular, long and bent flexible connexions feeding units are a prime source of noise. Badly-made units may leak high pressure air from their casings and burrs on poorly moulded nozzles may generate unexpected noise. If the ductwork system has not been sized to achieve a natural balance, excessive pressure will be absorbed at the unit damper, giving unacceptable noise.

2.11 Fan coil systems

FAN COIL UNITS

These are small air handling plants, usually located in the room being conditioned. They are simple in construction and are fed with chilled water and/or LTHW. Sometimes they are provided with electric re-heaters. Fresh air may or may not be supplied, the decision depending on the application and the economics. Historically, they have been in use all over the world since the early part of the 20th century. In the UK they have increasingly found favour for air conditioning office blocks, particularly since 1980 and, although originally intended for installation as free-standing, cased units, they are commonly also located above suspended ceilings and compete successfully with variable volume systems.

A fan coil unit comprises one or more fans, recirculating air from the room and blowing it over a cooler coil. The fans are driven by totally enclosed, single phase, permanent split capacitor motors, with absorbed powers in the range of 20–190 W. There are usually three possible running speeds, delivering airflow rates that are 100% at medium speed, about 83% at low speed and 118% at high speed. The driving motor absorbs about 75% power at low speed, 100% at medium speed and 140% at high speed. For a building of some size, it would be necessary to provide switchgear to give a random start for the fan coil units in the morning, when the air conditioning system was first switched on.

Sometimes two coils are provided, one for cooling and one for heating. The coils are made from copper tubes (typically 9–16 mm outside diameter) with aluminium plate fans. Chilled water flow rates for room fan coil units lie between 0.05 and 0.4 l s^{-1} with water pressure drops through the coil of about 5–25 kPa. Standard coils have two or three rows of tubes for cooling when used in two-pipe systems but, in four-pipe installations, the coil may have four rows, three for cooling and one for heating, sharing a common fin block. It is cheaper to make coils that are wide, rather than tall, with the further advantage of better condensate drainage down a shorter fin distance to the collection tray, if the coils run wet. Hence the face of the coil is rectangular, with a fairly large aspect ratio. A consequence of this is that multiple fans are used in parallel, to give a more uniform air distribution over the face of the coil. One or more driving motors maybe used, close-coupled to the shaft of the fans. The fans themselves are double inlet, with forward curved centrifugal impellers and airflow rates in the range 70–450 l s^{-1}, when running at medium speed.

THE SUPPLY OF FRESH AIR

As a means of air conditioning, particularly in tropical countries, fan coil systems have much in their favour, not least being the relatively small amount of time required for design. In the simplest and cheapest form of system the cooler coil in the unit receives chilled water at the lowest practical temperature and does both the sensible and latent cooling. Fresh air enters fortuitously through ill-fitting windows and condensate off the coil is drained to waste. A small improvement accrues if outside air, plus traffic noise and dirt, comes in through a purpose-made hole in the wall behind the unit, induced through the casing and mixed with recirculated air by the fan, prior to being blown over the cooler coil and delivered to the room.

A third possibility is to add a mechanical ventilation system that handles filtered and heated (as necessary) fresh air, ducted at low velocity to supply grilles located at high level on the corridor side of the room and blowing towards the windows. While this assists proper air distribution and removes noise and dirt from the fresh air, the fan coil units still run wet. In this connexion, if a changeover mode is used, wet and dirty unit coils may generate unpleasant smells, absorbed during the previous form of operation, as they dry out when hot water circulates.

The fourth and best option is to provide a cooler coil for dehumidification in the central air-handling plant and arrange for the fan coil units to run dry, doing only sensible cooling. The units will then cost more money because they will remove less sensible heat with the higher secondary chilled water temperature necessary to prevent latent cooling. This extra cost is partly offset by the sensible cooling contribution of the ducted air. As with the induction system, there is no need to install piping for condensate drainage. Any condensate formed during start-up, drains into an emergency collection tray beneath the cooler coil and is subsequently re-evaporated.

When fan coil units are installed at high level, above suspended ceilings, it is essential to ensure that the air distribution in the room is satisfactory and that any ducted fresh air is also properly admitted to the room.

The air supplied from the fan coil unit must be smoothly fed into a ceiling diffuser that will give draught-free, quiet air distribution. The used air must be extracted from the room in a proper fashion that does not result in any short-circuiting or noise. If there is any doubt about the air distribution, tests should be carried out to establish air temperatures, air velocities, and noise levels, either in the manufacturer's works or in a full-scale mock-up. The fan in the unit must be able to develop enough additional pressure to offset the resistance to airflow of any air distribution ducting and diffusers attached to the unit.

When a fresh-air ducted supply is used in conjunction with fan coil units located above a suspended ceiling, the fresh-air duct branch for each module should be extended to be close to the back of the fan coil unit (see Figure 3.1). The fresh air will then be handled as part of the fan coil unit duty and be properly distributed in the room. When this is done, the effect the fresh air has on the state of the air entering the fan coil unit must be considered and its influence on the capacity of the fan coil unit assessed.

HEATING

If a two-pipe fan coil system is used in the UK climate, a difficulty arises with heating. The options are:

(1) Use a changeover system. This is unsuitable for the UK climate and for most buildings generally because of the thermal inertia of the piping system and the building itself. This has been discussed in Section 2.10.

(2) Supply warmed air from an auxiliary mechanical ventilation system. This is also not suitable, for the following reasons:

  (a) Heat losses from the supply duct, even though it is lagged, will mean that the temperature of air delivered to rooms remote from the heater battery in the air handling unit will be too low. Attempting to deal with this by increasing the air quantity supplied is self-defeating and a waste of time.

  (b) It is not possible to balance a ducted supply air system perfectly. Plus or minus 10% in the supply of fresh air may escape notice but such a typical imbalance will give a significant variation in the heating capacity.

  (c) Air supplied from a grille at high level in the corridor wall (say) is really in the wrong place for single glazing. Air ought then to be delivered under the windows to offset downdraught, and so on.

  (d) If the supply air is delivered at high level at a temperature exceeding 35°C, stratification becomes a difficulty, with inadequate heating in the occupied part of the room.

(3) Use a four-pipe, twin coil, system. This is the correct choice for a fan coil system in a temperate climate.

FAN COIL UNIT CAPACITY

Most fan coil units are available for horizontal or vertical mounting and provided with a metal casing for immediate installation, or are uncased when they are fitted behind or above joiners' or builders' work panels. Manual selection for running the unit fans at high, medium or low speed is usual, and for applications in the UK it is customary to select units to match office loads at medium or low speed, but for hotel bedrooms selection is at low speed. Units in offices would normally be operated at a fixed speed after installation, without the option of an easy speed change, but in hotel bedrooms guests would have the chance to select a higher speed, giving greater-than-design capacity with a higher-than-design noise. In warmer climates a choice at high speed is often the rule, the greater noise levels being acceptable. As with cooler coils generally, total cooling is largely a function of entering wet-bulb temperature but when coils are chosen for sensible cooling only, the dry-bulb dominates. Commercial fan coil units, suitable for dealing with the loads in individual offices or hotel bedrooms, have cooling capacities from about 600 W to as much as 70 000 W, depending of course on the entering air state, with fans running at high speed and when fed with chilled water at entering temperatures between 11 and 4.4°C. Air quantities vary but one typical manufacturer offers four unit sizes with fans handling 100, 150, 200 and 300 $l\,s^{-1}$ of air.

Performances are usually shown in tables but catalogue information, particularly for sensible cooling, is seldom well presented. However, as with induction units, sensible cooling capacity can be regarded as proportional to the difference between the entering air dry-bulb and the entering chilled water temperature, for a given water flow rate and fan speed. Selection to give an exact match between sensible and latent loads and corresponding fan coil capacities is rarely possible and, as long as the sensible gain is dealt with in a satisfactory way, this need not matter very much because of the relative unimportance of humidity in human comfort. Figure 2.25 shows performance curves, based upon technical information extracted from a manufacturer's catalogue, for fan coil units handling 200 $l\,s^{-1}$; with the fan

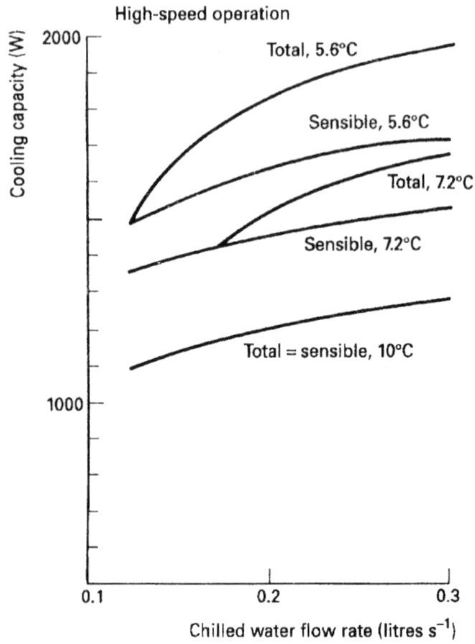

Figure 2.25 Performance of a typical fan coil unit, with a three-row cooler coil, for various entering water temperatures and an airflow duty of 200 $l\ s^{-1}$. State onto the fan coil unit = 22°C dry-bulb, 15°C wet-bulb.

running at high speed. The way in which the latent capacity falls off as the flow rate diminishes can be seen. For an entering water temperature of 7.2°C, for example, the sensible and total curves merge at a flow rate of about 0.17 $l\ s^{-1}$ when the thermal resistance of the water film within the tubes is great enough to ensure a surface temperature entirely above the entering dew-point. With a higher temperature, say 10°C, no latent cooling occurs at all for the entering air state of 22°C dry-bulb and 15°C wet-bulb (sling) quoted.

EXAMPLE 2.16

Select a fan coil unit to deal with the design heat gains to a west-facing module on an intermediate floor in the hypothetical office block. Assume a central mechanical ventilation plant delivers filtered, but uncooled outside air to the modules at a rate of 1.4 $l\ s^{-1}\ m^{-2}$ expressed at the outside state and allow a temperature rise of 0.3 K to cover supply fan power. Assume also that chilled water is available at 5.6°C.

Answer

The outside air delivered to the offices constitutes part of the load on the fan coil units. For the established outside and inside design states, caused in Example 1.9, the loads for a west-facing module on an intermediate floor are as follows.

Fresh air sensible load (by Equation (2.3)):

$$[(14.4 \times 1.4)/10^3] \times (28.3 - 22) \times [358/(273 + 28)] = 0.151 \text{ kW}$$

Fresh air latent load (by Equation (2.4)):

$$[(14.4 \times 1.4)/10^3] \times (10.65 - 8.3667) \times [856/(273 + 28)] = 0.131 \text{ kW}$$

Hence

Sensible gain at 1500 h sun time in July:	1394 W
Fresh air sensible gain:	151 W
Total sensible load on the fan coil unit:	1545 W
Latent heat gain at 1500 h sun time in July:	134 W
Fresh air latent load:	131 W
Total latent load on the fan coil unit:	265 W

The overall load on the fan coil unit is 1545 + 265 = 1810 W.

Figure 2.25 shows that, when running at full fan speed and delivering 200 l s^{-1} of air, a typical four-pipe fan coil unit, using three rows for cooling with chilled water flow rate of 0.3 l s^{-1} at a temperature of 5.6°C, has a sensible cooling capacity of 1720 W and a total capacity of 1980 W. By difference, the latent cooling capacity is 260 W. The capacity multiplier at medium speed is 0.85. Hence the capacities at medium speed will be: 1462 W, sensible, 1987 W total and 526 W latent, by difference. The sensible gain to be dealt with by the fan coil unit is 1545 W and so the unit will probably have to run at full speed during summer design weather but may be reduced to medium speed, with a lower sound power level, for a significant part of the time. The latent capacity of 260 W almost equals the calculated latent load of 265 W when the unit is running at full speed but with a reduced airflow over the coil, at medium speed, more dehumidification is done and the latent capacity of 525 W is rather more than the latent gain of 265 W. This is not uncommon and results in a lower humidity than intended in the conditioned space. There is very little effect on human comfort but, if the difference is excessive, the cooling load for the whole building may be increased.

The latent capacity of a fan coil unit is approximately proportional to the difference between the room dew-point and the chilled water flow temperature onto the coil. At the same time, the latent heat gain to the module, as given by Equation (1.12), is dependent on the moisture content in the room. Since moisture content is directly related to dew-point, the latent heat gains to a module can be expressed in terms of room dew-point. Hence the approximate humidity in the room can be determined.

EXAMPLE 2.17

For the case of Example 2.16, determine the approximate humidity in a west-facing module if the fan coil runs at medium speed.

Answer

The latent gain by Equation (1.2) and the latent gain due to the mechanical ventilation system, by Equation (2.4), is:

$$\begin{aligned} q_l &= [0.8 \times 0.5 \times 37.44\,(10.65 - g_r) + (2 \times 50)] + [(1.4) \times 14.4) \times (10.65 - g_r) \\ &\quad \times [856/(273 + 28)] = 14.976 \times (10.65 - g_r) + 100 + 57.332\,(10.65 - g_r) \\ &= 870.08 - 72.31\, g_r \end{aligned}$$

For a room condition of 22°C dry-bulb, 50% saturation, the dew-point is 11.3°C, the moisture content is 8.366 g kg^{-1} and the latent load on the fan coil unit is 265 W. This establishes point 1, in Figure 2.26. If the room dew-point is 5.6°C (equal to the chilled water flow temperature onto the fan coil unit) the corresponding room moisture content is found from psychrometric tables or a chart, to be 5.658 g kg^{-1} and, using the equation obtained above for the particular case being considered here, the latent heat gains are calculated as $870.08 - 72.31 \times 5.658 = 461$ W. This establishes point 2 in Figure 2.26. Joining points 1 and 2 gives the line representing the latent load on the fan coil unit. When the state of the air onto the fan coil unit is 22°C dry-bulb and 15°C wet-bulb (Figure 2.25) the room dew-point is 10.1°C and the latent cooling capacity at medium speed is 525 W. This establishes point 3 in Figure 2.26. When the room dew-point equals the chilled water flow temperature of 5.6°C, the latent cooling capacity of the fan coil unit in zero. This establishes point 4 in Figure 2.26. Joining points 3 and 4 establishes the line for the latent cooling capacity of the fan coil unit. The two lines, for latent load and latent capacity, intersect at a dew-point of about 8.7°C. At a dry-bulb temperature of 22°C the corresponding humidity is about 42% saturation, from psychrometric tables or a chart.

EXAMPLE 2.18

Select a fan coil unit to deal with the design heat gains to a west-facing module on an intermediate floor of the hypothetical office block assuming that a central mechanical system delivers dehumidified air at a state of 13°C dry-bulb, 7.702 g kg^{-1}. Secondary chilled water at 11°C is available. Refer to Example 1.11 and Figure 1.4, as necessary.

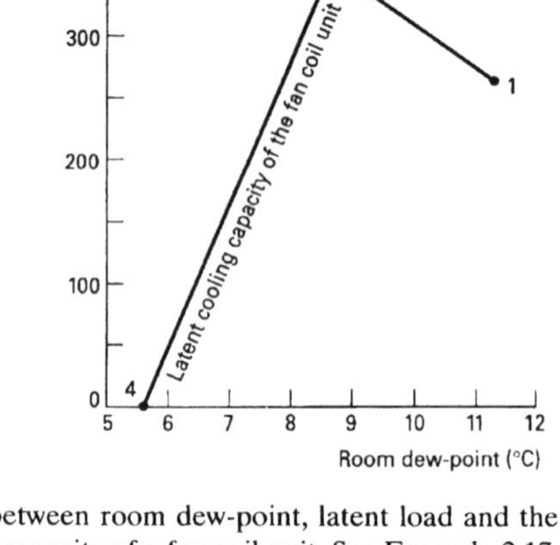

Figure 2.26 Relationship between room dew-point, latent load and the approximate latent cooling capacity of a fan coil unit. See Example 2.17.

Answer

To determine the cooling load in Example 1.11 for the whole of the hypothetical office block, when conditioned by a fan coil system, it was established that 46.071 $m^3 s^{-1}$ of air at 13°C and 7.702 g kg^{-1} must be delivered to the building, in order to deal with the design latent heat gains. There is a total of 864 modules in the entire building and hence the supply of air to each module is 46.071 × 1000/864 = 53.32 l s^{-1}, at 13°C.

The sensible cooling capacity of this air, using Equation (2.3), is 53.32 × (22 − 13) × 358/(273 + 13) = 601 W. Hence the sensible heat gain to be dealt with by the fan coil unit is 1394 − 601 = 793 W. Figure 2.27 shows that a size 3 unit, running at high speed, will give a sensible cooling capacity of about 793 W when the chilled water flow rate is 0.16 l s^{-1}. It is seen in Figure 2.28 that a fan coil unit running at high speed produces a sound pressure level that is about 5–8 dB louder than a unit running at medium speed, if the room effect is zero (see Section 7.8). A size 4 unit, running at high speed, has a sensible cooling capacity of about 1070 W with a chilled water flow rate of 0.16 l s^{-1}. Applying a multiplier of 0.75 gives a sensible cooling capacity of about 802 W when running at slow speed. This will give a quiet but more expensive unit. However, there is reduced scope for good capacity control by throttling the flow of chilled water: 0.16 l s^{-1} is at the bottom end of the flow rates for a size 4 unit. Under such circumstances, if the larger but quieter unit is used, the sensible cooling capacity might be increased by using a flow rate nearer to the middle of the range (with a better control potential) but then reduced to 793 W by choosing a higher chilled water flow temperature than 11°C. The sensible cooling capacity of a fan coil unit, for a given fan speed, is approximately proportional to the difference between the room dry-bulb and the chilled water flow temperature onto the unit. Such options have commercial consequences which must not be overlooked.

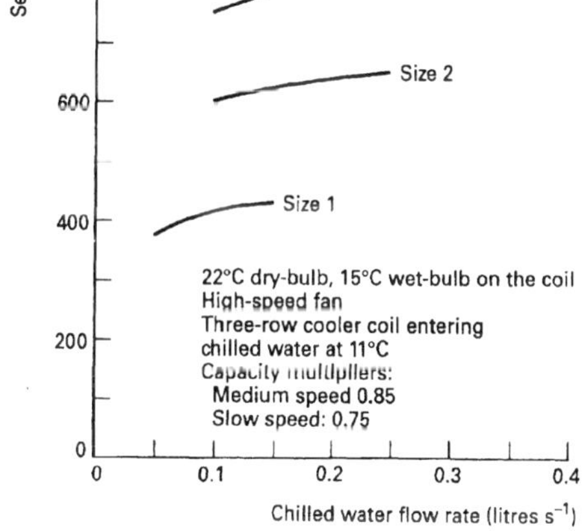

Figure 2.27 Typical sensible cooling performance of fan coil units.

Reference to Figure 4.7 shows the need to provide system pressure control when units are controlled by throttling water flow: the upstream pressure increases as the unit valves throttle. If this risk is ignored, the system will control poorly and be noisy at low duties.

With a four-pipe fan coil unit, the capacities of the cooling and heating coils are controlled in sequence. Hence re-heat never occurs. The heating capacity of the unit must be enough to deal with the heat loss and to warm the air introduced by the central mechanical system. The flow temperature of the LTHW supplied to the units is compensated against outside air temperature for normal operation but with intermittent system operation the flow temperature is elevated to provide the necessary boosted capacity in early morning heat up periods. During such periods the central air-handling plant would be off, with the dampers on its outside air inlet and discharge openings tightly closed.

EXAMPLE 2.19

Calculate the heating capacity required from a four-pipe fan coil unit for a module on an intermediate floor of the hypothetical building. Assume that the central mechanical air handling plant delivers 52.32 l s^{-1} of air at 13°C. Take the inside and outside temperatures in winter as 20°C and −2°C, respectively, and assume one air change per hour of natural infiltration. Refer to Section 1.2, as necessary, for details of the module.

Answer

Heat loss:	
Glass: 3.168 × 3.3 × (20 + 2)	230
Wall: 4.753 × 0.45 × (20 + 2)	47
Infiltration: (1.0 × 37.44/3) × (20 + 2)	275
Total heat loss:	552 W
Air heating: 53.32 × (20 − 13) × 358/(273 + 13)	467
Total heating capacity needed:	1019 W

Four-pipe systems having fan coil units with only a single coil, used for cooling or heating in sequence, are possible. They are not recommended: proper design is lengthy and hydraulic problems abound. *This form of four-pipe system should never be used.* There are also three-pipe systems which have a pair of flow lines, one for chilled water and the other for LTHW, supplying a single coil. Water leaving the coil is fed into one pipeline and returned either to the boiler or to the water chiller. The water returned to the plant is too warm for the chiller and too cold for the boiler. Mixing losses in the return line are very large and *three-pipe systems must never be used.*

NOISE

The noise mostly originates from the fan but sound power levels can vary according to the type of fan used, its bearings and mountings, the rigidity of the metal casing and the smoothness of airflow from the discharge grille. Manual changes in fan speed give variations in unit capacity according to the typical factors in Table 2.10, and refer to the units typified by Figures 2.25 and 2.27. It is possible to get sound power values from manufacturers and, when

Table 2.10 Typical variations in unit capacity with fan speed

Unit	Low fan speed		Medium fan speed	
	Total capacity	Sensible capacity	Total capacity	Sensible capacity
1	0.90	0.89	0.81	0.79
2	0.90	0.88	0.80	0.77
3	0.85	0.85	0.75	0.74
4	0.85	0.80	0.70	0.67

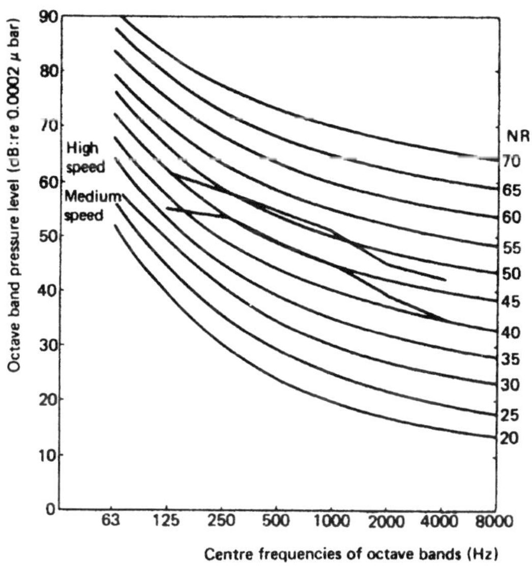

Figure 2.28 Noise ratings achieved with a size 3 fan coil unit in an office with zero room effect.

such information is used, the approximate acoustic performance of a new unit as shown in Figure 2.28.

CONTROLS

There are two basic ways of controlling unit capacity: cycling the unit fan and varying the chilled water flow rate. Cycling the fan from a thermostat behind the recirculation grille is the cheapest but least satisfactory method. Although adequate control is achieved by this method under tropical conditions, the variation in air movement and noise is seldom well-received in temperate climates. Water control, preferably by two-port modulating valves, is the best method although three-port valves also give good results.

Not much energy may be saved by varying the fan speeds. Shaded pole motors with low power factors are used and speed reduction in achieved by dropping the supply voltage across a resistance.

2.12 Chilled ceilings and chilled beams

The method of air conditioning by chilled ceilings has not proved particularly popular even though it is most economical to run, occupies no lettable floor area, requires no additional expenditure on joiners' work casings for terminal units, is silent in operation and includes an acoustic ceiling. Three possible explanations for this lack of application are: an apparently higher capital cost; some extra depth needed to accommodate the ceiling; and a comparative inflexibility in responding to partition rearrangement. However, insufficient credit is allowed for the acoustic ceiling and for the absence of joiners' work casings on terminal units. Secondly, as little as 200 mm is sufficient to accommodate the ceiling if, as is desirable, the auxiliary ventilation ducting is run above the ceiling in the central corridor, to feed side-wall grilles in the rooms. In any event, 200 mm is needed for a suspended ceiling with recessed light fittings. Thirdly, a suspended metal pan ceiling can carry chilled water pipes at 200, 300 and 450 mm centres, to offer a fair degree of rearrangement possibilities.

Originally, chilled ceilings were provided by pipe coils embedded in the soffit of concrete floor slabs during building construction and finished with an appropriate rendering or plaster. Although this approach permitted pipe centres as close as 100 mm, with greater cooling capacity, capital costs were high and it has been replaced by acoustic pan ceilings, of various types, clipped to pipes and suspended by short drop rods from the soffit of the slab. Figure 2.29 shows one such type of chilled ceiling and in Figure 7.5 the typical acoustic performance is shown for 600 × 600 mm panels with 25 mm of fibreglass laid above the pipe coils. Different makers offer various forms of chilled ceiling. The acoustic quality is given by perforations in the panels with a glass fibre blanket above the pipes, as shown in Figure 2.29, or glass fibre pads, encased in polythene bags, between the pipes. An auxiliary air handling plant is essential, to deliver filtered, cooled and dehumidified air through a low velocity ducting system to side-wall grilles or ceiling diffusers. Ceiling diffusers may not be suitable because the ceiling depth is not always enough to allow horizontally-flowing, ducted air to be turned smoothly into the diffuser necks and noise will result if there is an uneven distribution of air over the diffuser cones. The auxiliary air is enough for ventilation and,

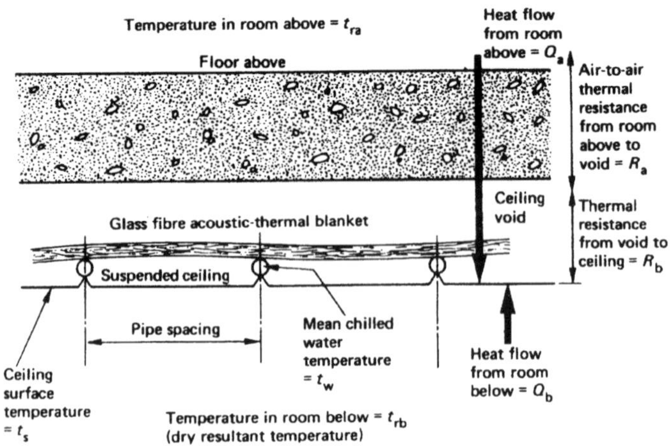

Figure 2.29 Section of one type of chilled suspended ceiling showing heat flows, thermal resistances and temperatures.

being mixed with recirculated air as necessary, deals with all the latent heat gains and some of the sensible heat gains in the treated rooms. The remainder of the sensible gain, and also any peripheral heat loss, is offset by the ceiling. Variations in sensible gain are met by thermostatic control over the flow rate of chilled water through the pipework, on a modular basis. For unfinned tube the thermal resistance of the inside water film is a comparatively small proportion of the overall, water-to-air, thermal resistance and it is, therefore, important that the water temperature is above the room dew-point at all times. Therefore, a secondary chilled water circuit is required and proportional plus integral plus derivative control must be exercised over its flow temperature.

For the best results, according to Jamieson (1963), the variable part of the heat gain should be dealt with by the thermostatically controlled capacity of the ceiling and the relatively constant or predictable portion of the gain by the constant or programmable capacity of the auxiliary air system. This is best illustrated for the case of a conditioned core area where the lights are on constantly and would be ideally covered by the cooling capacity of the air. The variable load is provided by the people and this is offset by the chilled ceiling. When using chilled ceilings for perimeter applications the part of the piping circuit nearest to the window wall can have a heating capacity. The piping circuit feeding the modular panels would be split so that the two or three pipes parallel to the wall are fed with compensated LTHW in sequence with chilled water, whereas the remainder of the piping receives chilled water only.

HEAT TRANSFER FROM CHILLED CEILINGS

Heat transfer from a ceiling is by both radiation, q_r (W m^{-2}), and natural convection, q_c (W m^{-2}). In a simplified, theoretical form these are expressed by the following equations

$$q_r = 5.67\varepsilon[(T_s/100)^4 - (T_{rm}/100)^4] \tag{2.14}$$

where ε is the emissivity of the emitting surface, T_s (K) is its absolute temperature and T_{rm} (K) is the mean radiant temperature of the room.

$$q_c = 1.3\ (t_s - t)^{1.25} \tag{2.15}$$

for the case of a horizontal surface looking downwards, where t_s is the surface temperature (°C) and t is the air temperature (°C).

EXAMPLE 2.20

Given that the surface temperature of a ceiling is 16°C, that its emissivity is 0.93, that the air temperature is 23°C and that the mean radiant temperature of the room is 21°C, determine the total, theoretical, thermal absorption by the ceiling, using Equations (2.14) and (2.15).

Answer

$$\begin{aligned} q_r &= 5.67 \times 0.93 \times [((273 + 16)/100)^4 - ((273 + 21)/100)^4] \\ &= -26.1 \text{ W m}^{-2} \end{aligned}$$

$$\begin{aligned} q_c &= 1.3 \times (16 - 23)^{1.25} \\ &= -14.8 \text{ W m}^{-2} \end{aligned}$$

Total thermal absorption = −26.1 − 14.8 = −40.9 W m^{-2}

Human comfort depends on air dry-bulb temperature, air movement, mean radiant temperature and, to a less extent, relative humidity. A chilled/heated ceiling offers opportunities for providing comfort by using a higher dry-bulb in summer and a lower dry-bulb in winter, than would be the case for systems having convective methods of heat transfer. When following this approach and using Equation (2.14) a problem arises in the use of mean radiant temperature. This is defined as the temperature of a notional, blackened, copper sphere that surrounds the point of measurement and radiates heat to the point at the same rate as the actual surfaces of the room do. Thus the value of the mean radiant temperature varies from place to place in a room and this poses a difficulty in choosing a value for T_{rm} when using Equation (2.14). Hence it is common to use the mean surface temperature of the room, T_{sm} (K), in Equation (2.14), instead of mean radiant temperature, because it does not vary with position in the room.

EXAMPLE 2.21

For the case of a west-facing module on an intermediate floor of the hypothetical office block, calculate the mean surface temperature at 15.00 h sun time in July, making use of the following data, as necessary and assuming that internal Venetian blinds are fitted and adjusted to exclude the entry of direct solar radiation.

Total internal surface area of a single office:	72.48 m^2
Window area:	3.168 m^2
Internal area of the outside wall:	3.072 m^2
Floor area:	14.4 m^2
Gross ceiling area:	14.4 m^2
Area of luminaire:	0.9 m^2
Net ceiling area:	13.5 m^2
Partition area (including door to corridor)	37.44 m^2
Volume of room:	37.44 m^3
Surface temperature of the wall:	22.0°C
Surface temperature of the partitions and floor:	23.0°C
Surface temperature of the luminaire:	30.0°C
Room air dry-bulb temperature:	23.0°C
Surface temperature of the chilled ceiling:	t_c
Heat transfer coefficient of the outside surface of the glass:	22.7 W m^{-2} K^{-1}
Heat transfer coefficient of the surface of the Venetian blinds:	7.9 W m^{-2} K^{-1}

The CIBSE Guide (1986) gives the total solar irradiation normally incident on a west-facing window at 15.00 h sun time in July as 555 W m^{-2} and, according to Pilkington (1991), the absorption coefficient for 4 mm clear glass fitted with internal Venetian blinds is 0.44. Hence, considering the window–blind combination as a whole unit (at a temperature t_b), and assuming that part of the heat absorbed by the whole combination is emitted to the room and the remainder is emitted to outside, the approximate temperature of the blinds, t_b (°C), can be calculated:

Table 2.11 A summary of chilled ceiling temperatures for Example 2.21

t_c	(°C)	10	11	12	13	14	15	16	17
t_{sm}	(°C)	21.3	21.5	21.7	21.9	22.0	22.2	22.4	22.6

$$0.44 \times 555 = 7.9 \times (t_b - 23) + 22.7 \times (t_b - 28)$$

whence $t_b = 34.7°C$.

The total internal surface area of the room is 72.48 m^2 and hence the mean surface temperature can be calculated:

$$\begin{aligned} t_{sm} &= [(3.168 \times 37.4) + (3.072 \times 22) + (14.4 \times 23) + (13.5 \times t_c) + (0.9 \times 30) + (37.44 \\ &\quad \times 23)]/72.48 \\ &= 19.39 + 0.19t_c \end{aligned}$$

Making assumptions for the ceiling temperature the consequent mean surface temperatures for the room are found, as Table 2.11 summarises.

For a typical room state of 22°C dry-bulb and 50% saturation the room dew-point is 11.3°C and, remembering that the ceiling temperatures are mean values this seems likely to rule out ceiling temperatures less than about 13°C. It appears that, with the above theory and assumptions, achieving a mean surface temperature less than 21.5°C is going to be difficult.

THE ACTUAL COOLING CAPACITY OF A CHILLED CEILING

The sensible cooling capacity of a chilled ceiling depends on several practical factors:

(1) The pitch of the pipes. The interval between pipes should normally not exceed 450 mm for cooling. With some makes of ceiling a pipe spacing of 600 mm will not give enough sensible cooling although it can be used for heating.

(2) Water velocity within the pipes. If this is too slow not only will the water temperature rise be large, increasing the mean temperature of the ceiling, but laminar flow will prevail within the pipes and heat transfer from the tube walls to the water will fall off dramatically. The mean water velocity should not be less than 0.3 m s^{-1} for sizes up to 500 mm diameter if air bubbles are to be conveyed, although 0.6 m s^{-1} is better. Velocities should not exceed 2.3 m s^{-1}, otherwise erosion of the pipe walls will occur, particularly at the heels of bends and if the water is very acidic or alkaline. A water temperature rise of between 2–8 K is possible and about 5 K is typical.

(3) The tightness of the grip between the pipes and the suspended ceiling panels.

(4) The material of the ceiling.

(5) The finish on its lower surface.

(6) The difference between the room temperature and the mean temperature of the ceiling surface. (It is not always clear from manufacturers' literature what is meant by room temperature.)

(7) The cooling capacity is modified by the airflow over the ceiling: side-wall grilles or diffusers blowing across the ceiling can increase convection by 10%.

In principle, the design procedure is as follows:

(a) Decide on a practical off-coil state, consistent with the type of water chiller to be used and the method to be adopted for controlling its capacity. Determine also the state of the air returned to the air handling plant for recirculation.
(b) Determine the minimum fresh airflow rate for summer design conditions.
(c) Calculate the auxiliary supply airflow rate to deal with the latent heat gains.
(d) Establish the psychrometry for the auxiliary airflow, verify that the cooler coil performance is practical by determining the contact factor and calculate its cooling load.
(e) Calculate the sensible cooling capacity of the auxiliary supply airflow rate and determine the sensible cooling load remaining for the chilled ceiling to deal with.
(f) Decide on the area of ceiling available in the module for the chilled ceiling.
(g) Refer to the manufacturer's data for the cooling capacities possible from the ceiling and follow the procedure recommended by the manufacturer for selecting a ceiling to deal with the remaining sensible heat gain.

EXAMPLE 2.22

Determine the necessary supply airflow rate of low velocity, ducted, auxiliary air and the sensible cooling capacity required from a chilled ceiling, for a west-facing module on an intermediate floor of the hypothetical building. Use the results of Examples 1.9 and 1.10, as appropriate. Take the supply fan total pressure as 0.625 kPa with the motor in the airstream to give a temperature rise of 0.75 K. Assume the supply duct heat gain causes a temperature rise of 2.25 K and take the total return air temperature rise to be 0.3 K. The refrigeration plant uses reciprocating compressors and produces chilled water at 6.5°C under design load conditions. Assume the minimum fresh air allowance is 12 l s^{-1} for each person and that there are two people present, requiring a total of 24 l s^{-1}.

Verify the psychrometry for the auxiliary ducted air supply. Decide on the area of ceiling possible for cooling and consider the possibility of selecting a ceiling based on the capacity shown in Figure 2.30.

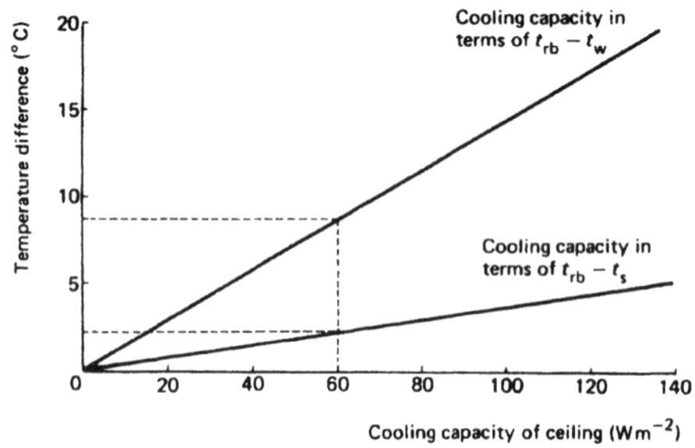

Figure 2.30 Typical capacity of an aluminium pan ceiling with chilled water pipes at 300 m centres. t_{rb} is the temperature in the room beneath the ceiling.

Answer

Outside design state: 28°C dry-bulb, 19.5°C wet-bulb (sling), 10.65 g kg^{-1}, 55.36 kJ kg^{-1}
Room design state: 22°C dry-bulb, 50% saturation, 8.366 kg^{-1}, 43.39 kJ kg^{-1}, 11.3°C dew-point.
Sensible heat gain (Example 1.9): 1394 W
Latent heat gain (Example 1.10): 134 W

(a) With a cooler coil having six rows of tubes, 315 fins per metre (8 fins per inch), 2.5 m s^{-1} face velocity and 1.0 m s^{-1} water velocity, the minimum practical off-coil state with a chilled water flow temperature of 6.5°C is 10.5°C dry-bulb, 10°C wet-bulb (sling), 7.448 g kg^{-1} and 29.33 kJ kg^{-1}.

The state of the air returned to the air handling plant is 22.3°C dry-bulb and 8.366 g kg^{-1}. Hence its enthalpy is determined from psychrometric tables or a chart as 43.70 kJ kg^{-1}.

(b) Assuming two people are present the minimum fresh air allowance is $2 \times 12 = 24$ l s^{-1}.

(c) The supply air temperature, t_s, is $10.5 + 0.75 + 2.25 = 13.5$°C. By Equation (2.4) the auxiliary ducted airflow rate needed to deal with a latent gain of 134 W is:

$$\dot{v}_{13.5} = [134/(8.366 - 7.448)] \times [(273 + 13.5)/856] = 48.9 \text{ l s}^{-1}$$

Percentage fresh air = (24 × 100)/48.9 = 49.1% = 50%.

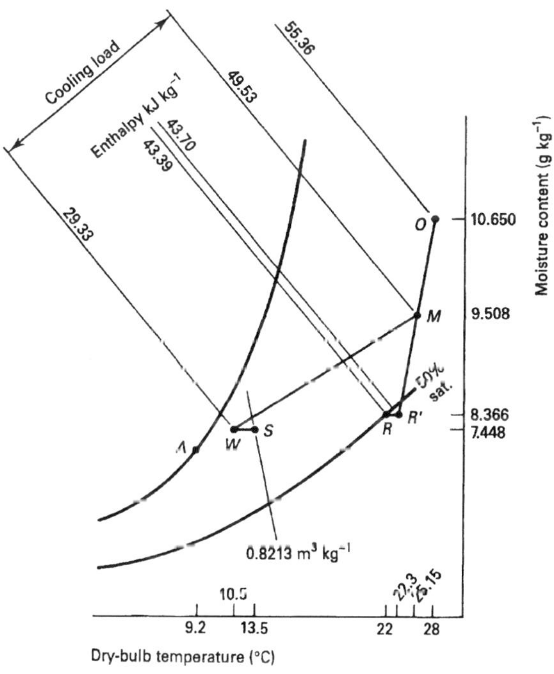

Figure 2.31 Psychrometry for the auxiliary ducted air supply in Example 2.22.

(d) Figure 2.31 shows a verification of the psychrometry. The apparatus dew-point, *A*, is found by the intersection of the process line, *M–W*, and the saturation curve, at a temperature of 9.2°C. The temperature of *M* is 25.15°C, midway between 22.3°C and 28°C, because 50% fresh air is used. Hence the contact factor, β, for the cooler contact is determined as

$$\beta = (25.15 - 10.5)/(25.15 - 9.2) = 0.92$$

This is a practical value for the sort of cooler coil specified. Noting that the auxiliary supply airflow is at state *S*, its specific volume is determined as 0.8213 $m^3\ kg^{-1}$ from psychrometric tables or a chart (Figure 2.31) and the cooling load of the auxiliary air supplied is calculated as

$$[48.9/(0.8213 \times 1000)] \times [49.53 - 29.33] = 1.203 \text{ kW}$$

(e) The sensible cooling capacity of the auxiliary ducted airflow is calculated from Equation (2.3)

$$\text{Sensible cooling capacity} = [48.9 \times (22 - 13.5) \times 358]/[(273 + 13.5)] = 519 \text{ W}$$

Hence the sensible cooling required from the chilled ceiling is 1394 – 519 = 875 W.
This represents 63% of the sensible gain. Referring to Example 1.9 it is seen that the gains from people, lights and business machines amount to 51% of the total and are likely to remain fairly constant. The remainder of the gains (glass, wall, infiltration and solar) are 49% and are likely to vary and to be matched by the ceiling, under thermostatic control.

(f) The maximum number of 300 × 300 mm ceiling tiles that could be accommodated in a single module is 8 × 20. A strip of half-tiles would have to be left to run adjacent to each of the four walls of the module leaving 7 × 19 as the maximum number of tiles, having an area of 11.97 m^2. From this the area of the luminaire must be deducted and the net area available for cooling is then 11.97 – 0.9 = 11.07 m^2.

(g) With a room dew-point of 11.3°C (at 22°C dry-bulb, 50% saturation) the lowest acceptable chilled water flow temperature at the ceiling is 11.5°C and, taking a water temperature rise of 3.3 K through a ceiling panel coil, the mean chilled water temperature is 11.5 + 1.65 = 13.15°C. Using the notation of Figure 2.30, $t_{rb} - t_w = 22 - 13.15 = 8.85$ K and the upper line in Figure 2.30 shows that the ceiling absorbs about 60 W m^{-2}. The lower line in the figure shows that $t_{rb} - t_s$ is equivalent to 2 K, so the mean surface temperature of the ceiling is about 20°C. Although lagging is provided above the pipe coils a ceiling also absorbs some heat from the module above. Assuming a thermal resistance (R_b) between the void and the ceiling of 0.4 $m^2\ K\ W^{-1}$ and a resistance (R_a) between the void and the room above of 0.5 $m^2\ K\ W^{-1}$ (which could be calculated) the heat absorbed by the chilled ceiling in the module beneath the floor considered is (22 – 10)/(0.4 + 0.5) = 2.2 W m^{-2}. This represents about 3.5% of the total cooling capacity in this case but it clearly depends on the values used for the two thermal resistances. Hence the total sensible cooling capacity provided by the ceiling is 62.2 W m^{-2}. The available area of ceiling for cooling is 11.07 m^2 and the sensible cooling capacity obtainable is 62.2 × 11.07 = 689 W. This is not enough to match the design sensible load of 875 W.

It would no doubt be possible to increase the auxiliary air supply rate but this would increase the proportion of the cooling capacity that was not under thermostatic control from room temperature and might prejudice good performance. The correct solution would be to reconsider the pitch of the pipes and perhaps choose 200 mm instead of 300 mm. Alternatively, another manufacturer's product, with a tighter grip between the pipes and the ceiling panels, might offer more cooling capacity and a better answer to the problem. Commercial considerations would dominate.

CHILLED BEAMS

These are available in various forms and the aim is not to use the whole ceiling, but to locate the cooling capacity parallel to the lines of the partitions, the window, or the corridor wall. A hollow sheet metal beam usually contains one or more pairs of finned tubes, running parallel to the major beam axis. The auxiliary air supply duct may run between the finned tubes and deliver air to the room in a way that induces airflow from the room or the ceiling void over the finned tubes. Sometimes the ducted air supply is separate and cooling is from natural convection over the finned tubes in the beam. Thermostatic control of room temperature is by means of two- or three-port motorised valves regulating the chilled water flow rate through the finned tube. With some arrangements, the surfaces of the sheet metal beam are also the walls of the auxiliary air supply duct and become cooled by direct contact with the airflow. These surfaces then give a small measure of radiant cooling in addition to the convective cooling provided by the finned tube and the ducted air supply. Cooling is largely by convection and sensible cooling capacities seem to be in the range of 20–270 W, per metre of beam length. As a generalisation, sensible cooling capacities appear to be about 60 W m^{-2}, referred to the floor area, but more is possible, depending on the temperature difference between the room air and the chilled water.

Beams may be mounted in a suspended ceiling or surface-mounted. Typical dimensions of the metal casings are 150–300 mm wide × 200–300 mm deep and a clearance of 50–100 mm is required above some types of beam. Beam lengths are available in suitable increments to total lengths of about 1.2–5.0 m.

HEATING BY CEILINGS AND BEAMS

The methods of heating already considered for induction and fan coil units apply equally here. The only correct system for the UK is four pipe.

A criticism of radiant heating from ceilings in the past has been that down draughts from single glazing are not dealt with in a satisfactory manner. An answer to this is that radiation from the ceiling warms the floor and the wall next to the window and this gives rising convection currents that deal with the down draught. In any case, the increased use of double glazing in the UK, with *U*-values of 3.3 W m^{-2} K^{-1}, or less, has reduced the likelihood of down draughts at windows.

CHILLED WATER DISTRIBUTION

Chilled ceilings and chilled beams must only provide sensible cooling. Consequently, a secondary chilled water circuit is essential, with proportional plus integral plus derivative control over the flow temperature. See Figure 4.7.

FREE COOLING

A supply of secondary chilled water has to be available to all air–water systems throughout 12 months of the year. This is because in modern offices there are often net heat gains from people, lights, business machines and solar radiation through windows, in winter as well as in summer, resulting in a cooling load for the building. The options for providing secondary chilled water are as follows:

(1) Run the refrigeration plant all the year round. This is certainly feasible, provided that the correct steps are taken to allow the cooling tower (or air-cooled condenser) to run safely in cold winter weather. This method consumes energy and increases running costs.
(2) Use the dehumidifying cooler coil in the air handling plant to chill the water. See Figure 4.7 and Section 2.10. If the compressor in the water chiller is switched off but the primary chilled water pump is kept running, cold outside air at a temperature of less than about 5°C, flowing over the outside of the coil can chill the water flowing inside the coil tubes to about 11 or 12°C, which is likely to be satisfactory for the secondary circuit. Protection against freezing is needed for the chilled water.
(3) Use a thermosyphon cycle. Pearson (1990) has shown that significant cooling can be obtained for much of the year in the UK by switching off the compressor of a refrigeration plant and allowing a natural circulation of refrigerant to chill the water flowing through the evaporator. For free cooling in cold weather, the compressor is switched off and chilled water flowing through a shell-and-tube evaporator is chilled by giving up heat to the liquid refrigerant, which is boiling at a lower temperature. The vapour by-passes the compressor and rises naturally to the condenser, at a higher level. This can be an air-cooled condenser or a water-cooled condenser, fed with water from a cooling tower handling outside air at a low wet-bulb, which is less than the temperature of the refrigerant vapour. The refrigerant condenses and flows by gravity back to the evaporator. The cycle repeats and is continuous. Cooling capacities as large as 30% of the design refrigeration duty have been claimed and it is said that the compressor can be switched off for a substantial part of the year in the UK.

Exercises

1 Calculate the maximum cooling load for the hypothetical office block (see Section 1.1), assuming it is conditioned by a high velocity, perimeter-induction system with a low-velocity extract system and that its major axis is aligned east-west. Take the moisture content in August as 9.65 g kg^{-1}.
(*Answer* 1432 kW, 102.3 W m^{-2})

2 Using Tables 2.5 and 2.6, design a VAV ductwork distribution system alternative to that in Example 2.10(a) for the hypothetical office block. Assume that (a) the two service areas for ducts, etc. are at the two short, windowless ends of the building and (b) the two service areas are within the main body of the building, on its major axis, each positioned one-quarter of the way from the short end wall nearest to it.

3 Select fan coil units, using Figures 2.25 and 2.27, to match the loads of eastern modules in the hypothetical office block (a) with chilled water at 5.6°C at the units with 1.4 l s^{-1} m^{-2} of fresh air being supplied by a mechanical ventilation system and (b) with chilled water at 11°C at the units to do sensible gains only, 46.5 l s^{-1} of dehumidified air at 13°C being supplied to each module, by an auxiliary air system.

4 An acoustically hard office has the following room effect:

Mid octave-band frequency (Hz)	125	250	500	1000	2000	4000
Room effect (dB)	−1	−1	0	−1	−1	−1

The office is air conditioned by four VAV terminals, each handling 80 l s^{-1} with a duct pressure of 0.75 kPa under design load conditions but reducing to 20 l s^{-1} with 0.25 kPa for minimum sensible heat gains. Making use of Table 2.7 and a standard sheet for NR values (Figure A.2), determine the NR value likely in the room for (a) design conditions and (b) when the sensible gains are at a minimum.
(*Answer* (a) NR 51; (b) NR 26)

5 (a) A room which is treated by a constant-volume, all-air system has sensible heat gains of 2040 W by transmission, 1800 W from people, 4320 W from lights, 10710 W by solar gain through glass and latent gains of 1340 W, based on 28°C dry-bulb, 19.5°C wet-bulb (sling) outside and 22°C dry-bulb, 50% saturation inside. The treated room is 374.4 m^3. The supply temperature is 14°C dry-bulb, the allowance for supply fan power and duct gain is 0.5°C and 10% by mass of the air supplied is from outside, the remainder being recirculated from the treated room. Using a psychrometric chart determine the supply air quantity and specific volume, the design states on and off the cooler coil, the design cooling load and the air change rate.
(b) Determine the outside air temperature at which the refrigeration plant can be switched off. What is the on-coil state immediately prior to switch-off? Assume that the inside state is 20°C dry-bulb, 34% saturation in winter at switch-off, that fixed proportions of fresh and recirculated air are handled and that the differential on the switching thermostat is ±1°C. Ignore solar heat gains in winter but assume full gains from people and lights. Assume the outside state is saturated at switch-off.
(*Answer* (a) 1.891 $m^3 s^{-1}$, 0.824 $m^3 kg^{-1}$, on-coil: 22.6°C dry-bulb, 15.9°C wet-bulb, off-coil: 13.5°C dry-bulb, 12°C wet-bulb, 24.06 kW of refrigeration, 18.2 air changes per hour; (b) 6.5°C; 18.6°C dry-bulb, 11.2°C wet-bulb)

6 A condensing set using R22 has a performance as follows when condensing at 40°C and using four cylinders:

Suction temperature (°C)	−10°	−5°	0°	+5°
Suction pressure (kPa)	219	261	309	362
Cooling capacity (kW)	57.9	77.3	99.8	126.5

The set is coupled to a direct-expansion cooler coil which cools air from 17°C wet bulb to 10°C wet-bulb, for which the calculated duty is 100 kW of refrigeration. Plot the characteristic of the condensing set for four and two cylinders in coordinates of temperature (abscissa) and cooling capacity (ordinate). Also plot the characteristics for the cooler coil and determine the evaporating temperature and duty for four cylinders. In addition, determining the evaporating temperatures and duties for two cylinders when the entering wet-bulb falls to 13°C. Allow 1°C as corresponding to the pressure drop in the suction line.
(*Answer* Evaporating temperature = 1°C, duty = 100 kW; evaporating temperature = 4°C, duty = 57 kW)

Notation

Symbols	Description	Unit
c_d	Coefficient of discharge	—
c_p	Specific heat capacity	J kg^{-1} K^{-1} or kJ kg^{-1} K^{-1}
c_s	Coefficient of discharge through a supply grille	—
d	Relevant linear dimension, or	m
	depth of a grille	m
f	Free area ratio of a grille	—
g	Moisture content, or	kg kg^{-1} or g kg^{-1}
	acceleration due to gravity	m s^{-2}
g_r	Room air moisture content	kg kg^{-1} or g kg^{-1}
g_s	Supply air moisture content	kg kg^{-1} or g kg^{-1}
H	Stack height	m
h	Height of a door, or	m
	enthalpy of air	kJ kg^{-1}
h_a	Air film heat transfer coefficient	W m^{-2} K^{-1}
$h_{\text{off coil}}$	Enthalpy of the air leaving a cooler coil	kJ kg^{-1}
$h_{\text{on coil}}$	Enthalpy of the air entering a cooler coil	kJ kg^{-1}
k	Thermal conductivity	W m^{-1} K^{-1}
n	Fan speed	rev min^{-1}
p	Static pressure	Pa or kPa
p_{at}	Atmospheric (barometric) pressure	kPa
p_{ss}	Saturation vapour pressure	kPa
p_{tF}	Fan total pressure	Pa or kPa
Q_s	Airflow rate through the open area of a supply grille	m^3 s^{-1}
Q_t	Airflow rate trying to enter the upper half of a doorway	m^3 s^{-1}
q_c	Heat transfer by convection	W m^{-2}
q_r	Heat transfer by radiation	W m^{-2}
R_a	Air-to-air thermal resistance between a room and the void under the floor slab but above the suspended ceiling	m^2 K W^{-1}
R_b	Air-to-air thermal resistance between the void above a suspended ceiling and the ceiling surface	m^2 K W^{-1}
r_m	Metal thermal resistance of a cooler coil	m^2 K W^{-1}
r_w	Water film thermal resistance of a cooler coil	m^2 K W^{-1}
S	Sensible/total heat transfer ratio	—
T	Absolute temperature	*K*
T_o	Outside air absolute temperature	K
T_{rm}	Mean radiant temperature	K
T_r	Absolute air temperature inside a building	K
T_s	Absolute temperature of a surface or,	K
	absolute supply air temperature	K
T_{sm}	Absolute mean temperature of the surfaces in a room	K
t	Temperature or surface temperature	°C
t_{ao}	Outside air temperature	°C
t_p	Primary air temperature	°C
t_r	Room temperature	°C

t_{ra}	Temperature in a room above a chilled ceiling	°C
t_{rb}	Temperature in a room beneath a chilled ceiling	°C
t_s	Supply air temperature	°C
t_w	Mean chilled water temperature in a chilled ceiling or entering secondary chilled water temperature	°C
U	Thermal transmittance coefficient	W m^{-2} K^{-1}
u	Air velocity	m s^{-1}
u_o	Outside wind speed	m s^{-1}
u_s	Air velocity through the open area of a supply grille	m s^{-1}
u_t	Velocity of airflow trying to enter a doorway	m s^{-1}
v	Specific volume	m kg^{-3}
$v_{off\ coil}$	Specific volume of the air leaving a cooler coil	m^3 kg^{-1}
$\dot{v}_t$	Volumetric airflow rate at temperature t	m^3 s^{-1} or l s^{-1}
W_f	Fan power	W or kW
w	Width of a door	m
(Nu)	Nusselt number = $h_a d k^{-1} \theta^{-1}$	—
(Pr)	Prandtl number = $c_p \mu k^{-1}$	—
(Re)	Reynolds number = $u d \rho \mu^{-1}$	—
β	Contact factor of a cooler coil	—
ε	Emissivity of a surface	—
η	Fan total efficiency, as a fraction	—
θ	Temperature difference	K
ρ	Air density	kg m^{-3}
ρ_o	Outside air density	kg m^{-3}
ρ_r	Air density inside a building	kg m^{-3}
ρ_s	Supply air density	kg m^{-3}
μ	Percentage saturation, or absolute viscosity	% N s m^{-2}
ϕ	Angle between an airstream and the vertical	degrees

References

ASHRAE, *Handbook and Product Directory*, Systems, 1966.

Keith Blackman, *Centrifugal Fan Guide*, Keith Blackman Ltd, 1986.

BS 848, Fans for general purposes, Part 1, Methods of testing performance, 1980.

BS 6583, Methods of volumetric testing for rating of fan sections, in central station air handling units (including guidance on rating), 1985.

CIBSE Guide, Fig C1.1, Variation of barometric pressure with altitude, p. C1-3, 1986a.

CIBSE Guide, Psychrometric properties at non standard barometric pressures, p. C1 2, 1986b.

CIBSE Guide, Properties of humid air, pp. C1-1–C1-66, 1986c.

CIBSE Guide, A2, Weather and solar data, Table A2.27, 1986d.

Jamieson, H. C. and Calland, J. R., The mechanical services at Shell Centre, *JIHVE*, **31**, 1, 1963.

Jones, W. P., *Air Conditioning Engineering*, 4th edn, pp. 4–34, 130, Edward Arnold, London, 1994.

Martin, P. L. and Curtis, D. M., *M–C Psychrometric Charts for a Range of Barometric Pressures* (in SI units), Troup Publications, 1972.

Pearson, S. F., Thermosyphon cooling, *Proc. Inst. of Refrig.*, 6-1, 1989–90.
Pilkington, Glass and transmission properties of windows, Pilkington Glass Ltd, February, 1991.
Shataloff, N. S., High velocity dual duct systems for multi-zone installations, *Air Conditioning, Heating and Ventilating*, August, 1964.

3

Applications

3.1 Principles

Since most plants operate at partial load for the majority of their life, the essential feature of good application is to choose a system with a capacity that can be satisfactorily controlled to match the full range of load variation expected, within the limits of the design brief. The nature of the load is also a very important factor. For example, a theatre auditorium with a large population is better suited by an all-air system that can easily provide the vast quantity of fresh air needed than by an air–water system which probably cannot. The size of the application also has a bearing on the choice of systems, for example small loads are dealt with more cheaply by direct-expansion, all-air systems than by water-chillers with air–water systems. In addition, it is important when a system serves different areas that all the rooms in the group treated should have similar patterns of use, so that the plant can be programmed to operate in an economical manner.

When choosing a system it is vital to select plant components that are suitable for each other and mutually compatible. It would be useless to couple a sophisticated and expensive automatic control system with a commercial, comparatively crude plant, for a tightly controlled, industrial application. The outcome would be a poor performance that is often wrongly blamed on the controls. The quality of the plant controlled must be on a par with the quality of the automatic control system if the best is to be obtained from both. By similar reasoning, it is uneconomical to select expensive plant and refined controls for a commercial, comfort conditioning application that could well be dealt with by equipment of lesser quality and price. The design calculations that precede system selection and the commissioning that follows its installation should be above reproach. Finally, the system must be commissionable, otherwise it will never work properly, no matter how appropriate otherwise for the job.

3.2 Office blocks

Air conditioning in office blocks has provided the stimulus for system design and application over many years. Dirty, noisy, urban centres with high-rise, greenhouse buildings, have been where the market is and also where the capital has been available. Consequently a very large number of air-conditioning systems have been developed for this application and, therefore, which system to use is open to the designer's choice. The aesthetics of architectural inspiration and whim have given some variety in the design of window-wall facades and in the exploitation of the available site plan to maximise the lettable floor area produced for a given expenditure of capital, within the restraints of statutory obligation. However, two classes of office building have generally emerged: the narrow block, usually rectangular in form and having a central corridor 1.5 m wide, bounded by two usable peripheral strips of about 6 m width inboard from the windows: and, to a lesser extent in the UK, the deep plan building, often lacking an internal light well.

The first, narrow plan, epitomises the speculative development, with concrete floor slabs, fairly extensive glazing and curtain walling. The modular construction concept involved has been associated with the provision of demountable internal partitions that allow the tenant to have many individual offices and to change the size and arrangement of them during the term of the tenancy, at his or her own, expense. Such a design aim has provoked the development of office air-conditioning systems that can offer flexibility in control and performance to cater for partition rearrangement, and has tended to require a controllable air-conditioning terminal in each module. Architectural and development philosophy has also influenced the size and cooling capacity of the terminal units themselves. For example, there is a small amount of evidence that a modular width of about 2.4 m is most economical in terms of unit selection, and capital cost, but this depends to a large extent on the amount of glazing and the level of illumination. The building regulations in the UK have exerted a very strong influence on building design and on the energy consumed by the mechanical and electrical services in commercial buildings, in recent years. This has been notable in a progressive reduction in the acceptable *U*-values for walls, roofs and windows, with a consequent effect on the design of air-conditioning systems and on their thermal and electrical loads under design conditions. The incorporation of an atrium in many building designs has also played a part in system development.

Latent gains in an office are relatively trivial with fairly good building construction in the UK, although this is not always the case in a tropical environment where a high outside moisture content and a poor building construction may give a latent load by infiltration, which dominates plant selection and performance. In Britain, the small latent heat gains derive, in almost equal proportions, from natural infiltration and the sparse population. Sensible gains come from air-to-transmission (T), people (P), electric lighting (L), solar gain through glass (S) and business machines (M). Example 1.9 showed these to be of the order of 7% for T; 13% for P; 17% for L; 42% for S; and 21% for M. Transmission gains and losses are regarded as being predictable since they are related to outside air temperature and consequently are often offset by compensated heating, for example with the independent perimeter heating system sometimes adopted for VAV schemes. Gains from lights are frequently fairly stable, once patterns of usage for the building have been established, and with block switching, it ought to be possible to cut back zoned, compensated heating schedules to conserve energy, at least for part of the day, although this is not very often done. People and business machine gains also follow a common behaviour, since the use of the latter is related to the presence of the former. They should, therefore, have similar diversity factors applied to them according to CIBSE (1992) and throughout the working day they ought not to present big variations in load. It must not be concluded from this that there is any case for abandoning individual thermostatic control for the modules. Generally speaking zone control, that is controlling all the units on one face of a building from one thermostat or from several averaging thermostats, has proved a failure. Solar gains are unpredictable in the UK and, for many buildings, significantly large. The designer must make sure that windows are adequately shaded as it is not possible to be comfortable in direct sunlight within a building, regardless of the temperature maintained by the air-conditioning system.

In the past, the two-pipe, non-changeover, perimeter induction system (see Section 2.10) has been a commercial solution for air-conditioning office buildings and, in its four-pipe version, with two coils, has given a very good performance. This system is now obsolescent, partly because the rather high static pressure needed to produce the induction effect helped to give high fan total pressures (about 2 kPa) with a consequent high electrical energy consumption. The variable air-volume system (Section 2.6), in its various forms, has displaced the perimeter induction system but the four-pipe fan coil system, with two coils in each unit and auxiliary mechanical ventilation (Section 2.11), has become a first choice for office blocks. More recently, principally because it appears to be a system that is economical in its use of electrical energy, the chilled ceiling with an auxiliary mechanical ventilation system, has made some headway (Section 2.12). Various sorts of chilled beams are also available (Section 2.12) but these provide most of their cooling by convection and some use a high static pressure at the terminal beam to produce an induction effect, with obvious consequences for the fan total pressure. Water loop, air conditioning/heat pump units (see Section 2.3 and Figure 2.6), with auxiliary mechanical ventilation systems, have been used extensively, being particularly suitable for refurbishing office buildings.

More recently, the availability of efficient speed control methods (using inverters) for refrigeration compressors, the development of motorised electronic expansion valves, and the use of microprocessors has led to the use of so-called, cassette units, with variable refrigerant flow (Section 2.3 and Figure 2.5). These seem to offer a commercial solution for small office buildings. Dual duct systems have generally trailed behind the field, for fairly obvious reasons. On an *ad hoc* basis, particularly for smaller buildings, individual offices and in the tropics, through-the-wall room air-conditioning units have been extensively and successfully used for individual offices. All these systems offer individual thermostatic control, usually on a modular basis, but not all provide the same flexibility in responding to load variations and in coping with partition rearrangement (see Chapter 2). Perimeter systems, generally, are best suited to dealing with a peripheral strip of 4.5 m width inwards from the window-wall, up to a maximum of 6 m.

For occupied core areas beyond such treated peripheral strips, the standard treatment in the past used to be a low or medium velocity, constant volume, reheat system with one or more reheaters per floor, the number used depending on the number of tenancies anticipated by the letting arrangements. In more recent years, variable air volume or fan coil systems have been applied to the core areas in deep plan buildings with the advantage of control and flexibility referred to and with considerable success. Floor area cannot be conveniently occupied by any system intended for a core and so where fan coil units have been chosen for such applications they are located at high level, which then causes difficulties of access and maintenance. The fan coil system is not intended for treating core areas and should not be so misapplied in this way unless air distribution and air temperatures have been properly tested in a full-scale mock-up.

For refurbished offices, air conditioning has been successfully provided by three systems: water-loop air conditioning/heat-pump units, four-pipe fan coil systems, and cassette units, all with auxiliary mechanical ventilation.

For air conditioning in an office block, or anywhere else, to be a success a source of heat is required and this means that the boilers must be able to run throughout the summer as well as the winter. The only exception to this is in hot climates where buildings never suffer a heat loss, although it must be remembered that desert environments and locations at high altitudes have night-time temperatures that often sink to very low values and heating is still needed for comfort.

3.3 The atrium in buildings

An atrium is a link between the working environment and the outside world and is for people to use and enjoy. The concept of the atrium dates to classical Greek and Roman times but, in its modern idiom, it is a glass-covered courtyard within a building, often graced with plant life, fountains and well-designed lighting arrangements, both natural and artificial. Such aesthetic aims are not always easily achieved because of the very necessary restrictions imposed by fire protection, smoke control, and the escape of people from the building.

PLANTS IN THE ENVIRONMENT OF AN ATRIUM

Three phenomena in the life of plants dictate the steps to be taken for the provision of a suitable environment:

(1) *Photosynthesis*. Carbon dioxide is absorbed from the atmosphere by the plant during the day, combined with water through the root system, and turned into carbohydrates for growth purposes and into chlorophyll for the absorption and decomposition of carbon dioxide.
(2) *Photomorphogenesis*. This is the response of the plant to the spectrum of light in the atrium that is its source of energy for continued development and growth. Most light is absorbed by plants in the blue and red ends of the spectrum with a dip in the middle frequencies. The shape of the leaves and the form of the plant itself are affected by the spectrum: tungsten filament lamps are not really suitable because of the large red component in their spectrum but fluorescent lighting does provide an acceptable source of light.
(3) *Photoperiodism*. This describes the period of daylighting preferred by a plant for optimum growth. It varies with the species.

Proper lighting is the single most important environmental factor. Plants like to be illuminated from above and the varieties most likely to succeed in an atrium are those accustomed to a shady habitat, preferring comparatively low levels of illumination. As a general level, 600–700 lux may be suitable but some plants can tolerate as little as 500 lux. In parts of an atrium, daylight gives higher levels than this for some of the day but artificial lighting is likely to be necessary as well. Ultra-violet light is bad for plants and clear or tinted glass will filter out most of this while still admitting an adequate amount of the visible spectrum. Special "grow" lamps are not needed in an atrium.

A period of darkness is necessary for plants to thrive and hence security lights ought not to be switched on over planted areas during night-time. Plants do not like draughts and this should be remembered when designing the air distribution system for the atrium. Except for tropical plants a high relative humidity is unnecessary. (Tropical plants are not suitable anyway because they also require a higher temperature than would be associated with comfort conditions in the UK and with energy conservation.) Plants suitable for an atrium are of the hardy or half-hardy types. These do not like excessive solar radiation, preferring diffused daylight from above. It is difficult to obtain flowers that bloom regularly from plants grown in an atrium, because the absence of seasonal climatic changes inhibits the normal flowering cycle. The temperate-zone plants selected for an atrium should be able to tolerate a wide range of temperatures, from about 7°C to over 30°C and, since a high humidity is not necessary, only a minimal air treatment is needed for the plant environment. Transpiration by the plants involves the liberation of some water vapour during the daytime when photosynthesis is occurring and, although this is not likely to amount to much for the comparatively small amount of greenery in an atrium, there is a slight chance of condensation on the glazing in the coldest weather. The rate of photosynthesis and the production of water vapour falls at night-time to roughly 15% of the peak occurring at mid-day. With the much reduced intensity of illumination at night in an atrium the risk of condensation becomes very small. Any fancied risk that may remain is easily dealt with by fitting fan coil units to sweep the glass or providing finned-tube perimeter heating. To economise on the consumption of thermal energy these should be controlled in a manner related to the risk of condensation. Decorative fountains are not likely to cause condensation on the glazing in an atrium. First, the degree of contact between any water droplets and the air is very small and, secondly, the fountains should be switched off at night-time when outside air temperatures are at their lowest.

GLASS IN AN ATRIUM

The choice of glass depends on three factors: the behaviour of glass in a fire, the transmission of solar radiation and mechanical strength.

Ordinary window glass, plate glass or float glass, is a fused mixture of the various oxides of aluminium, calcium, silicon and sodium, cooled from a molten state without crystallisation, and fed with additives to give a colourless appearance. The softening point is about 595°C and the melting point about 1130°C. Thermal radiation from a flame at a typical temperature of 2000°C is readily transmitted through glass and some is absorbed. The middle part of the pane increases in temperature but the periphery of the glass, attached to the frame, is cooler. After two or three minutes the temperature difference from the centre to the perimeter reaches about 40 K and the glass breaks and falls from the frame.

To make toughened glass (termed tempered glass in the USA), the surface of ordinary molten glass is rapidly chilled while its centre stays hot, putting the glass surfaces into a state of compression and the centre into tension. Since bending the glass requires the surface to be de-compressed before it enters tension, higher stresses can be tolerated before the glass fractures. The temperature difference across the pane, caused by thermal radiation from a high-temperature flame, can rise to 250 K before fracture occurs and this may take as much as 10 minutes. The glass then fractures into multiple small pieces which may remain interlocked and not fall out of the frame for perhaps a further 20 minutes.

The commonest form of wired glass comprises a 13 mm square mesh of wires, electrically welded at the intersections and embedded in clear plate glass. It has a 1-hour fire rating. When irradiated from a high-temperature flame its behaviour is similar to that of toughened glass but the wire mesh retains the fractured pieces together until the glass reaches its softening temperature and slumps out of the frame. The time taken for this to occur depends in the size and type of frame: according to BS 476 Part 8 (1972), as long as 90 minutes can elapse before failure, on occasions. Horizontal or inclined glass would fall earlier.

Laminated safety glass consists of two layers of float or plate glass, held firmly together by an interlayer of polyvinyl butyral with a melting point of about 90°C. The adhesive property of the interlayer retains the integrity of the glass after fracture, when the temperature difference from the centre of the glass to the frame reaches 40 K. After this its behaviour is similar to that of ordinary clear glass. Other proprietary forms of laminated glass are available, with claims of up to 90 minutes of fire insulation.

Polycarbonate sheet is a plastic material, regarded as the toughest known at ordinary temperatures but, according to BS 476 Part 7 (1971), it is hot-workable at 150°C, although meeting the Class I standard for flame spread. This implies softening at a temperature much lower than the value of 595°C for ordinary glass. It also appears to be combustible if enough heat is supplied to it and it emits some toxic gas when burning. Its maximum operating temperature is 120°C. As regards mechanical strength, although polycarbonate is the toughest, the choice for roof lights would be toughened, laminated or wired glass, as approved by the local fire brigade. Of these types, wired glass would be favoured, because of its ability to retain glass fragments after fracture.

Virtually all the ultra-violet radiation is removed from the solar spectrum by clear glass. The transmission of the visible and thermal parts of the spectrum depends on the angle of incidence between the incident ray and the normal to the glass, the type of glass, and its thickness. For angles of incidence less than 40° the value of the transmission coefficient is almost constant, at about 0.87 for clear float or plate glass, 4 mm in thickness.

NATURAL ILLUMINATION

Various techniques (such as horizontal louvres adjacent to the window lintel to reflect direct sunlight onto the nearby ceiling) have been adopted to provide the diffused, natural illumination that plants prefer. There is no doubt that the cost of electric lighting can be significantly reduced if glazed atrium walls and roofs are designed to diffuse direct sunlight and give acceptable natural lighting, not only in the atrium itself but also in the offices and other rooms overlooking the atrium.

CONTROL OF SOLAR HEAT GAINS

It is well established that people sitting in direct sunlight, coming through unshaded windows, will feel uncomfortable after a short time. Air conditioning cannot prevent this happening. All roof glass, and all windows facing in directions other than NNW, N or NNE, must be protected with some sort of adequate solar control. The options for solar control are: external shading, internal shading, heat-reflecting glass, or heat-absorbing glass.

External shading is the best protection. Nearby tall buildings will cast moving shadows over the glazed parts of the atrium roof and walls but the changing pattern of external shading they provide can be predicted by computer analysis. It may be possible to augment such natural shading by using purpose-made, fixed shades but unless it is clearly shown that such shading is acceptable for the comfort of the occupants, motorised shades, preferably louvres, will be needed, with adequate access for regular maintenance.

For internal shading, blinds of various sorts have been used but, to be fully effective, they should be white and reflective with easy access for adjustment by the occupants to exclude the direct rays of the sun. The penetration of direct solar radiation into the atrium and its subsequent incidence on the windows of rooms overlooking the courtyard should be investigated to establish the need for shades on such inward-facing windows.

Heat-reflecting glass will be effective if its total shading coefficient does not exceed 0.27. The shading coefficient is defined as the ratio of the total thermal transmittance through a particular type of glass and shade, to the total thermal transmittance through ordinary, clear, 4 mm glass. It is to be noted that ordinary, clear, 4 mm glass, with internal, white, reflective, Venetian blinds, drawn to exclude the direct rays of the sun, has a total shading coefficient of 0.53. This is acceptable and the reason is that the blinds transmit virtually none of the direct solar radiation which, being from a surface temperature of 6000°C at the sun, causes discomfort. The radiation from the drawn blinds is from a surface at a temperature of about

40°C, which is acceptable. On the other hand, the small amount of direct radiation transmitted through heat-reflecting glass does cause discomfort and hence its acceptable shading coefficient must not exceed the much lower figure of 0.27.

Heat-absorbing glass is not effective, by itself, in reducing the transmitted direct solar radiation to a tolerable level and internal shades are invariably also required. When such additional shades, of any sort, are used, reference must be made to the manufacturers of the heat-absorbing glass to establish that any thermal stresses set up in the glass will be acceptable. Pilkington (1980) gives guidance on this.

MECHANICAL SYSTEMS FOR TREATING AN ATRIUM

Natural ventilation has been used in the past, with mixed results. Mechanical ventilation is a possibility that might be considered, principally to satisfy the requirements for smoke exhaust but, if tolerable, controllable conditions are desired for human comfort, air conditioning is necessary.

The three matters to be considered are the usage of the space, the environment for the plants, and smoke control. There is no need to provide individual thermostatic control in multiple rooms and hence a low velocity, constant volume supply and extract system is the answer, having a cooler coil controlled in sequence with a heater battery, subject to high-limit humidity over-ride. See Section 2.4.

One technique used in the past was to arrange for the spillage of used air from the adjoining offices into the courtyard. To maintain a balanced airflow for the building the amount spilled was the difference between the supply and extract air quantities for the offices, hence corresponding to the minimum fresh-air quantity. This would not be satisfactory for other than a very transient occupancy and would not be acceptable if the atrium was an escape route.

The atrium has also been used as part of the return air system, it being argued that the presence of the vegetation in the atrium improved the quality of the air returned to the air-handling plant for recirculation. Problems arose with negative air pressures in the atrium and the plants seemed to require an excessive amount of watering.

The correct technique is to provide separate supply and extract air-handling plants for each atrium, independent of the rest of the systems in the building. There is then full scope for providing conditions to suit the needs of the plants or the people and satisfy the requirements of the local fire officer. The system can be engineered to cope with the use of the building, the proper control of conditions in the atrium and the need for energy conservation.

In general, the supply of conditioned air should be where the people are. There is no point in attempting to condition the entire large volume of the atrium. However, fan coil units or finned tube, with a compensated supply of LTHW and independent thermostatic control from local temperature, may be desirable, to deal with down draught or the remote possibility of condensation, in winter.

An atrium can be used for more than just a walk-way. It may form part of the commercial and social activity of the building and have a degree of permanent occupancy throughout the hours that the building is in use. The air-conditioning system for the atrium must then be designed to suit the needs of the occupants, giving them sufficient fresh air and comfortable conditions of temperature, air movement and humidity. If there is commercial use, individual tenants would normally be expected to make their own arrangements for their own premises. Such usage of an atrium is at the discretion of the local authority in the UK and if commercial use were not allowed the atrium would become merely a communication space with transient occupancy. The fresh-air requirement would be much less and a comparatively wide swing in temperature could be acceptable, consistent with the needs of the plants.

In the event of a fire, the fire brigade may require the system dealing with the atrium to be entirely under their control for smoke exhaust or other purposes. The necessary fire dampers and fire-resistant materials should therefore be included in the system. The fans and their electric driving motors must be able to withstand the anticipated temperatures when operating in an emergency.

It is not possible to predict the attitudes that may be taken by local authorities in respect of smoke control and the escape of people, except to say that human safety will be the overriding factor. Knight and Jones (1995) give some guidance on fire protection and smoke control. The attitude of the local authority to smoke control will depend on whether the atrium is an escape route and this will influence the treatment provided. Whatever the view of the local authority and the fire brigade it is possible to design and control a system to suit.

3.4 Hotels

The treatment for a hotel falls very clearly into two parts: that needed for the bedrooms and that necessary for the public rooms, the load characteristics between the two being very different.

In the bedrooms the load is somewhat similar to that in an office, with windows tending to be smaller and perhaps with external shading if balconies are provided. The electric lighting will probably be a good deal less, although some or all of it will be from tungsten lamps. A coloured television set is standard in modern hotels, liberating about 400 W when switched on. The design of bedrooms is usually for two people, twin single beds, or twin double beds, being the most common furnishing. So sensible heat gains are in the region of 60–80 W m^{-2} of floor area, say about 1500 W per bedroom. Some rooms are likely to be occupied, with the television set on, at the time of peak afternoon sensible heat gain, although part, or all, of the lighting may be off. The standard fresh-air allowance for a double bedroom is 25 l s^{-1}, on the assumption that this is exhausted to waste through the bathroom mechanical extract ventilation system.

Several systems of air conditioning have been used, with varying degrees of success. In a tropical environment a through-the-wall room air-conditioning unit can provide the welcome relief so necessary in a hot climate, without its excessive noise being regarded as a nuisance, but a more critical guest in a higher class hotel, or in Europe or America, might not be satisfied. Two-pipe fan coil units, frequently floor-mounted, are also used, in both temperate and warmer climates. They are operated as a changeover system if there is any need for winter heating and the changeover problems (Section 2.10) either do not exist or at least do not present a serious problem. With some designs, the fresh air is cooled and dehumidified by a central plant and delivered to the corridors, to find its own way through grilles or under-cut doors into the bedrooms for exhaust to waste via the bathroom extract ventilation system. With bathrooms generally located close to the entrance from the corridor the amount of fresh air supplied to the bedrooms is often only nominal. Corridors also tend to be over-cooled. Two-pipe and four-pipe induction systems have been used to condition hotel bedrooms, with mixed success. The induction system is wrongly applied in these instances as units cannot be turned off and, except in the four-pipe case, there is not a quick enough response to the capacity changes sometimes demanded by the guest.

Water-loop air conditioning/heat pump units have also been used successfully in the UK, backed up by an auxiliary mechanical ventilation system providing filtered, and, when necessary, tempered, fresh air that is not cooled in summer, the terminal units in the bedroom doing all the cooling needed. Units are usually located above the suspended ceiling in the bathroom, as is the fan coil unit shown in Figure 3.1. Although much quieter than through-the-wall room air-conditioners, these units tend to be a little on the noisy side when the refrigeration compressor runs, unless encased in a special acoustic enclosure.

There is no doubt that the correct choice for hotel bedrooms is the four-pipe, twin-coil, fan coil system with an auxiliary, low velocity, mechanical supply of filtered, cooled and dehumidified fresh air, ducted to the vicinity of the fan coil units and subsequently removed through the bathroom extract ventilation system. Figure 3.1 shows a typical method of installation. The fan coil units can run at low, medium or high speed, by manual choice, with their selection made to cope with maximum design heat gains at low speed to give a noise level below NR 30 and as near as possible to NR 25. The unit can be switched off by

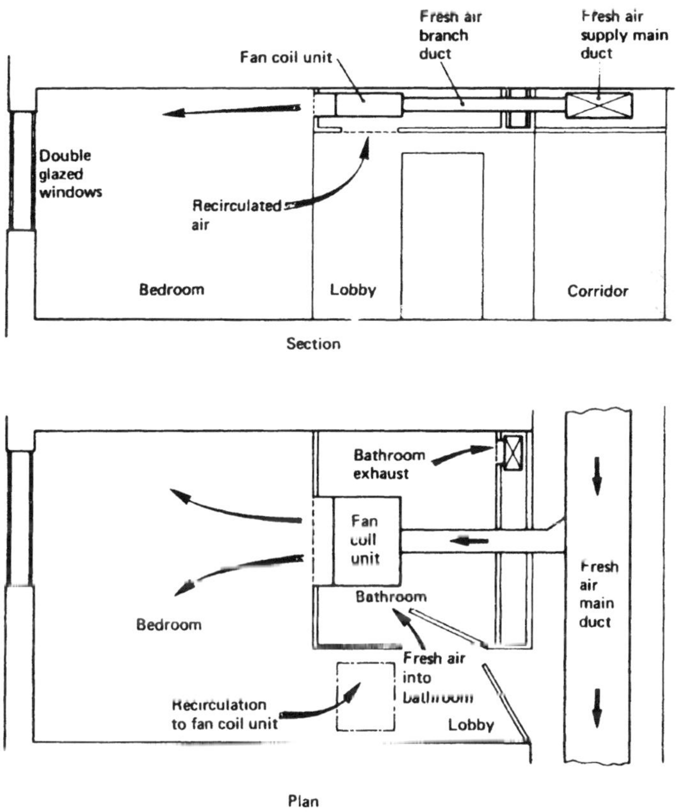

Figure 3.1 Section and plan of a typical four-pipe, twin coil, fan coil unit installation. Fresh air goes from the bedroom through the lobby into the bathroom. A condensate collection tray under the fan coil unit must be provided and drained to waste in the plumbing system.

the management when the room is unlet and the hotel guests can switch it off themselves if they wish. Furthermore, increased cooling or heating capacity can be rapidly obtained by switching to medium or even high-speed operation if the penalty of a higher noise level is accepted. This facility of quick capacity change by switching fan speeds, coupled with the constant, simultaneous availability of full cooling and full heating by the four-pipe distribution to the twin coils, means that a wide range of settings is possible on the room thermostat and the system response to set point alterations or load fluctuations is very rapid. The whims of most room occupants can be accommodated, although if all guests chose to run their units at full capacity at the same time the central refrigeration or boiler plant would experience some difficulty.

EXAMPLE 3.1

Suppose that the hypothetical office block described in Section 1.2 is used as the bedroom block of a hotel. Each bedroom is of similar construction and dimensions to a module in the hypothetical block except that it is 3.6 m wide and contains only one window, with an area of 3.168 m^2, which is double glazed to reduce the intrusion of traffic noise and is fitted with a Venetian blind, between the panes of glass. The *U*-value of the glazing is 3.3 W m^{-2} K^{-1} and the *U*-values of the wall and roof are each 0.45 W m^{-2} K^{-1}. Each bedroom is 3 m deep with a further 3 m to cover the lobby and the bathroom.

The central corridor is 1.5 m wide and lifts, escape staircases, and all vertical services ducts, pipes and cables, are accommodated at the ends of the block, as shown in Figure 1.1. Assume that these two vertical services blocks do not affect the heat flow into or out of the main body of the building. Each bedroom is for two people and contains a television set that liberates 400 W and electric lighting that emits 300 W. There is no natural infiltration. The air-conditioning system is by means of a four-pipe, twin-coil, fan coil system with an auxiliary air supply to offset the latent gains and to provide for the fresh-air needs. The air supply state is 13°C dry-bulb, 7.702 g kg^{-1}. The minimum fresh-air rate is 25 s^{-1} for each bedroom. Auxiliary air at a rate of 1.4 l s^{-1} m^{-2} is supplied to the corridor. The auxiliary air system is at low velocity. Allow 2 K temperature rise for supply fan power and duct gain and 0.2 K rise for the extract fan power. The motor in each fan coil unit liberates 115 W of waste heat into the bedroom. Referring to Examples 1.9 and 1.11 as necessary and using CIBSE data from Tables A.5 and A.9, calculate the following:

(a) The sensible heat gain at 1500 h sun time in July for a west-facing bedroom.
(b) The latent heat gain for a bedroom.
(c) The supply airflow rate necessary to deal with the latent heat gain in a bedroom.
(d) The sensible cooling capacity of the ducted auxiliary air supply to the bedroom and the sensible heat gain to be dealt with by the fan coil unit in a west-facing room.
(e) The total supply airflow rate of the ducted auxiliary air system for the whole building.
(f) The fresh air supply rate to the bedrooms, the corridors and the whole block.
(g) The cooling load for the whole block at 15.00 h sun time in July.

Answer

Reference to Example 1.11 will give details of the relevant sol-air temperatures for the calculation of the heat flow through the opaque parts of the structure.

(a) Design sensible heat gains at 15.00 h sun time in July

	W
Glass: $3.168 \times 3.3 \times (28 - 22)$:	63
Wall: $8.712 \times 0.45\ [(26 - 22) + 0.2 \times (18 - 26)]$:	9
Solar: $0.39 \times 0.80 \times 270 \times 3.168$:	267
People:	180
Lights:	300
Television:	400
Total sensible heat gain:	1219 W

This represents a specific sensible gain of $1219/(3.6 \times 6) = 56.4$ W m^2, over the floor area of the bedroom, lobby and bathroom.

(b) Latent heat gain from people: $2 \times 50 = 100$ W.

(c) By Equation (2.4) the airflow supply at 13°C and 7.702 g kg^{-1} is

$$\dot{v}_{13} = [100/(8.366 - 7.702)] \times [(273 + 13)/856] = 50.3\ \mathrm{l\ s^{-1}}$$

This is about twice the minimum fresh air rate so it would be reasonable to use 50% fresh air and 50% recirculated air as the auxiliary air supply to the bedrooms.

(d) By Equation (2.3)

Sensible cooling capacity of the auxiliary air supply to one bedroom =
$(50.3 \times (22 - 13)) \times 358/(273 + 13) = 567$ W.

Hence

Sensible duty for the fan coil unit = $1219 - 567 = 652$ W.

(e) Corridor area = $12 \times 1.5 \times 86.4 = 1555.2$ m^2
Supply rate to the corridors = $(1.4 \times 1555.2)/1000 = 2.18$ m^3 s^{-1}
Half of this will be fresh air.
The number of modules in each long face = $(86.4 \times 12)/3.6 = 288$
Total number of bedrooms = $2 \times 288 = 576$
Total number of people = $2 \times 576 = 1152$
Supply rate to the bedrooms = $(50.3 \times 576)/1000 = 29.97$ m^3 s^{-1}
Total auxiliary supply airflow rate for the whole block = $2.18 + 29.97 = 32.15$ m^3 s^{-1}

(f) Taking the fresh air proportion as 50% of the total, the fresh air supply is 14.985 m^3 s^{-1} to the bedrooms, 1.09 m^3 s^{-1} to the corridors and 16.075 m^3 s^{-1} to the whole building.

(g) Noting that the areas of glass and wall for a particular bedroom are 3.168 m^2 and 8.712 m^2, respectively, the cooling load for the entire building at 15.00 h sun time in July is calculated as follows:

1. Sensible heat gain (Q_s). Diversity factors have to be applied to the heat gains from people, lights and television sets. There is no guidance on this but it is suggested that 0.2 be allowed for both people and television, with 0.1 for lighting.

		W
East glass:	24 × 12 × 3.168 × 3.3(28 – 22)	18 065
West glass:	24 × 12 × 3.168 × 3.3(28 – 22)	18 065
North wall:	(13.5 × 3.3 × 12) × 0.45 × [(23.5 – 22) + 0.2(20 – 23.5)]	192
South wall:	(13.5 × 3.3 × 12) × 0.45 × [(26 – 22) + 0.2(18 – 26)]	577
East wall:	(24 × 12 × 8.712) × 0.45 × [(26 – 22) + 0.2(28 – 26)]	4 968
West wall:	(24 × 12 × 8.712) × 0.45 × [(26 – 22) + 0.2(18 – 26)]	2 710
Roof:	(86.4 × 13.5) × 0.45 × [(31 – 22) + 0.77 (48 – 31)]	11 595
East glass (solar):	(24 × 12 × 3.168) × 0.58 × 0.8 × 154	65 195
West glass (solar):	(24 × 12 × 3.168) × 0.39 × 0.8 × 270	76 859
People:	0.2 × 1152 × 90	20 736
Lights:	0.1 × 576 × 300	17 280
Television:	0.2 × 576 × 400	46 480
Total sensible heat gains at 15.00 h in July:		282 722 W

Over the treated floor area of 13 997 m^2, this represents a specific sensible heat gain of 20.2 W m^{-2}.

2. Latent heat gain (Q_1). There is no infiltration so the latent gain is from people only and equals 0.2 × 1152 × 50 = 11520 W.
3. Fresh air load (Q_{fa}). The fresh air must be supplied whether the people are present or not. The total fresh air supply is 16.075 m^3 s^{-1} and, if this is expressed at the supply state of the auxiliary air, namely 13°C dry-bulb and 7.702 g kg^{-1}, the specific volume (from CIBSE psychrometric tables or chart) is 0.820 m^3 kg^{-1}. The outside design state at 15.00 h sun time in July is 28°C dry-bulb, 19.5°C wet-bulb (sling), for which the enthalpy is 55.36 kJ kg^{-1}. The design room state is 22°C dry-bulb, 50% saturation, for which the moisture content is 8.366 g k g^{-1}. There is a temperature rise of 0.2 K across the recirculated air fan and, by Equation (1.14), the enthalpy at 22.2°C and 8.366 g kg^{-1} is 43.59 kJ kg^{-1}. Alternatively, the enthalpy could have been determined by interpolation in psychrometric tables or, less accurately, read from a chart. Hence the fresh air load is

   $$Q_{fa} = (16.075/0.820) \times (55.36 - 43.59) = 230.74 \text{ kW}$$

4. Supply fan power and duct heat gain (Q_{sf}). This is calculated using Equation (1.16) as follows:

   $$Q_{sf} = 32.15 \times 2 \times 358/(273 + 13) = 80.49 \text{ kW}$$

5. Power of fan coil unit driving motors (Q_{tfp}). There are 576 bedrooms with one fan coil unit liberating 115 W in each, when switched on. The policy of the hotel regarding switching fan coil units off, in the absence of guests from a bedroom, would have to be determined. In this case it is supposed that a diversity factor of 0.2, as used for people, can be applied to the fan coil units. Hence the load is determined:

   $$Q_{tfp} = 0.2 \times 576 \times 115/1000 = 13.25 \text{ kW}$$

6. Recirculated fan power load (Q_{ra}). Only 50% of the auxiliary air is recirculated so the cooling load attributable to this is calculated by Equation (2.3) as

$$Q_{ra} = 0.5 \times 32.15 \times 0.2 \times 358/(273 + 13) = 4.02 \text{ kW}$$

7. Summary:

1. Sensible heat gain	282.72
2. Latent heat gain	11.52
3. Fresh air load	230.74
4. Supply fan power and duct gain	80.49
5. Fan coil unit motors power	13.25
6. Recirculated fan power	4.02
7. Refrigeration load	622.74 kW

This is at 15.00 h sun time in July, in the UK.

For the bedroom block as a whole it is possible, but not certain, that the maximum cooling load will occur during the period 14.00–17.00 h sun time in July. A calculation survey using a computer will resolve this but the real problem facing the designer is in the allowance he or she must make for diversity factors on lights, television and people.

In this total cooling load the elements subjected to virtually guessed diversity factors, add up to 109.3 kW or about 18%. This comparatively small percentage reduces the risk associated with the guesses. For the hotel as a whole, the cooling load in the public rooms is very significant and it is not likely that these areas will be densely occupied at 15.00 h sun time in July. A better time for assessing the maximum cooling load for the bedroom block might be 13.00 h sun time, to coincide with a peak of activity in the public rooms. In Exercise 1 this can be calculated as 454 kW, making the same assumptions for diversity.

The public rooms in a hotel comprise reception area and foyers, dining rooms, bars and a banqueting suite. The policy of many hotel developers is not to provide areas that fail to yield a revenue, so residents' lounges are rare in modern hotels, and, for the same reason, foyers and reception areas often include shops. Hotel designs vary enormously but a common practice is to house the public rooms in a three- or four-storey podium and to have a tower block of bedrooms above it, sometimes with a penthouse suite or a restaurant at the top.

The obvious characteristic of public rooms is a high population, so all-air systems are the correct choice for air conditioning. Three types are possible selections: multizone, constant volume reheat and, rarely, double duct. The multizone system is the most popular because it takes full account of cooling load diversification and is comparatively cheap in first cost if the plants can be located fairly close to the areas treated in order to minimise duct runs. The double duct system might provide a good solution to a small part of the public areas if there exists a case for multiple, thermostatic control, say for a restaurant with alcoves for semi-private dining parties, but its use is very rare because of its higher capital cost. The constant volume reheat system is really only acceptable for a single area where one heater battery could be controlled in sequence with the cooler coil, so that reductions in the heat gains would not be cancelled with reheating, otherwise no advantage could be taken of load diversity and the installed refrigeration capacity and capital cost would be large.

It makes sense to select a multizone unit to deal with a group of rooms sharing a common pattern of usage. Thus one unit would be chosen to treat the three or four bars in the hotel, another to condition the group of dining rooms, a third for the banqueting suite – which is often one very large room with one or more demountable partitions – and a fourth for the foyers and reception area. The first three of these groups would probably require to be conditioned at different times and the last might need air conditioning continuously. The administrative offices sometimes receive a supply of conditioned air from the reception plant and achieve individual thermostatic control with fan coil units. Sometimes they are only heated, with radiators, if they have openable windows. Shops can be treated in a similar way or merely provided with a supply of chilled water and be expected to make their own arrangements for conditioning. Another approach is to fit extract fans at the back of the shops which gives them second-hand air conditioning by drawing air in from the treated foyer or reception area. (This is done with shopping centres.)

Noise in public areas is never as critical as in the bedrooms but acceptable levels must be provided, i.e. NR 45 or NR 40, or lower, depending on the use of the space. Some hotels have conference facilities and in these cases the auditoria must have separate plants, using all-air systems that are acoustically compatible.

The location of plant has an important bearing on capital cost and the acoustics. The aim must be to minimise duct runs and avoid noise and vibration causing a nuisance within the treated areas or outside to the occupants of the adjoining buildings, although the latter may be less of a problem in the Middle East and the Gulf where noisy air-cooled condensers are the norm and are acceptable. For the typical tower block/podium arrangement mentioned, the air-handling plant for the bedrooms and possibly the cooling towers would be on the roof of the tower, the multizone units for the public rooms would be on the roof of the podium, suitably disguised, or possibly in plant rooms inside the podium block itself, refrigeration and boiler plant would be in the basement. An alternative site for the cooling towers, giving shorter runs of cooling water piping, would be on the podium roof but screening the towers acoustically and visually, without affecting the airflow they need for their proper performance, can introduce problems.

The proportions of the public areas allotted to the various functions (bars, restaurants, and so on) differ among hotels, as also do the standards adopted for lighting and population densities. Furthermore, the ratio of bedroom area to public area depends on the commercial policy followed by the hotel operating group. It is consequently not easy to generalise about cooling loads in hotels in the way that one can about office blocks (see Table A.10). A similar situation exists for estimating the boiler power and often the best policy for a designer is to seek the advice of the chief engineer of the hotel group.

The results of a survey of nine air-conditioned hotels in the UK, completed in the 1970s, are shown in Table 3.1. The average ratio of private to public areas is 3.3 and the mean specific refrigeration load is 129 W m^{-2}. Hotel number 9 is not far from the average and its design values of population density and power liberated by the electric lighting are given in Table 3.2, although the diversity factors applied to them are not known. The lighting is largely by tungsten filament lamps in the public areas.

With appropriate assumptions a cooling load for the complete, hypothetical hotel can now be estimated. The aim is to determine the maximum coincident sum of the cooling loads in the private and public areas and the proper way to do this is by calculating the combined load at hourly intervals, using a computer and making assumptions for diversity factors for lights and people. Without doing this though, the peak load in the public rooms can be judged to occur at lunch time, or at dinner time in the evening, say at 2 pm or 9 pm, clock time, respectively. Consider a possible lunch time load at 13.00 h sun time, for which the calculations in Exercise 1 yield a cooling load of 454 k W of refrigeration for the bedroom block, based on an outside enthalpy of 47.99 kJ kg^{-1} in July, at this time, determined from Equation (1.1), and an outside state of 26.5°C and 8.366 of kg^{-1}. Deciding on diversity factors for the population in the public rooms is difficult. If 2.87 is assumed to be the ratio of private

Table 3.1 Comparison of air-conditioning loads in hotels in the UK

Hotel	Public area (m^2)	Bedroom area (m^2)	Ratio of bedroom area to public area	Total cooling load (kW of refrigeration)	Specific refrigeration over the total treated floor area ($W\ m^{-2}$)
1	8500	15000	1.65	1758	78
2	1500	9340	6.23	2271	210
3	1000	3000	3.00	774	194
4	4425	13150	2.97	2532	144
5	2690	14950	5.56	1758	100
6	2700	5000	1.85		
7	2600	8000	3.08		
8	1000	2060	2.06		
9	1640	5290	3.23	1125	162

Table 3.2 Typical approximate heat gains for a hotel at 1300 h sun time in July in the UK

	People					Lighting		
	Floor area (m^2)	Design population	Diversity factor	Sensible gain (kW)	Latent gain (kW)	Design load ($W\ m^{-2}$)	Diversity factor	Sensible gain (kW)
Bars	1220	678	0.6	36.6	20.3	75	0.8	73.2
Restaurants	1317	693	1.0	62.4	34.6	75	0.8	79.0
Banqueting	634	334	1.0	30.1	16.7	150	1.0	95.1
Reception	293	39	0.2	0.7	0.4	40	0.8	9.4
Committee room	49	(20)	0	0	0	–	0	0
Foyers	585	162	0.2	2.9	1.6	75	0.8	35.1
Administration	439	86	0.2	1.5	0.9	40	0.8	14.0
Cloakrooms	341	19	0.2	0.3	0.2	75	1.0	25.6
Total	4878	2011		134.5	74.7			331.4

to public areas then the public area associated with a bedroom block of about 14 000 m^2 is 4878 m^2. If the restaurants are full (diversity factor = 1.0), the bars will be less full (diversity factor = 0.6, say) and the other public areas comparatively empty (diversity factor = 0.2, say). The banqueting suite, on the other hand, could well be fully occupied, but the committee room can be assumed to be empty, any occupants being at lunch. A similar line of reasoning cannot be easily applied to the lighting and it is possible that in some hotels the lights will be permanently on. However, at lunch time in July it is reasonable to suppose there is some natural illumination through windows and that some of the lights will, therefore, be off intentionally, or off for maintenance. Hence, a factor of 0.8 can be assumed for all lights, except in the cloakroom and the banqueting suite, both of which can be assumed here to be internal rooms and have a diversity factor of 1.0. Allowing 90 W for sensible heat gains and 50 W for latent heat gains from people and taking the appropriate values from Table 3.3, the loads are as shown in Table 3.2.

Table 3.3 Design values for population density and lighting power for a hotel in the UK

	Proportion of the total public area (%)	Design population density (m^2 per person)	Design lighting power ($W\ m^{-2}$)
Bars	25	1.8	75
Restaurants	27	1.9	75
Banqueting	13	1.9	150
Committee room	1	2.4	75
Reception	6	7.5	40
Foyers	12	3.6	75
Administration	9	5.1	40
Cloakrooms	7	18.3	75

These figures must not be regarded as typical but only as an indication of possibilities.

EXAMPLE 3.2

Making use of the loads established above, for people and lights in the public areas and 454 kW of refrigeration for the bedroom block (Exercise 1), make an estimate of the cooling load for the whole hypothetical hotel building at 13.00 h sun time in July. Assume the public areas are in a two-storey podium of plan dimensions 87 × 28 m, located beneath the bedroom block and with its major axis pointing north–south. Take the building construction to be similar to that of the bedrooms and assume 40% double glazing on its two long faces, with internal Venetian blinds, not between the panes of glass. Allow 16 $l\ s^{-1}$ per person as a ventilation rate in the bars, restaurants and banqueting suite and 12 $l\ s^{-1}$ elsewhere. There is no natural infiltration.

Answer

From Table A.9 the outside air temperature at 13.00 h in July is 23.0°C, to which a correction of 3.5 K must be added because the design outside temperature at 15.00 h is 28°C, whereas the value at 15.00 h given in Table A.9 is 24.5°C. Hence t_{13} = 26.5°C (with an outside moisture content of 10.65 $g\ kg^{-1}$). The relevant sol-air temperatures are determined from Table A.9, taking the wall surfaces to be light-coloured and the roof surface to be dark in colour. The total treated floor area of the podium is 2 × 87 × 28 = 4872 m^2 and this is close to the total area of 4878 m^2 given in Table 3.2 for a typical hotel at 13.00 h sun time in July in the UK. Hence use the heat gains quoted in Table 3.2 for the public rooms in this example. The heat gains are now calculated as follows:

	kW
Glass, east: $[(0.4 \times 87 \times 6.6) \times 3.3 \times (26.5 - 22)]/1000 =$	3.41
Glass, west: $[(0.4 \times 87 \times 6.6) \times 3.3 \times (26.5 - 22)]/1000 =$	3.41
Wall, north: $(28 \times 6.6 \times 0.45) \times [(23.5 - 22) + 0.2\,(15.5 - 23.5)]/1000 =$	−0.01
Wall, south: $(28 \times 6.6 \times 0.45) \times [(25 - 22) + 0.2\,(15.5 - 25)]/1000 =$	0.19
Wall, east: $(0.6 \times 87 \times 6.6 \times 0.45) \times [(26 - 22) + 0.2\,(15.5 - 26)]/1000 =$	0.29
Wall, west: $(0.6 \times 87 \times 6.6 \times 0.45) \times [(26 - 22) + 0.2\,(15.5 - 26)]/1000 =$	0.29
Roof: $(87 \times 28 - 86.4 \times 13.5) \times 0.45 \times [(31 - 22) + 0.77\,(35 - 31)]/100 =$	6.90
Subtotal:	14.48 kW

Glass solar gains, east: [(0.4 × 87 × 6.6) × 0.95 × 0.91 × 172/1000] =	34.15
Glass solar gains, west: [(0.4 × 87 × 6.6) × 0.74 × 0.91 × 248/1000] =	38.36
People:	134.50
Lights:	331.40
Total sensible gain:	552.89 kW
This is rounded off to 552.9 kW.	
Total latent gain:	74.7 kW

The contribution of the supply and extract fan powers can be approximately assessed by assuming that low velocity, all-air systems have been installed, which take full account of diversity in the heat gains. In other words, constant volume re-heat systems have not been used. If we assume a temperature rise of 2 K to account for supply fan powers and duct gains, and 0.2 K for the extract fan powers, this could lead us to a psychrometric consideration similar to that shown in Figure 1.4, with a supply air temperature of 13°C. On this basis the total supply air quantity for the podium block of public rooms is given by Equation (2.3) as

$$\dot{v}_{13} = [552.9/(22 - 13)] \times [(273 + 13)/358] = 49.1 \text{ m}^3 \text{ s}^{-1}$$

Hence, by Equation (2.3):

Supply fan power and duct gain: (49.1 × 2 × 358)/(273 + 13) = 122.9 kW

Extract fan power: (49.1 × 0.2 × 358)/(273 + 13) = 12.3 kW

The fresh air allowance is 16 l s^{-1} for each person in the bars, restaurants and banqueting suite. Referring to Table 3.2 and allowing a diversity factor of 1.0 this amounts to [(16 × 678) + (16 × 693) + (16 × 334)]/1000 = 27.28 m^3 s^{-1}. For the remaining rooms, the total population is 306, assuming that the committee room is not in use and its plant is switched off. The fresh air supply to these rooms is (12 × 306)/1000 = 3.67 m^3 s^{-1}. The total fresh air quantity handled by the various plants dealing with the public rooms is 27.28 + 3.67 = 30.95 m^3 s^{-1}.

The recirculated air enthalpy was calculated as 43.59 kJ kg^{-1} in Example 3.1. The outside state in this example is 26.5°C dry-bulb and 10.65 g kg^{-1}. The enthalpy can be calculated by Equation (1.14), or determined from psychrometric tables or a chart, as 53.81 kJ kg^{-1}. Using a specific volume of 0.820 m^3 kg^{-1} at the supply state (see Figure 1.4) the fresh air load is determined as

$$(30.95/0.82) \times (53.81 - 43.59) = 385.7 \text{ kW}$$

Summary

The cooling load for the complete hotel at 13.00 h sun time in July in the UK is calculated as follows:

Sensible heat gain	552.9
Latent heat gain	74.7
Fresh air load	385.7
Supply fan power and duct gain	122.9
Extract fan power	12.3
Subtotal	1148.5
Bedroom block at 13.00 h sun time in July	454.0
Total refrigeration load	1602.5 kW

This represents a specific cooling load of 1 602 500/(14 000 + 4878) = 84.9 W m^{-2}, which is at the bottom end of the values in Table 3.1.

3.5 Residences and apartments

In a hot climate air conditioning is common and, in some instances, essential for living quarters, particularly for bedrooms where cooler conditions assist sleep and discourage insects. The approach has been to install through-the-wall, room air-conditioning units (see Section 2.3) on an *ad hoc* basis to meet the needs of individual rooms or dwellings. More refined treatments have been adopted for houses, generally using constant volume all-air systems with a single direct expansion cooler coil and a remote air-cooled condenser. One or more reheaters (or sequence heaters) give temperature control, although a single thermostat in the corridor with one heater battery is often enough. Such systems are also used in apartment blocks, the advantage being in operation when a separate system is used for each apartment but the disadvantage is in accommodating the air-cooled condenser in a place with access to a plentiful supply of outside air. A variation for apartment blocks is to provide water-cooled units with multiple cooling towers on the roof and vertical flow and return cooling water mains to feed the self-contained units in each apartment. A third possibility is to install a central water-chilling plant and to distribute chilled water to each apartment for its air-handling unit. At the luxury end of the market the best approach is to treat the apartment block as a high-class hotel and use a four-pipe, twin-coil, fan coil system with a common, low-velocity ventilation plant feeding filtered, dehumidified, tempered (in winter) air to each flat. Enough fresh air is delivered in this way to give either 25 l s^{-1} for each person normally present or enough to make good the mechanical extract ventilation, whichever is the greater. Such mechanical extract is essential and should take the form of 25 l s^{-1} from the bathroom and 20 air changes per hour in the kitchen. The fan coil units would be selected to meet the design heat gains and losses when running at slow speed and although the cooler coils would normally do sensible cooling only, condensate drain lines must be fitted and connected into the plumbing system, to cater for the odd occasion when many more than the design population is present. It is also desirable to provide smell removal, which can be done by a comparatively small unit, comprising an activated carbon and an electrostatic filter, handling a proportion of the recirculated airstream. A system for a luxury apartment must have: more than enough capacity to meet the design loads; quietness (NR 20 is the aim); a quick response to a manual resetting at a thermostat; plenty of fresh air; and system reliability. Each room, except the corridor, bathroom and kitchen, should have independent thermostatic control, in contrast to the simpler systems mentioned previously, where an entire house of two or three bedrooms, etc., is controlled on an average basis from a single thermostat in a corridor.

In a temperate climate and in the UK, the same principles apply but the case for air conditioning is more difficult to make for this sort of application. The most compelling argument is for dwellings in urban environments where noise and dirt must be kept at bay with double glazing. Air conditioning then becomes a serious consideration, to prevent excessive rises in internal temperature when the windows must be kept shut. However, sensible gains are often less than for offices, with a much lower level of illumination and a smaller proportion of glazing in the walls. Mechanical ventilation, to give an adequate air change rate of, say, 10 per hour, is then a possible alternative. It costs a good deal less than air conditioning to buy and operate, and may give acceptable results although the client must expect an increase in room temperatures above the value outside during part of the summer.

3.6 Shopping centres

The design of air-conditioning systems for shopping in the UK is subject to two constraints:

(1) The economic pressures imposed by the need for the developer to pitch the amenities at a level that will command a rent local traders are willing to pay, whilst providing him or her with a satisfactory return on capital.
(2) The desires of the local fire officer to ensure the safe escape of the people present in the event of a fire.

Shopping-centre design has tended to follow an established pattern. Central malls are bounded by shopping units and are connected at large squares which become the focal points for aesthetic display, with fountains, exotic lighting, sunshine roofs and sometimes areas for rest fitted with benches and tables and provided with a light refreshment service. Most shopping centres are single-storey or have double-height malls with shop units opening directly onto them at ground level and further shops at the first floor, fronting onto a balcony. Sometimes, however, the development is over several conventional storeys, with malls and shop units on each. For a shopping centre to be viable it is necessary that several nationally-known chain stores take up important positions in the development, smaller concerns then being encouraged to cluster around them.

Lighting levels in the malls are at 200–400 lux but in the shop units they vary, with 500 lux being an average. It is common practice to include one or two supermarkets in a development and these have lighting levels that can be much higher, perhaps up to 2000 lux. The contrast ratio, shops to malls should not be greater than 5:1. The lighting will be provided in a variety of ways in the shops. Food shops, newsagents, chemists and the like will be lit by fluorescent tubes but boutiques, restaurants, and so on, will have spot lighting and tungsten filament lamps. Fluorescent tubes could produce 200 lux by liberating about 10 W m^{-2} of floor area, and 500 lux by 20 to 25 W m^{-2}. Tungsten lamps will dissipate, on average, five times these amounts.

There is little well-established information on population densities in malls and shops. A study by Martin (1970) of the pattern of customer movement in a department store suggested 2.5–1.0 m^2 per person over 60% of the gross floor area, the remainder being counter space. This corresponds to 4.3–1.7 m^2 per person over the gross area. Some supermarket operators use 3 m^2 per person. For the malls there is even less information but Stinson (1970) suggests 10 m^2 per person. Population densities vary enormously, Saturday being busier than any other day, with a peak at Christmas. A reasonable design basis is 3.3 m^2 per person for shop units and 10 m^2 per person for the malls.

The CIBSE guide recommends that 16 l s^{-1} of fresh air be supplied for each person when there is some smoking and 24 l s^{-1} each when there is heavy smoking. Taking a population distribution in a mall of 10 m^2 per person these fresh airflow rates correspond to 1.6 and 2.4 l s^{-1} m^{-2}, respectively. This is less than the usual supply airflow rate to the malls of 5 or 6 l s^{-1} m^{-2} and hence the air handling unit treating the malls can be selected and operated economically to provide adequate quantities of fresh air.

A conventional design approach, commonly adopted in the past and still in use in the UK and elsewhere, did not attempt to calculate the supply air quantity needed to offset the sensible heat gains in the malls. Instead, roof top units supplied air to the malls at a rate of 5 or 6 l s^{-1} m^{-2}, referred to the floor area. At the back of each shop unit an extract fan, sometimes two-speed, drew air from the mall and discharged it to waste. The shops, therefore, got second-hand air conditioning and were warmer than the malls in the summer by about

2 K or more, depending on the activity in them. The tenant in the shop provided his or her own heating, usually electrical, but there was seldom much need for heating apart from early in the day because of the heat gain from lights, appliances and people. With this approach, one or more roof-top units, each comprising a direct, gas- or oil-fired heater battery, direct-expansion air cooler coil, in-built air-cooled condensing set and centrifugal supply fan (see Section 2.5) are mounted over the malls. They generally deliver air through a low-velocity ducting system to diffusers or side-wall grilles and handle an adjustable mixture of fresh and recirculated air. Extract is locally through a grille in the ceiling of the mall and it is commonly arranged to halve the amount of fresh air in winter, for economy.

This design technique appears to work quite well in the UK and in other parts of the world, where the climate in summer is not extreme. One important point, that cannot be over-emphasised, is the necessity of having conventional doors at the ends of the malls, as it is impossible to prevent the entry of cold draughts by attempting to pressurise the malls or by fitting an air curtain (see Section 2.9). Door heaters are desirable, even so, at the doors, to temper the inrush of cold winter air when they are opened to let people pass through.

Shop units are now generally fitted with sprinkler systems but malls are not, unless the mall is used for commercial or other purposes involving the presence of combustible materials.

Although extract fans at the back of shop units are still used they are out of favour and the common approach is to extract air through the roof of the mall. An exception might be when the contents of the shop make it essential that a separate extract system is provided. An alternative, better, smoke-relief system from shops, that could be acceptable, is a central, ducted, mechanical extract installation which, in the event of a fire, automatically closes all the grilles except in the shop where smoke is detected. The central fan then gives increased smoke exhaust from the single shop. Another possibility is to distribute clean, cooling water from a central cooling tower and plate heat exchanger complex to the shop units, in addition to conditioning the malls. The tenant then installs his or her own heating and water-cooled air-conditioning unit. Although cooling water distribution is cheaper than chilled water distribution, because no lagging is required, this approach does not seem to have been extensively adopted.

Large shops, department stores and supermarkets in shopping centres are invariably air conditioned by their own, independent systems and this must be remembered when planning the treatment for the rest of the centre.

The behaviour of fire and the production of smoke in shopping centres has been well studied by Butcher (1987), Morgan (1976), Hinkley (1971) and others and clear conclusions reached. Because the quantity of combustible material in a shop is large compared with that in a mall it is more likely that a fire will start in a shop, where a layer of hot gas and smoke will rapidly form beneath the ceiling and spread laterally until it reaches the walls of the shop, when the layer will thicken until it reaches the top of any opening through which it can escape. If there are no openings, or if they are too small, the layer of smoke will thicken downwards to below head level in the shop. Except for large shops, greatly exceeding 1000 m^2 in floor area, the time taken for this to happen is seldom more than a minute or two. When hot, smoky gases escape through openings in the shop front into the mall, they rise and form a layer beneath the ceiling and then rapidly advance, at 1 m s^{-1} or more, towards the ends of the mall. When the end of the mall is open the smoke flows out at high level but local draughts and wind return large amounts back to the mall, where the lower levels are now filled with smoke returning with the cold air inrush to feed the fire. If the ends of mall are closed, the smoke doubles back on itself, again filling the lower reaches of the mall. Therefore, whether the ends of the mall are open or closed, it is rapidly clogged with smoke.

The aim of smoke control is to keep the mall free of smoke so that it can act as an escape route for people. Because of the speed with which smoke spreads, fusible links do not act quickly enough to be of any value in smoke control and smoke detectors must be used to sound an alarm. In the mall itself, the ceiling should be subdivided by downstanding screens into multiple smoke reservoirs, each of 1000 m^2 maximum area. The screens should be reasonably airtight, although small openings around pipes where they pass through the screens are acceptable, and not more than 60 m apart and of at least 1 m depth. Ideally, the screen should extend downward for at least one-third of the way from the ceiling to the floor to give a clear distance of 3 m from the underside of the reservoir to the floor of the wall. It is essential that each smoke reservoir has a natural smoke vent in the ceiling that can be opened to the atmosphere in the event of a fire. This is done automatically or manually when the fire alarm sounds. Mechanical smoke relief systems are not liked because of their dependence on an electrical power supply and the risk of malfunction. One solution is to use natural ventilation openings that are held in the closed position by compressed air, against springs. The exhaust vents then fail safe to the open position.

3.7 Supermarkets

There are three prime considerations when determining the heat gains: population density; electric lighting; open refrigerated cabinets.

Figures on population densities in chain stores and supermarkets are not too well-established but one by Martin (1970) suggests 1.7–4.0 m^2 per person over the gross floor area (Section 3.6). Doone (1971) suggested 3 m^2 per person over the gross floor area. The density varies with the front third having the heaviest load of people. With a conditioned state of 22°C dry-bulb and 50% saturation the heat output per person would be 100 W sensible and 60 W latent.

Since the object of a supermarket is sales, lighting to enhance the attractiveness of the goods on display is most important and illumination levels, mostly by fluorescent tubes, are in the range 1000–1500 lux. For a typical lighting level of 1400 lux, the net gain to the conditioned space could be in the range 45–80 W m^{-2}, depending on the type of luminaire installed.

There are two types of open, refrigerated display cabinet: that with a condenser at the bottom of the cabinet and that with a remote condenser, outside the conditioned space. In the former all the power used by the compressors is an extra load on the room and there is no benefit from the heat absorbed by the frozen food in the cabinets themselves. Stores with this type of display cabinet do not suffer from the underheating sometimes experienced by those having the other type. The second type of open refrigerated cabinet has a big impact on the air-conditioning load. All the heat gain to the cabinets is from the conditioned space and is, therefore, a credit, reducing the sensible heat gain because it is ultimately rejected at the remote condensers. This effect, plus the latent cooling also done at the cabinets, is very significant and must be taken into account when calculating the heat gains, the cooling load and the sensible/total ratio of the cooler coil performance in the central air-handling plant. Complaints sometimes arise because cold air spills out of the cabinets and stratifies above the floor, to a depth of about 1 m. Air distribution can be arranged to deal with this effect, if extract grilles are located at low level, or even in the floor in front of the cabinets. Refrigerated cabinets with remote condensers remove heat from the sales area 24 hours per day, 365 days of the year, regardless of the room temperature; clearly, underheating is sometimes a difficulty, at unexpected times.

It is important to maintain humidity at or below 50% to reduce the latent removal at the cabinets which results in a need for frequent defrosting, shortening the life of the product in the cabinet. High limit humidity control can thus be desirable and, if the sensible/total ratio of the cooler coil condition line is small, i.e. the slope of the condition line from the 'on' to the 'off' states is steep, this may be incompatible with maintaining a required room temperature of 20–22°C, unless reheat is used. There is then a good case for using the condenser reheat. The simplest way of doing this is by directing hot gas from the compressor to the reheater instead of to the condenser, as required. If this is done the need to stabilise condenser pressure remains, as even when the reheater is being used it may not be able to reject enough heat. A more expensive approach that removes some of the refrigerant pressure control and oil return problems is to use a water-cooled condenser. When this is adopted, a plate heat exchanger is essential to ensure a supply of clean, warm, cooling water from the condenser to the reheater, dirty cooling water being circulated to the cooling tower on the other side of the heat exchanger when it cannot reject enough heat through the heater.

Overall cooling loads vary from 90 to 200 W m^{-2} of the total sales floor area, depending on the illumination level and the type of refrigerated cabinet used.

All-air systems of the constant volume, reheat type, preferably using variable proportions of fresh and recirculated air as dictated by the outside air state and an economical operation of the refrigeration plant, should be used. It is highly desirable to design the system to be as simple as possible to operate because the staff of the supermarket are not technically competent to do more than switch it on and off. With this in mind it is common to use air-cooled condensers, obviating the problems of corrosion, scaling and water treatment that would otherwise occur. It is also desirable to adopt heat reclaim techniques, if these are simple, to give economical system operation. Most air-handling plants and air-cooled condensing sets are roof-mounted, proper attention being paid to vibration isolation for the dynamic loads involved (see Sections 7.25–7.27) and the risks of a noise nuisance to neighbouring properties.

Air distribution by low-velocity ducting should be arranged to deal with the possible spillage of air from refrigerated display cabinets, but sensible gains are not usually big enough to make the use of ceiling diffusers or side wall grilles or slots difficult. Because most of the load is near the entrance, 50% of the total supply air quantity should be delivered to the front third of the sales area. The entrance itself should be provided with a door heater to mitigate the worst effects of infiltration as customers pass through the doors in winter. More air should be supplied than is extracted, to achieve a small but unquantifiable positive air pressure within, and so discourage infiltration.

EXAMPLE 3.3

A supermarket of dimensions 39 m long, 17.5 m wide, 3.785 m floor-to-ceiling height, is to be air conditioned at 22.5°C dry-bulb, 50% saturation when the outside state is 28°C dry-bulb, 19.5°C wet-bulb (sling). The structural and solar heat gains amount to 25 kW, the population density is 3 m^2 per person (each emitting 100 W sensible and 60 W latent) and the illumination level is 1000 lux, obtained by the dissipation of 60 W m^{-2}. Open, refrigerated, display cabinets absorb 30 kW, of which 15% is latent, from the sales area and reject it to air-cooled condensing sets located outside the conditioned space. Natural infiltration amounts to one air change per hour. Determine the supply air quantity and the cooling load if an all-air system is to be used with low-velocity air distribution.

Answer

The heat gains can be calculated and are summarised as follows:

	W
Structural and solar	25 000
People: (39 × 17.5/3) × 100	22 750
Lights: (39 × 17.5) × 60	40 950
Infiltration (Equation (1.9)): (1 × 39 × 17.5 × 3.785/3) × (28 – 22.5)	4 736
Total gross sensible heat gain:	93 436
Credit from refrigerated cabinets: 0.85 × 30 000	25 500
Net total sensible heat gain:	67 936
Latent gain from people (Equation (1.11)): (39 × 17.5/3) × 60	13 650
Latent infiltration gain (Equation (1.10)): (0.8 × 39 × 17.5 × 3.785) × (10.65 – 8.632)	4 170
Total gross latent heat gain:	17 820
Credit from refrigerated cabinets: 0.15 × 30 000	4 500
Total net latent heat gain:	13 320

Sensible/total ratio: 67 936/(67 936 + 13 320) = 0.84

A low velocity supply air distribution system is likely to have a fan total pressure of about 625 Pa and, assuming the driving motor is in the airstream, there will be a temperature rise of 1.2 K for each kPa of fan total pressure, giving an increase of 0.75 K across the supply fan. If the plant room is fairly close to the food hall the supply duct runs will be short and a reasonable assumption for the rise due to duct heat gain is 0.75 K, giving an overall rise of 1.5 K. An examination of the psychrometry shown in Figure 3.2, with the design room ratio line having a slope of 0.84, suggests that a supply air temperature of 13°C would be suitable, if we assume that an off-coil dry-bulb temperature of 11.5°C will be compatible with a contact factor of approximately 0.82 to 0.92 and a four- or six-row cooler coil in the air handling unit. Using Equation (2.3) the supply air quantity is calculated as

$$v_{13.5} = [67.936/(22.5 - 13)] \times [(273 + 13)/358] = 5.713 \text{ m}^3 \text{ s}^{-1} \text{ at } 13°\text{C}$$

The corresponding air change rate is (5.713 × 3 600)/(39 × 17.5 × 3 785) = 8 per hour, which is well within the capacity of conventional air distribution methods.

According to the CIBSE Guide (1986), a suitable fresh air allowance, when there is no smoking, would be 8 l s^{-1} for each person, namely, (39 × 17.5/3) × 8 = 1820 l s^{-1} which, if increased to 1904 l s^{-1}, is one-third of the supply air quantity. If, to discourage entering draughts, 1 $\text{l s}^{-1}\text{ m}^{-2}$, namely 39 × 17.5 × 1 = 682 l s^{-1}, is allowed to exfiltrate through the entrance, then the amount of air extracted mechanically is 5713 – 682 = 5031 l s^{-1}. The amount of fresh air handled is 5713/3 = 1904 l s^{-1} and the amount recirculated is thus 5713 – 1904 = 3809 l s^{-1}. Hence the amount discharged to waste under summer design conditions is 5031 – 3809 = 1222 l s^{-1}. All the above air quantities are expressed at 13°C dry-bulb, for convenience.

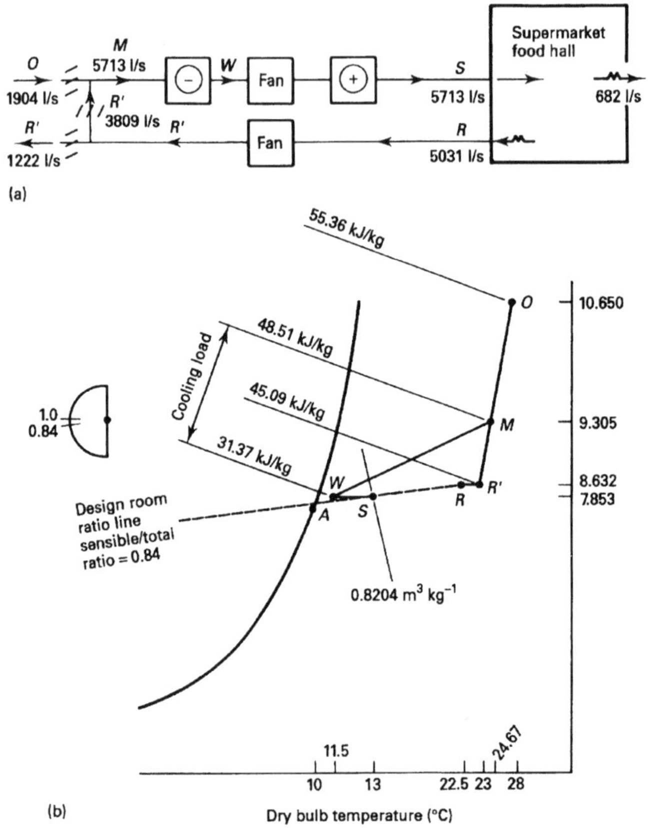

Figure 3.2 Plant and psychrometry for Example 3.3.

A low-velocity extract air system is likely to be working at a fan total pressure of 0.5 kPa and, assuming that the driving motor is outside the airstream, the temperature rise due to fan power is 1 K per kPa of fan total pressure, namely, 0.5 K. Heat gain to the recirculated air ducting is of no consequence because of the very small temperature difference across the duct walls. The recirculated air state, *R′*, is therefore 23°C dry-bulb, 8.632 g kg^{-1} and 45.09 kJ kg^{-1} by interpolation in CIBSE psychrometric tables, or from Equation (1.14) or, less accurately, from a psychrometric chart. The mixture state, *M*, onto the cooler coil, is determined as:

$$t_m = (0.333 \times 28) + (0.667 \times 23) = 24.67°C$$

$$g_m = (0.333 \times 10.65) + (0.667 \times 8.632) = 9.305 \text{ g kg}^{-1}$$

$$h_m = (0.333 \times 55.36) + (0.667 \times 45.09) = 48.51 \text{ kJ kg}^{-1}$$

The moisture content of the air supply state, *S*, to deal with the latent heat gains is determined from Equation (2.4) as

$$g_s = 8.632 - (13\,320/5713) \times [(273 + 13)/856] = 7.853 \text{ g kg}^{-1}$$

Hence the off-coil state, *W*, is established from psychrometric tables or a chart as 11.5°C dry-bulb, 7.853 kg^{-1} and 31.37 kJ kg^{-1}. Reference to tables or a chart also shows that the specific volume at the supply state, *S*, is 0.8204 $\text{m}^3\ \text{kg}^{-1}$. Hence the cooling load can be calculated using Equation (1.17)

$$\text{Cooling load} = (5.713/0.8204) \times (48.51 - 31.37) = 11.9.4 \text{ kW}$$

This should be checked using a different method (see Section 1.9)

	kW
Sensible heat gain:	67.936
Latent heat gain:	13.320
Fresh air load: (1.904/0.8204)(55.36 – 45.09)	23.835
Supply fan power (by Equation (2.3)): (5.713 × 1.5 × 358)/(273 + 13)	10.727
Recirculated fan power: (3.809 × 0.5 × 358)/(273 + 13)	2.384
Cooling load:	118.202

This is about one per cent different from the other answer.

Plotting the states shown in Figure 3.2 on a psychrometric chart allows the apparatus dew-point, *A*, to be determined at 10°C. The contact factor, β, is then calculated

$$\beta = (24.67 - 11.5)/(24.67 - 10) = 0.90$$

This is a practical value and would mean using a six-row cooler coil with a face velocity of about 2.5 m s^{-1} and tubes having 315 fins per metre (8 fins per inch).

3.8 Department stores

A department store comprises a complex of different sales areas in which lighting levels vary considerably, population densities are different and internal heat gains from various appliances are possible. The air-conditioning treatment should therefore be subdivided among several plants, the duty of each being related to the load and application of the area treated. Although floor-mounted unitary equipment has been used in the past, it is not popular with store operators because of the valuable sales floor area occupied. Population densities tend to be high and it follows that all-air systems are favoured, generally located on the roof, but occasionally mounted at high level, above suspended ceilings in the sales area itself. A plant mounted at high level in an occupied area is difficult to get at and will not receive proper maintenance. It is much better to seek an outside location with easier access.

Although high-velocity air distribution has been used to help accommodation of the ductwork, noise problems become a risk and the unavoidable, higher fan total pressures cause increased running costs and greater energy consumption. It is better to distribute air at conventional, low velocities, if at all possible. Department store layouts may be changed, from time to time, so the duct and plant design should be conservative, with enough flexibility and capacity built into them to permit any future reallocation of air quantities necessary for load revisions. The modern trend is to use packaged air-handling plants and this sets a practical limit on the area treated by one plant, which should correspond to a maximum supply air quantity of about 12.5 $m^3 s^{-1}$. Final air distribution can usually be through conventional slots, diffusers and grilles. It is a good policy to use adjustable diffusers and grilles that allow modification of the air pattern, should this be necessary, to counter complaints of draughts or lack of air movement, particularly after a departmental layout rearrangement. Although several reheaters can be used with one air-handling plant, it is better to split the duty among several plants and have only one heater battery for each. In this way every plant can be economical in running cost, its heater battery being controlled in sequence with its cooler coil to conserve energy. With sequence control it is necessary to introduce a high limit humidistat, set at 55%. Upon rise in humidity the heater battery is overriden but allowed to continue controlling temperature as a reheater, the cooler coil coming on to secure some dehumidification. When the humidity fails the system is permitted to return to its normal mode of control. Multizone systems have been used with success but variable air volume systems require a cautious approach because of the reduction in the fresh air quantity that may accompany a throttled air supply. Variable proportions of fresh and recirculated air should be provided for and it should be arranged that the balance of the supply and extract quantities between departments and from floor-to-floor is such as to avoid excessive air movement along escalators and through doorways.

Central boiler plant and water-chilling plant, located at basement level, are needed to distribute LTHW and chilled water to air-handling units. While air-cooled condensers are simple to operate it is possible that the refrigeration load may be too big and cooling towers will be wanted.

In broad terms, the major sources of heat gain are the electric lights and the people. Some of the lighting will be by fluorescent tubes but some is likely to be by tungsten filament lamps with as much as five times the dissipation of heat for the same illumination. An overall figure of 60 W m^{-2} has been suggested by Murphy (1970) with average population densities of 3 m^2 per person in the basement sales area, 4 m^2 per person on the ground floor and 6 m^2 per person on upper floors, compared with those given earlier by Martin (1970) of 1.7–4.3 m^2 per person.

Certain areas have unique problems that merit particular attention. In a beauty salon the gains are from dryers and other appliances, in addition to lights and people. One person should be allowed for each basin and dryer, plus one member of staff for every four customers. A negative pressure is necessary in the salon to discourage smells from spreading to other parts of the store and for this reason a separate air-handling plant must always be used for this area. Fitting rooms are usually located around the perimeter of the sales floor and require supplementary heating, in particular if the floor overhangs the street or the car park and so gives an extra heat loss. Alterations rooms have higher heat gains because of steam irons used for pressing, and exhaust in their vicinity is sometimes a good policy, even to the extent of using hoods. At the same time, women sitting down at sewing machines must not feel draughts. Television and hi-fi display areas often suffer very heavy heat gains – a colour television set can liberate 500 W – so a proper allowance must be made, based on the number of sets on view and a diversity factor that is best agreed with the store operators. Local exhaust ventilation may be considered to reduce the worst effects of the heat gains. Lamp sales areas also have exceptionally high heat gains with radiant components that are not directly dealt with by air conditioning. Again, local exhaust may be useful. Entrances must receive extra heating to combat infiltration. Door heaters are the minimum provision and in some instances air curtains have been successful over much of the year (see Section 2.9), enabling the doors to be wide open at times, encouraging the entry of customers. Where possible, without compromising the design commercially, 10 W m^{-2} may be added to the heat gains to cover an additional and unpredictable use of appliances over and above the other heat gains.

Fresh air allowances should follow the recommendations of the CIBSE guide.

3.9 Kitchens and restaurants

The essential principle of air distribution in kitchens and restaurants is to arrange for airflow from the restaurant into the kitchen in order to avoid the retention of food smells in the restaurant and the spread of odours to adjacent premises. The essential difference in treatment is that restaurants can be air conditioned to a comfort standard, whereas kitchens cannot. In kitchens there is a high proportion of radiant heat gain and the best that can be done is to cool the supply air to about 12 or 14°C, thus improving temperature conditions of as high as 45°C to about 30°C.

Restaurant loads are predominantly from people, with a population density of 1.8–2.0 m^2 per person and electric lighting, largely from tungsten sources, up to 75 W m^{-2}. Odours from food, people and smoking must be diluted by using recommended fresh air supply rates, according to the CIBSE guide (1986) of at least 8 l s^{-1} for each person. The type of restaurant influences the choice of inside design conditions, lower temperatures (22°C dry-bulb) being chosen for establishments with longer term occupancies. In this connexion, it is worth noting that because of the high latent heat gain from people the room ratio line may have a steep slope, and maintaining 50% humidity with 22°C dry-bulb may not be a practical proposition with commercially available cooler coils and refrigeration plant. In such cases up to 60% is tolerable and should be the design aim; except during peak occupancy the humidity will be much lower than this. If a bar forms part of a restaurant, extract ventilation should be concentrated over the bar, to remove local smells and tobacco smoke.

Kitchens should have at least 20 air changes per hour of extract ventilation, removed mostly through hoods over the cooking equipment. Exhaust air quantities through hoods should be in accordance with the recommendations of the CIBSE (1986) with an absolute minimum air velocity of 0.3 m s^{-1} over the face of any hood. Hoods are ineffective with velocities less than this. Just as there is some extract from parts of the restaurant, such as the bar, remote from the kitchen, so there is often a need for the supply of air to parts of the kitchen, to supplement the air drawn from the restaurant. Such air must be filtered and tempered in winter and it should be cooled in summer as suggested, if this can be afforded. There should be no attempt to specify a condition to be maintained in the kitchen but it is most important to ensure that the balance of airflow is from the restaurant into the kitchen.

Air change rates in commercial kitchens can be very high indeed and it is then worth considering the use of proprietary forms of kitchen exhaust hood that provide local supply, specifically for the hood, thus easing the problems of supply air distribution and saving energy.

3.10 Auditoria and broadcasting studios

Auditoria may be roughly classified as cinemas, theatres and concert halls, in terms of the importance of noise control and the attention that must be paid to this in the design of the air conditioning system. The sound production system in cinemas is sufficiently loud to make noise control of least importance among the three but in theatres audibility at all parts of the auditorium is critical and there must be no intrusive noise from the mechanical services. In concert halls audibility is even more important and background noise of any sort is unacceptable. The ASHRAE guide (1995) quotes noise criteria that may be interpreted as NR 30 for cinemas and NR 20 for theatres and concert halls.

The major component of the load is from people in all these cases and it is evident that all-air systems are required. Low-velocity air distribution should be used because of the low noise levels necessary and because there is usually enough space to run ductwork. It is suggested by ASHRAE (1995) that 2–3 m^2 per person be allowed for population density in the lobby spaces and 0.75–1.0 m^2 per person in the auditorium, referred to the total area including the aisles if a seat count is not possible. In cinemas and theatres the lights are off or dimmed, except when cleaning takes place and it is recommended by ASHRAE (1995) that 5–10% of the installed lighting power be taken for design purposes, coincident with the population load. Electric lighting is seldom a significant proportion of the heat gain in an auditorium of this sort, although in a convention hall this might not be the case.

The situation is quite different on the stage where lighting presents a most complex problem. Tungsten lights located around the proscenium can be dealt with by extract fittings that remove 40–50% of the heat liberated and return it to the central plant for discharge to waste or recycling, as appropriate. Over the stage itself the presence of movable scenery and the need for spot lighting raise difficulties; on the other hand, theatre-in-the-round, introduced as a break away from traditional stage/audience configurations, may have all its lighting in the auditorium, aiming largely at the stage and representing loads as high as 200 kW for a seating accommodation of 1400 according to Thornley (1969).

In assessing structural heat gains, advantage can be taken of stratification in high auditoria, if the air supply grilles are 1.5 m or more below the depth assumed to be occupied by the stagnant air. Such stratified air is assumed to suppress the natural convection component of the inside surface heat transfer coefficient, allowing heat gain by radiation only. This is then usually taken as 33% of the total structural gain.

An excellent survey of mechanical services in auditoria by Thornley (1969) shows that there are at least six effective possibilities for air distribution:

(1) Downward – air enters from ceiling diffusers and is extracted beneath the seats. This gives draughtless air movement but advantage cannot be taken of stratification. Supplementary heating is needed at the sides, where there is a heat loss and near emergency exits, in winter.
(2) Upward – air is supplied beneath the seats and extracted at high level. This is the best arrangement, in theory, because cooling air is delivered at the place where there is the major source of heat gain, i.e. from the people. Air has been successfully introduced in this way through perforations in the pedestals supporting the seats and also through specially designed perforated diffusers in the risers of the seating tiers. Careful design is essential and this technique of supply should never be used without a preliminary, full-scale, mock-up and a proper programme of test for the air distribution and the air temperatures involved. Draughts at the ankles are the risk and this is exaggerated bv a large cooling temperature difference, room-to-supply.
(3) Front-to-rear – long-throw grilles are needed and it is difficult to select these for a throw exceeding 10 m without producing too much noise. Taking advantage of stratification is possible if the supply grilles are sufficiently below the level of stratified air. Some variation of temperature and freshness is not unlikely over the length of the auditorium, in the direction of air throw.
(4) Front-of-stage-to-rear – air is blown from beneath the front of the stage to extracts at the rear of the hall. Similar considerations to (3) prevail. There must be a large distance

between the stage and the first row of seats, if draughts are to be avoided there. A variant used in some of the early cinema installations was to extract the air on each side of the screen. When this is done supplementary heating is needed at the sides, where there is a winter heat loss.

(5) Rear-to-front – long-throw grilles are needed, with possible noise problems, and supplementary winter heating.

(6) Side-wall supply air with high level extract – this is a cheap arrangement in terms of ductwork but air distribution may not be adequate for the rear of the auditorium or the balcony.

A general comment on air distribution is that auxiliary supply and extract may be needed with any of the above systems at the back of the auditorium, above and beneath the balcony, to deal with vitiated, warm air that may stratify. It is sometimes claimed that air distribution over the large plan area of the auditorium can be helped by relying on the influence of the extract openings. However Jamieson (1959) has shown that it is only the momentum of the supply air that is significant in giving good air distribution and the location of the extract openings is irrelevant in this respect.

The presence of the fly tower over the stage of a theatre introduces a large stack effect that is often apparent by the curtain billowing when it is lowered. There are generally side stages and a rear stage in opera houses, and these require air conditioning or ventilation as well as having problems of air distribution.

The local authority, also the licensing authority in the UK, the building regulations, the fire brigade and other municipal influences, must be consulted when the mechanical and electrical services are being designed. Significant restraints are often imposed by these authorities. For example, the local authority in London may require a minimum fresh air provision of 28 m^3 per hour for each person in a place used for public entertainment, but this may be reduced to 14 m^3 per hour if air conditioning is installed and the humidity does not exceed 55%. The same authority may require that, with downflow air systems, the flow must be reversed, i.e. supply grilles becoming extract grilles, and vice-versa, in the event of a fire, to remove the smoke and so help people to escape from the auditorium.

Theatres and opera houses usually have 'crush bars' which are very densely occupied for fifteen minutes or so, once or twice during a performance, i.e. in the intervals. Heat gains from people are then very high indeed and local, temporarily boosted extract may be necessary to supplement the air conditioning. Where a concert hall or theatre forms part of a larger complex, incorporating restaurants, large foyers, bars, and so on, separate plants should be adopted for areas having similar patterns of usage, as for the public room in hotels (Section 3.4). The location of plant, particularly refrigeration machines and boilers, merits special care. On no account should they be in places where noise and vibration can be transmitted to the auditorium.

The inside design conditions chosen for the auditorium, or other areas where long-term occupancy is the rule, should be those normally adopted, for example 22–23°C dry-bulb and up to 55–60% humidity, in the UK. In entrance foyers, in the UK, it may be reasonable to design for a temperature half-way between that in the auditorium and that outside, bearing in mind that outside temperatures will fall as the time of the evening performance approaches.

Pre-cooling is sometimes advocated for the two or three hours preceding a performance to reduce the size of the installed cooling capacity necessary, but this does not always give good results, since temperatures at the start of a performance are too low and they rise to uncomfortably high values by its end. According to Thornley (1969), installed cooling capacities seem to be an average of 15–25 kW of refrigeration per 100 seats in European auditoria.

Television and broadcasting studios are very special cases indeed. Quite apart from the need to select quiet-running plants, provided with adequate silencers and mounted on properly selected vibration isolators, the air distribution from the supply grilles or diffusers may cause far too much noise. The problem is made worse by the very high lighting loads, 300 W m^{-2} being common in the UK, and as much as 600 W m^{-2} in Europe. The studios often have very large floor-to-ceiling heights. There may be a subceiling, acting as a walk-way for maintenance and access to the lights supported from it. The space above this can be 5–6 m high and the space in the studio beneath it as much as 12 m. Extreme care must be taken with the design of the air distribution system. Air has been successfully supplied at high level above the suspended walk-way, to spread evenly over it and diffuse downward through it into the studio, giving 20–25 air changes, over its height of 12 m. All supply and extract ducting must be acoustically lined, minimum 50 mm thick, in addition to the silencers required at the plant. A maximum velocity is 2.0 m s^{-1} through the free area of grilles or slots and velocities in the ductwork must not exceed 3.0 m s^{-1}.

A silencer must be provided on the supply side of the air handling plant, positioned so that no noise from the plant can outflank it. A silencer must also be fitted on the extract side of the air handling plant, again with due precautions about flanking noise. Terminal silencers are usually needed at each supply air terminal and at each extract point. Steps are necessary to prevent noise break-in through the duct walls, often taking the form of 1 mm steel duct walls bonded to 0.75 mm of lead, due regard being paid to the increased weight of the installation and the consequences of this on the building structure. It is essential that the air-conditioning system is isolated from the structure of the building and this involves the use of spring hangers for the ductwork, properly loaded and installed. The supply and extract diffusers and ducting must not touch the ceilings or rest of the building structure and properly selected and fitted gaskets are used as necessary.

Duct branches from the mains to the diffusers must be long in order to give smooth airflow into the diffuser necks and it is recommend that they should be at least 1.2 m in length. It is not possible to balance the diffusers by using dampers because of the noise they generate. If dampers are essential, they should be on the upstream side of the supply terminal silencers and the downstream side of the extract terminal silencers. It follows that the duct mains are sized with very low pressure drops in order to make the system virtually self-balancing. The cones in the diffusers may themselves generate unacceptable noise and in this case the cones have to be removed, with due consideration paid to the consequences of this on the air distribution in the broadcasting studio.

For the smaller size of studios it is desirable to use direct-expansion air cooler coils in the air handling plant with remote air-cooled condensers, properly fitted with purpose-made silencers and acoustic louvres. Larger studios are likely to need water chillers, probably water cooled, but whatever form the refrigeration plant assumes careful attention to the treatment of noise and vibration is vital. A measure of standby-by is necessary because of the sensible heat gains in the studios and, in the case of the supply and extract fans, this can be 100% by the provision of belted, running, stand-by driving motors (see Section 7.27).

For an air-conditioning system to be acoustically successful in an exacting application such as a concert hall or a broadcasting studio, it is essential that the mechanical services are designed, installed and commissioned in full consultation with the acoustic consultant and that the design and construction of the building itself deals with the risks of flanking noise paths and the transmission of sound and vibration through the structure.

3.11 Museums, art galleries and libraries

Books, paintings and some other valuable articles are often made of hygroscopic materials that expand and contract with variations in the ambient humidity. Such movements eventually cause cracks and deterioration. Since humidity is temperature-dependent it follows that both should be controlled if the objects stored or on display are to be protected. Furthermore, the air-conditioning system must run continuously and must also have adequate stand-by features. Although natural daylight is preferred for viewing paintings, spotlights are sometimes used in art galleries, and these impose local radiant heat gains that exacerbate the thermal and hygroscopic movements referred to. Enclosing articles of value in glass-fronted cases is not always an answer as the cases breath as they undergo thermal expansion and contraction, possibly made worse by spotlights. One solution is to enclose very valuable objects in hermetically sealed cases filled with helium. Another approach is to feed a supply of conditioned air to each case and so keep it at a slight positive pressure in relation to the ambient air and in a controlled condition.

Apart from structural gains the load is from people and lights, the population density varying from an average of 10 m^2 per person to a possible occasional peak of 2 or 3 m^2 per person for special exhibitions, or for a social function. Lighting is unlikely to exceed 500 lux, produced by fluorescent tubes liberating 15–27 W m^{-2} of floor area.

Atmospheric impurities in the outside air, notably SO_2, and even SO_3, are a danger to leathers, paper, textiles, wood and paint. Smoking is seldom permitted but, if allowed, will also damage these materials and many others. Filtration should, therefore, be aimed at removing these undesirable contaminants from both the outside and recirculated air.

All-air, low-velocity systems are the natural choice and if control over both temperature and humidity is demanded then constant dew-point plants with reheat are required. In many cases it is sufficient to keep the humidity reasonably constant at any value in the comfort zone, without paying too much attention to the precise value selected. Thus, 40% ± 5%, or 55% ± 5%, might be acceptable. It should be remembered that the first step to good control over humidity is good control over temperature.

3.12 Swimming pools

A swimming pool is intended for the bathers and their comfort is best served, in the UK, by the following temperatures:

Childrens' teaching pool or leisure pool	28–29°C
Pool for the general public	27–28°C
Pool for the elderly, disabled, or infirm	up to 30°C
Competition swimming	26–27°C

The air temperature should be one degree above the water temperature and the humidity is usually between about 50–60%. Air-conditioning a swimming pool hall to the comfort standards commonly accepted in this country for fully-clothed people is clearly out of the question. If the comfort of the spectators is a matter of concern they should be screened from the pool hall by vapour-tight glass partitions and conditioned by a separate, all-air plant. The pool hall itself must be ventilated and the following considerations apply in doing this.

Table 3.4 Standard dimensions for diving pits

Diving board		Approximate size of pit			Approximate water content
Type	Height (m)	Length (m)	Width (m)	Depth (m)	(m^3)
Spring	1	9.0	5.0	3.0	136.380
	3	10.5	7.0	3.5	227.3
Fixed	5	12.0	7.6	3.8	318.22
	10	15.0	9.0	4.5	590.98

THE SWIMMING POOL COMPLEX

Pools in the UK are commonly built to three standards:

(1) County – 25 × 12.8 m, surface area 320 m^2, content about 640 m^3
(2) National – 33.3 × 12.8 m, surface area 426.24 m^2, content about 910 m^3
(3) Olympic – 50 × 17 m, surface area 850 m^2, content about 2270 m^3.

The depth at the shallow end is 1 m and a diving pit is assumed at the other end. The dimensions of a diving pit depend on the type of diving board and its height. Standard dimensions according to Doe *et al.* (1967) are given in Table 3.4. When two boards, side-by-side, are used, the width is increased by 50%.

Leisure pools have no standard size or shape and are provided with many additional features, such as flumes, spas, wave machines, river rides, water mushrooms, etc. which add difficulty to a determination of the wetted surface area when calculating the evaporation rate.

Teaching pools are much smaller, 12.8 × 7.3 m, containing about 60 000 l and of a depth 600–700 mm. The rest of the accommodation generally provided includes changing rooms, toilets, showers and precleansing foot baths, cafeteria, administrative offices and a plant room. Terraced seating is the rule along one side of the pool, to deal with 200–300 people but sometimes as many as 2000 people can be seated, along both sides of the pool.

MECHANICAL VENTILATION

Mechanical ventilation is essential and must run continuously. Its purpose is to keep the humidity within the pool hall to an acceptably low level in both summer and winter, to minimise the extent of condensation forming on cold surfaces in winter, to minimise the rise in air temperature in summer, to dilute the concentration of objectionable odours and to provide an adequate supply of fresh air. It is also often used to offset the winter heat loss. A study by Doe *et al.* (1967) of the consequences of adopting different ventilation rates indicates that 15 l s^{-1} m^{-2} of wetted surface area is not enough to prevent humidity from becoming too high at certain times of the year nor enough to prevent condensation streaming down single glazed windows. It is recommended that 20 l s^{-1} m^{-2} of wetted surface area be supplied. This higher rate deals with the problem of high humidities for most of the year in the UK, except perhaps for the warmest summer weather. Condensation will still occur in winter on single glazing. There is also a chance that condensation may occur on double glazing and on metal window frames without a thermal break, at certain times of the year, depending on the activities in the pool and the number of bathers, both of which affect the wetted surface area, the evaporation rate and hence the relative humidity.

It is impossible to calculate the evaporation rate, W, in kg s^{-1}, or the latent heat emission, q_e, in W m^{-2}, from the wet surfaces in a pool hall with any certainty. For airflow across a simple horizontal water surface Thiesenhusen (1932) gives the rate of evaporation as

$$q_e = (0.0885 + 0.0779v)(p_w - p_s) \tag{3.1}$$

where v is the air velocity over the surface of the water in m s^{-1}, p_w is the saturated vapour pressure in Pa exerted by the water in the pool at its temperature t_w, and p_s is the partial pressure of the water vapour in the air above the pool surface in Pa. The equation is based upon a ducted airflow across an open surface of water in a pan, with controlled and measured conditions of temperature, humidity and air velocity. When using the equation for a swimming pool there is some doubt as to the air velocity to choose, particularly so because of the activity of the bathers. In terms of comfort alone, 0.15 m s^{-1} is reasonable. If the water temperature is taken as 28°C and its saturated vapour pressure as 3780 Pa, the above equation simplifies to

$$q_e = 379 - 0.1\,p_s \tag{3.2}$$

When using Equations (3.1) and (3.2) experience has shown that the surface area to be used is the pool surface plus 20%, to take account of the wetted surrounds and this has yielded satisfactory results for the design and operation of ventilation systems for swimming pool halls. However, the equations have been criticised in the past because they ignore the number of bathers present, their wet surfaces, and the influence the activity of the bathers may have on the evaporation rate from the water. Furthermore, no account is taken of the additional wet surfaces caused by features such as flumes, wave machines, etc. in leisure pools. It would be possible to estimate the surface areas of bathers by using an equation proposed by Du Bois and Du Bois (1916)

$$A_D = 0.202\,m^{0.425}h^{0.725} \tag{3.3}$$

where A_D is the Du Bois surface area of a body in m^2, m is the body mass in kg and h is the body height in m. It might also be possible to estimate the wet surface areas of the various water features, given their dimensions.

More commonly, two equations, due to Biasin and Krumme (1974) which purport to account for the presence of the bathers, have been used. These were originally formulated in non-SI units but, adopting SI units, their equation for occupied pools becomes

$$W_0 = [3.278 + 4.157\varepsilon(p_w - p_s)]\,A_p \times 10^{-5} \tag{3.4}$$

where W_0 is the evaporation rate of water in kg s^{-1}, ε is a dimensionless, normalised activity factor that depends on the number of bathers and the pool surface area (see Figure 3.3), and is defined by

$$\varepsilon = (n_n\,\Delta\theta))/(A_p\,n_p) \tag{3.5}$$

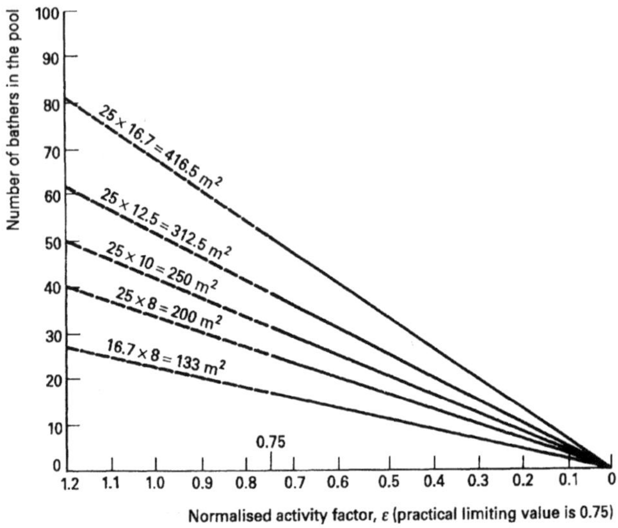

Figure 3.3 Normalised activity factor for evaporation from indoor swimming pools, for use with Equation (3.4). After Biasin and Krumme (1974).

in which n_n is the number of people arriving at the pool per hour, $\Delta\theta$ is the average length of stay of a bather in the pool and n_p is the number of bathers present per unit area of pool surface. In Equation (3.4) p_w is the saturation vapour pressure corresponding to the water temperature of the pool in kPa, p_s is the partial pressure of the water vapour of the air in the pool hall in kPa and A_p is the surface area of the pool in m^2. The evaporation rate can easily be turned into a latent heat gain to the pool hall by multiplying by the appropriate latent heat of evaporation (Table 3.5).

For an unoccupied pool, the evaporation rate, W_u in SI units is

$$W_u = [2.188 \times (p_w - p_s) - 1.639] A_p \times 10^{-5} \quad (3.6)$$

The units of vapour pressure are again in kPa.

The work by Biasin and Krumme was based on measurements in an apartment block for an actual, rectangular swimming pool having dimensions 12.5 × 5.12 m, giving a total pool surface area of 64 m^2. The pool hall was treated by a ventilation system handling 444.4 $l\,s^{-1}$ using a heat pump. This gives a specific ventilation rate of 6.94 $l\,s^{-1}\,m^{-2}$, which is a good deal less than the figures commonly used in the UK and mentioned earlier. The authors of the work stress that it is only possible to apply their equations to pools of different sizes if the pools have the same geometrical proportions and if the Reynolds number for

Table 3.5 Latent heat of evaporation

Temperature, °C	27	28	29	30
Latent heat of evaporation, kJ, kg^{-1}	2437.1	2434.7	2432.4	2430.0

the air circulation in the pool hall is the same. The authors do not quote the Reynolds number for their system and the pool hall. Further restrictions regarding the validity of their equations, when applied to pools of different sizes, are that the supply air temperature should be the same as the air temperature in the pool hall and the expression or measurement of humidity and temperature in the pool hall should be at a height of 1.7 m above the pool surface (or at the same height above the surrounding floor, adjacent to a wall). It was considered that the maximum value of ε could lie in the range 0.68 to 0.75 when the equations were extrapolated to other pools but that values of ε exceeding 0.75 had little meaning because the evaporation rate from the pool did not increase as ε rose above this value.

EXAMPLE 3.4

(a) Calculate the total evaporation rate from the pool and the total latent gain to the hall, for a pool of 416.5 m^2 area containing 20 bathers, using Equations (3.2) and (3.4). Refer to Figure 3.3 for a value of ε to be used in Equation (3.4). The pool water is at 28°C and the air in the pool hall is at 29°C dry-bulb and 55% saturation.
(b) Calculate the evaporation rate when the pool is not occupied, by means of Equation (3.6).

Answer
(a)

Vapour pressure at 29°C dry-bulb and 55% saturation, p_s, = 2.243 kPa (CIBSE tables)

Vapour pressure of water in the pool, p_w, at 28°C = 3.779 kPa (CIBSE tables)

Latent heat of evaporation at 28°C = 2434.7 kJ kg^{-1} (Table 3.5)

By Equation (3.2), assuming that the wetted surface area is 20% greater than the pool area, the specific latent heat gain to the pool hall by evaporation is

$$q_e = (379 - 0.1 \times 2243) = 154.7 \text{ W m}^{-2}$$

Total latent heat gain = (154.7 × 416.5 × 1.2)/1000 = 77.32 kW

Total evaporation rate = 77.32/2434.7 = 0.0318 kg s^{-1}

Specific evaporation rate = 0.0318/416.5 = 0.0000764 kg s^{-1} m^{-2}

Normalised activity factor, ε (see Figure 3.3) = 0.28

By Equation (3.4) the total evaporation rate

$$W_0 = [3.278 + 4.157 \times 0.28 \times (3.779 - 2.243)] \times 416.5 \times 10^{-5} = 0.0211 \text{ kg s}^{-1}$$

Specific evaporation rate = 0.0211/416.5 = 0.0000507 kg s^{-1} m^{-2}

Total latent heat gain = 0.0211 × 2434.7 = 51.37 kW

(b) When unoccupied, the total evaporation rate, W_u, by Equation (3.6) is

$$W_u = [2.188(3.779 - 2.243) - 1.639] \times 416.5 \times 10^{-5} = 0.007171 \text{ kg s}^{-1}$$

$$\text{Total latent heat gain} = 0.007171 \times 2434.7 = 17.46 \text{ kW}$$

At night, when the pool hall is not in use, a cover should be placed over the pool to restrict the evaporation rate to virtually zero.

Reeker (1978) made measurements of evaporation rates in three indoor swimming pools, equipped with mechanical ventilation systems using heat pumps, and found a wide difference in the results obtained with a dependence on the frequency of use of the pool and the relative humidity in the pool hall. It was concluded that the use of an unfavourable calculation method could lead to an installation having unnecessarily large energy losses, without improving the conditions in the pool hall. Typical measured evaporation rates were 0.000017–0.000022 kg s^{-1} m^{-2} for still water surface conditions and 0.000022–0.000036 kg s^{-1} m^{-2} for water surfaces in motion.

The answers in part (a) of Example 3.4 gave an evaporation rate of 0.0000764 s^{-1} m^{-2}, by Equation (3.2) and 0.0000507 kg s^{-1} m^{-2}, by Equation (3.6), suggesting that the use of Equation (3.6) gives results that are not unreasonable.

Dressing rooms must be ventilated by a separate plant using 100% fresh air and giving ten air changes per hour. Less than this rate is poor practice and six air changes per hour is totally inadequate for dealing with odours and helping to keep the floor dry. Pool halls and dressing rooms should be maintained at a slightly negative air pressure to discourage the migration of smells to adjoining premises.

The performance of a swimming pool hall in humidifying the ventilating air passing through it can be regarded as similar to that of an inefficient, steam-pan humidifier (see Figure 3.5).

EXAMPLE 3.5

(a) A swimming pool has a pool surface area of 416.5 m^2 and is maintained at a temperature of 28°C. The air in the pool hall is at 29°C when it is –1°C saturated outside. Determine the humidity in the pool hall if there are 40 bathers and the ventilation system handles 100% fresh air at rates of 15 and 20 l s^{-1} m^{-2}. Make use of Equation (3.4) and compare with the results achieved by Equation (3.2).

(b) Given that the heat loss from the pool hall is 40 kW and that this is to be offset by warming the fresh air delivered from the ventilation system, calculate the necessary supply air temperature for a rate of 20 l s^{-1} m^{-2}, referred to a wetted surface area of 1.2 × the pool area.

Answer

(a) Assuming for convenience that the air supplied is expressed at 29°C the two proposed rates of fresh air supply are 15 × 416.5 = 6247 l s^{-1} and 20 × 416.5 = 8330 l s^{-1}. The fresh air supplied has a moisture content 3.484 g kg^{-1} (at –1°C saturated). By Equation (2.4) its latent cooling capacity, when handling 15 l s^{-1} m^{-2}, is

$$Q_{lc} = [6.247 \times (g_r - 3.484) \times 856]/(273 + 29) = 17.71g_r - 61.69 \text{ kW} \tag{3.7}$$

For 20 $l\ s^{-1}\ m^{-2}$ the latent cooling capacity is

$$Q_{lc} = [8.33 \times (g_r - 3.484) \times 856]/(273 + 29) = 23.61g_r - 82.26\ \text{kW} \quad (3.8)$$

From Figure 3.3, the normalised activity factor is 0.58 and, from CIBSE psychrometric tables, the saturation vapour pressure, of the pool water at 28°C, p_w, is 3.779 kPa. By Equation (3.4), the evaporation rate from the pool to the air in the pool hall is

$$w_0 = [3.278 + 4.157 \times 0.58 \times (3.779 - p_s)] \times 416.5 \times 10^{-5} = 0.05160 - 0.01004\,p_s\ \text{kg s}^{-1} \quad (3.9)$$

Taking the latent heat of evaporation (Table 3.5) to be 2434.7 kJ kg^{-1} at 28°C, the latent heat gain to the pool hall is

$$Q_e = [0.05160 - 0.01004\,p_s] \times 2434.7 = 125.63 - 24.45\,p_s\ \text{kW} \quad (3.10)$$

Assuming two different values for the percentage saturation in the pool hall, say 40% and 60%, with a dry-bulb temperature of 29°C, allows the vapour pressure of the air in the pool hall, p_s, to be determined and hence the latent gain to the pool hall for the two assumed values of percentage saturation and the latent cooling capacity for the two ventilation rates:

				Latent cooling capacity (kW)	
			Latent heat gain	Ventilation rate ($l\ s^{-1}\ m^{-2}$)	
			Equation (3.10)	15	20
μ (%)	g_r (g kg^{-1})	p_s (kPa)	(kW)	Equation (3.7)	Equation (3.8)
40	10.29	1.641	85.51	120.5	160.7
60	15.43	2.442	65.92	211.6	282.0

If Equation (3.2) is used for the latent gain to the pool hall by evaporation, instead of Biasin's Equation (3.4), we have the following.

At 40% saturation and 29°C in the pool hall, $q_e = 379 - 0.1 \times 1641 = 214.9$ W m^{-2} and the total latent gain to the pool hall is $(214.9 \times 1.2 \times 416.5)/1000 = 107.4$ kW.

At 60% saturation and 29°C in the pool hall, $q_e = 379 - 0.1 \times 2442 = 134.8$ W m^{-2} and the total latent gain to the pool hall is $134.8 \times 1.2 \times 416.5/1000 = 67.4$ kW.

Figure 3.4 illustrates these results. The humidity prevailing would appear to be in the range from 30 to 40% saturation, depending on the assumptions made and the equations used. It is clear that calculations of the latent gain to the pool hall by evaporation cannot be accurate because of the uncertainty about the activity in the pool and the wetted surface areas. On the other hand, the latent cooling capacity of the air supplied is based on the use of Equation (2.4) which is fundamentally correct although, of course, the air distribution arrangements in the pool hall will have an influence. Both Biasin and Krumme (1974) and Reeker (1978) found that considerable variations of humidity occurred in occupied swimming pool halls.

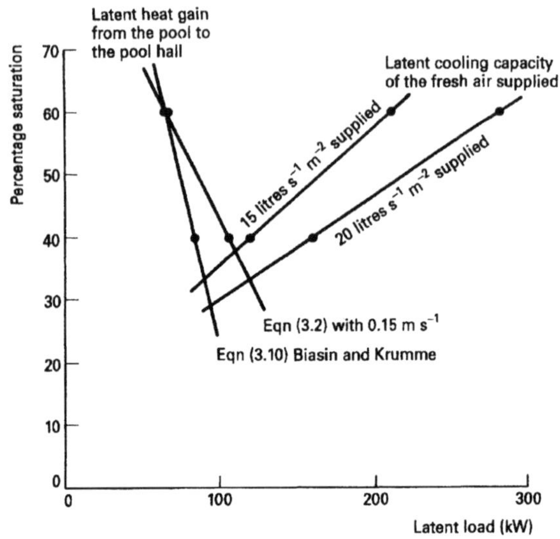

Figure 3.4 The results of Example 3.5.

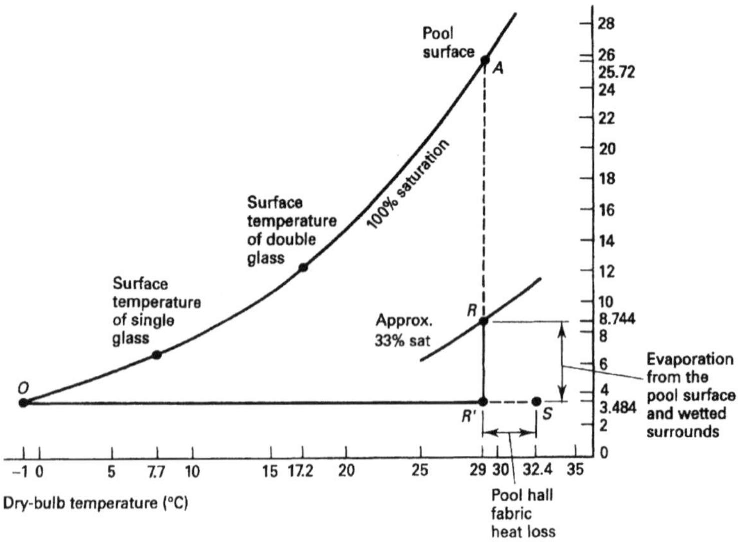

Figure 3.5 Psychrometry of pool hall ventilation in Example 3.5.

(b) The rate of supply of fresh air = 1. 2 × 416.5 × 20 = 9996 l s^{-1}, which is approximately 10 m^3 s^{-1}, expressed at 29°C. By Equation (2.3) the supply air temperature necessary to offset a heat loss of 40 kW is

$$t_s = 29 + (40/10) \times [(273 + 29)/358] = 32.4°C$$

Figure 3.5 illustrates the psychrometry. Fresh air is warmed from −1°C (state *O*) to 32.4°C (state *S*) and supplied to the pool hall, offsetting the heat loss of 40 kW and falling to 29°C (state *R′*). The air also flows over the pool surface and, if this were infinitely long, would attain 29°C at state *A*. In fact, it is humidified by evaporation from the pool and the wetted surrounds only to state *R*, at 33% and 29°C. It can be inferred that the pool and its surrounds have a humidifying efficiency of [(8.744 – 3.484)/(25.72 – 3.484)]100 = 24%, for a relative velocity of 0.15 m s^{-1}.

Although using recirculated air is a possibility it is generally considered wiser to use 100% fresh air, except, perhaps, when heat pumps are used. Energy can be reclaimed from the exhaust air by the use of run-around coils and transferred to the incoming fresh air (see Section 9.2).

CONDENSATION

Taking the inside surface resistance of glass to be 0.123 m^2 K W^{-1} and that outside to be 0.055 m^2 K W^{-1}, it can be estimated that the surface temperature of single glazing is 27 – [0.123/(0.055 + 0.123)](27 + 1) = 7.7°C. With an air gap having a resistance of 0.178 m^2 K W^{-1} a similar calculation shows that the room-side surface temperature of double glazing is 17.2°C. Single glazing will stream with moisture all the time in design winter weather, particularly at night-time when outside air temperatures will drop below −1°C. Even if double glazing is used, condensation must be expected on the frames, if they are metal and do not include a thermal break. Condensation on roof lights and their frames, and even on roof beams too if they are steel and can reach a low temperature by conduction to outside, is also a possibility. This sort of condensation may escape notice and result in unexpected corrosion if a suspended ceiling conceals the roof. When suspended ceilings are fitted it is recommended that a branch duct should supply fresh air to the ceiling void, so pressurising it and discouraging the entry of humid, corrosive air from the pool hall.

AIR DISTRIBUTION

Common sense should prevail in the location of supply openings, whose positions are much more critical for good air distribution than are those of the extract grilles. One of the best methods is to blow air upwards at the side walls and windows, particularly the latter where cold down draughts will be a problem otherwise.

HEAT RECOVERY AND HEAT PUMPING

A pair of eight-row, run-around coils, with 320 fins per metre and a face velocity of about 2.5 m s^{-1} in the fresh air and discharge air ducts will have a heat transfer effectiveness of about 50%. If such coils are used it is recommended that the coil in the discharge air duct should have copper tubes and fins, electro-tinned after manufacture. Some consideration should also be given to extra protection for the framework, which is normally made of galvanised mild steel. Thermal wheels will give much larger heat transfers and efficiencies, but at greater capital cost. It is not uncommon for the metallic fill of one type of thermal wheel to corrode quite rapidly, when handling swimming pool hall exhaust air. A further and most useful possibility for heat reclaim according to Braham (1975) is to use plate heat exchangers to transfer heat from the waste water leaving the showers to the cold feed for the HWS (hot water service) vessels.

Heat pumps have been used successfully and economically for the mechanical ventilation of swimming pools. Figure 3.6 shows a schematic for a typical plant arrangement and Figure 3.7 shows the related psychrometry for winter operation. In the simplest application there are no run-around coils (shown by dashed lines in Figure 3.6) and no mixing duct connection and dampers (R2b in Figure 3.6). A direct-expansion air cooler is located in the extract air duct from the pool hall and cools and dehumidifies the air from state *R* to state *W*. The air then flows through the extract fan, its temperature rising from t_W to $t_{W'}$ because of the fan power, before discharging to waste. One hundred per cent cold fresh air is introduced, filtered and passed over the finned tubes of the condenser, where the heat taken from the extract air together with the power absorbed by the compressor is rejected and warms the air supplied to the pool hall to offset the sensible heat losses. A re-heater, fed with LTHW is usually fitted, to support the condenser heater when necessary, but an auxiliary condenser is seldom needed, any surplus heat, not required to warm the air supplied, can be rejected to the pool water through a heat exchanger. A desuperheater is often provided in the hot gas line to warm the HWS feed. A boiler is necessary to produce LTHW for the auxiliary heater battery, when needed, and to provide most of the heat for the HWS.

Theoretically, when a heat pump is used like this in winter it is operating at very advantageous conditions: the air onto the evaporator in the extract duct is at a virtually

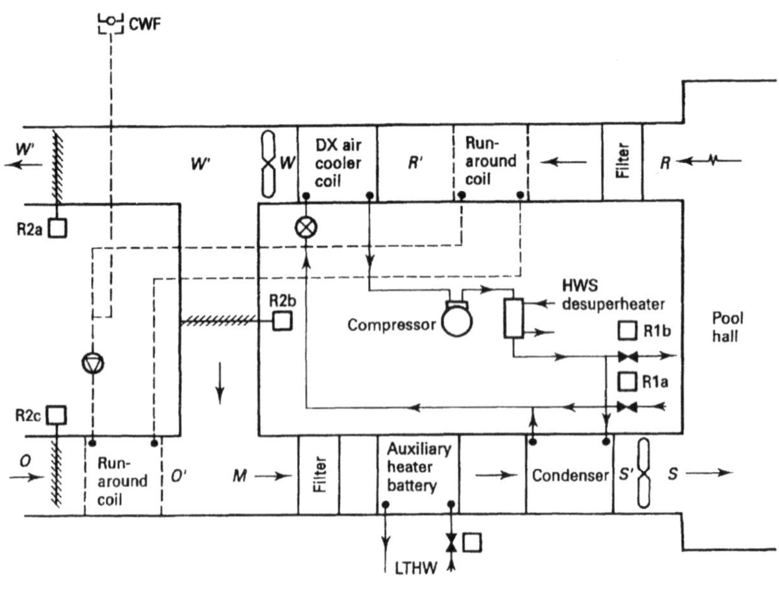

Figure 3.6 A schematic of a ventilation system for a swimming pool, using a heat pump.

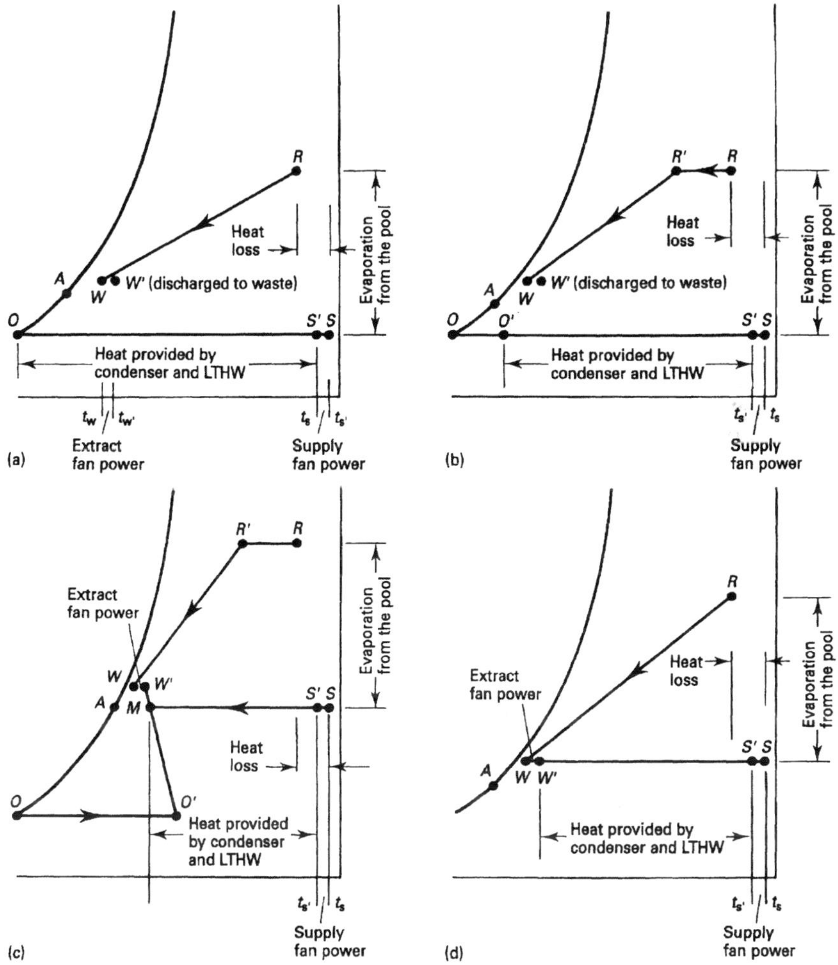

Figure 3.7 The psychrometry for winter operation of the plant schematic in Figure 3.6. (a) The simplest case: 100% fresh air in winter, without run-around coils. (b) 100% fresh air in winter, with run-around coils: *R–R′* and *O–O′*. (c) Minimum fresh air in winter with recirculation and run-around coils. The temperature rise across the fresh air run-around coil (*O–O′*) is larger than across the extract run-around coil (*R–R′*) because the fresh airflow rate is less. (d) 100% recirculated air.

constant state and hence the plant is running at a full load for most of the time with a high coefficient of performance, probably at about 5:1 for heat pumping. In practice, there are load variations because conditions change in the pool hall and the state of the air entering the fresh air duct varies. Loads will also change if variable proportions of fresh and recirculated air are handled. This would be to meet the varied needs of swimmers and spectators, to provide a small negative air pressure in the pool hall to discourage the migration of humid air to other parts of the building, and for operational economy. Proper control is essential to secure efficient operation and to minimise running costs. Using two compressors may also be desirable to optimise system performance under reduced load conditions and to provide a measure of stand-by.

The installed heat pumping load can be reduced by adding heat recovery devices, such as run-around coils (shown by dashed lines in Figure 3.6) to the extract and supply air-streams. Motorised dampers are essential so that the air-handling system can run with 100% recirculation to save thermal energy when the pool is shut down at night-time. However, corrosion is a considerable risk, particularly with steel ductwork, and it is suggested by Kay (1994) that, for pools treated by chlorine, as much as 30% of the air handled may be recirculated but if ozone is used for pool water treatment the figure may be as much as 80%. Plastic materials and builders' work have been successfully used for ducting. Operating costs may also be minimised if variable proportions of fresh and recirculated air are used under automatic control, as expedient with seasonal changes throughout the year.

BOILER POWER

The initial heating up of the water in the pool does not coincide with the demands for heating, ventilating and HWS. A reasonable basis for estimating the boiler power is to sum the heating requirements occurring simultaneously, as follows: fabric heat loss + fresh air load for ventilation + $^1/_7$ K h^{-1} for evaporative losses from the main pool + $^2/_7$ K h^{-1} for the evaporative losses from a teaching pool + the HWS load.

3.13 Bowling centres

For this competitive market, air conditioning has seldom been adopted in the UK, ventilation sufficing, but in warmer climates it is essential for the comfort of the spectators and bowlers. The major feature of the heat gain is the presence of large numbers of people in only a small part of the floor area, the lanes being unoccupied and, therefore, neither ventilated nor air conditioned, although air is sometimes exhausted above the pin setting machinery. Peak loads occur between 6 pm and 11 pm on weekdays and in the afternoon on public holidays and at weekends. Fresh air allowances should be generous to counter odours and smoking and 16–24 l s^{-1} per person is recommended.

Because only the area up to the foul line is treated, it is common to fit a curtain wall, sometimes transparent, coming down part of the way from the ceiling over the foul line to assist the separation between the conditioned and non-conditioned areas. It is important that this should not obscure the view of the spectators or the bowlers.

Although packaged units have been used, in the more competitive situations, the best air distribution and the best results are achieved with all-air systems, probably multi-zone to cater for the ancillary areas such as bars, cafeterias, and so on, and low velocity ductwork. Air is commonly supplied from a duct running parallel to the foul line and blowing air from double deflection grilles at a height of no more than 3.7 m from the floor, towards the bowlers and spectators. No more than two lanes should be dealt with by one grille. Air is exhausted at the back of the spectator area and recirculation can be used provided that precautions are taken that it is not from contaminated areas such as the cafeteria or from the bar, where smells and smoke are likely to be pronounced. According to Valerio (1959) it appears that in the United States, air change rates over the occupied area are between 10–15 per hour and the supply air rate is between 215–270 l s^{-1} for each lane. Cooling loads lie in the range from 3.8 to 6.4 kW of refrigeration per lane.

3.14 Clean rooms

The presence of moisture, dirt and impurities in certain machinery, assembly and production work spoils the high quality of the finished product and it becomes necessary to control the contamination by arranging for the work to be done in a clean room, or at a clean work station. Contamination control is best achieved in a clean room by adhering to the following six principles:

(1) Preventing particulate matter from entering the room
(2) Continuously purging the room of particles by providing a high air change rate
(3) Preventing particles from settling onto the product
(4) Restricting the generation of particulate matter within the room by the people present and their activity
(5) Arranging for the components dealt with, and the people handling them to be thoroughly cleaned, prior to entering the room.
(6) Constructing the room from materials that do not contribute to contamination.

Although well-designed clean room systems should prevent the entry of particles exceeding 0.5 μm in the diameter, the main source of contamination is the activity of the work people themselves. Initially clean rooms were air conditioned using conventional methods of air distribution with a high efficiency particulate air filter – the HEPA filter – introduced in about 1950. The HEPA filter is defined as having an efficiency of at least 99.97%, on a volumetric basis, for particles of 0.3 μm diameter, using the dioctyl phthallate (DOP) test. Clean rooms conditioned in this way were often unsuccessful in keeping the particle count down to the desired level because of the turbulent nature of the air distribution which did not provide a self-cleaning facility. Small particles became permanently trapped and circulated in eddies in some parts of the room. It was also found that low levels of contamination were only possible after the activity in the room had ceased and the conditioning system run for some further considerable time. A resumption of the activity immediately raised the contamination level. It was clear that a satisfactory level of contamination control depended on more than the quality of the filtration and that a new approach to air distribution was needed. So-called laminar flow air distribution was introduced in 1960 and successfully achieved the low particle counts desired by providing a supply airflow without turbulence that gave a self-cleaning function and flushed out dust particles from all parts of the room.

There is also an ultra low penetration air (ULPA) filter that has been introduced more recently. This is defined as having a minimum atmospheric dust spot efficiency of 99.999% for particles of 0.12 μm diameter.

POLLUTION SOURCES

A clean room or a clean device is an enclosed space in which the contamination of polluting particles is controlled to a specified standard and in which: the distribution of airflow follows a proper pattern, temperature and humidity are controlled to within specified tolerances, noise and vibration are limited, and electro-magnetic, radio frequency, and electrostatic fields are dealt with as necessary.

The sources of pollution in a clean room are:

(a) External. These arise from the provision of fresh air by the air-conditioning system, by the infiltration of dirty air through the fabric of the envelope where pipes, ducts and conduit penetrate it, by the entry of people from adjoining, dirtier spaces, and by the inflow of components or materials related to the processes carried out. Supply air filtration should deal with the first of these but, nevertheless, there appears to be a small correlation between dirty outside conditions and the indoor pollution level. Proper sealing around pipes, ducts and conduit at the entry and exit points through the building envelope is essential. A pressure gradient between the cleanest areas within the building and the outside, graded according to the classes of cleanliness of the rooms en route, can deal with the infiltration of air that accompanies the passage of people, components and materials through the building from the exterior.

(b) Internal. The greatest source of particulate contamination is from the people working in the clean room. Particles are being shed continuously from their bodies and clothing at a rate of between several thousand and several million per minute. Their actual working activities, even writing, also shed particles. These pollutants are kept under control by the provision of special clothing and by arranging for laminar downflow air distribution, which washes the particles from the surfaces of bodies and clothing and carries them away from the product being made. The airflow is not laminar in the true, dynamical sense of the word and is more correctly called uni-directional airflow. However, the term laminar has found common usage in this form of air distribution. Robots working in clean tunnels are sometimes used but these also shed particles from bearings, seals, and so on.

The internal surfaces of the clean room are additional sources of pollution, shedding particles and gassing out with the passage of time. Even the process itself can be a source of contamination, as also may be the materials provided for it.

STANDARDS OF CLEANLINESS

Particles of size greater than 5 μm tend to settle fairly quickly but it is important to control cleanliness generally for sizes much smaller than this, because a small particle can short-circuit a pair of conductors in an integrated circuit on a chip, rendering it useless. The smallest dimension in a micro-circuit is termed the line width and it is suggested by ASHRAE (1995) that particles having a maximum dimension of as little as 10% of the line width may result in a circuit failure. With circuit line widths approaching 0.25 μm, particles having a size of 0.025 μm are likely to be of concern.

The standards adopted for clean rooms throughout most of the world are those developed by the Federal Bureau of Standards in USA and embodied in Federal Standard 209E (1992). The class of cleanliness in a clean room was, and still is, described in the old, foot pound second units, by the number of particles of a specified size, per cubic foot of air in the room. Thus, Class 100 meant that the particle count should not exceed 100 per cubic foot, of size 0.5 μm or larger. A metric version of this, established in Federal Standard 209E, is in terms of the allowable number of particles per cubic metre, with class numbers prefixed by the letter M. Thus, Class M3.5 (corresponding to Class 100) defines a particle count not exceeding 3530 per cubic metre, of size 0.5 μm or larger (see Table 3.6).

Limits are given for particles of other sizes and a formula is quoted for establishing the allowable number of particles per cubic metre for metric class numbers:

$$p = 10^{M}(0.5/d)^{2.2} \tag{3.11}$$

where p is the number of particles per cubic metre, M is the numerical part of the metric

Table 3.6 Metric clean room classes according to Federal Standard 209F (1992)

		Maximum number of particles equal to or greater than 0.5 μm	
Old class name	Metric class name	per cubic foot	per cubic metre
1	M1.5	1	35.3
10	M2.5	10	353
100	M3.5	100	3530
1000	M4.5	1000	35 300
10 000	M5.5	10 000	353 000
100 000	M6.5	100 000	3 530 000

class name and *d* is the particle size in microns. Thus, to define a metric class M2 in terms of particles of size 0.3 μm

$$p = 10^2(0.5/0.3)^{2.2} = 308 \text{ particles per cubic metre}$$

The Federal Standard lists classes from M1 (10 particles per m^3 at 0.5 μm) to M7 (10^7 particles per m^3 at 0.5 μm) and covers the acceptable limits of particle sizes for 0.1, 0.2, and 0.3 μm, as well as 0.5 μm. The number of particles at 5.0 μm is given for the less clean classes of M4–M7. Federal Standard 209E does not specify downward air velocities, as was the case with earlier Federal Standards, but 0.45 m s^{-1} has become widely used for laminar downflow rooms to classes 100 (M3.5), or better. It is more difficult to maintain clean standards in large rooms than in small spaces, although Classes 100 000 and 10 000 have been maintained for large production halls in the aerospace industry. Classes 100, 10 and 1, in the semiconductor industry for the fabrication of integrated circuits, are only possible in small spaces.

In the UK and, to a lesser extent, abroad, BS 5295 (1989) has been used. The class of cleanliness is described by letters from C to M (excluding I) with A and B not described, presumably to allow for future, cleaner standards. BS 5295 gives figures for particle contents per cubic metre similar to those in Federal standard 209E, for particle sizes of 0.3, 0.5, 5.0, 10.0 and 25 μm. The maximum treated floor area in m^2 for each sampling position (when verifying performance) is given and the pressure differences between classified areas and unclassified areas, and also between a classified area and an adjoining area of lower classification, are quoted. The pressure in a clean room depends on the difference between the air extracted from the processes and the make-up air provided. When considering this it is important to appreciate that the airflow rate extracted through process equipment is set by the manufacturers and cannot be changed without risks to safety. Pressure differences are generally between 10 and 15 Pa and should be as small as possible: if they are too large, eddy currents at wall outlets may cause vibration problems, according to ASHRAE (1995).

Figure 3.8 compares the clean room classes of Federal Standard 209E (1992) and BS 5295 (1989).

Achieving a standard is not merely a matter of good filtration. Close attention must also be paid to: the pattern of airflow in the room, the quality and cleanliness of the materials of construction of the room and the services, commissioning, and maintenance. The installation must be proved and the standard achieved demonstrated by an actual particle count. Particle counts at regular intervals are subsequently necessary to verify that the correct class is being maintained.

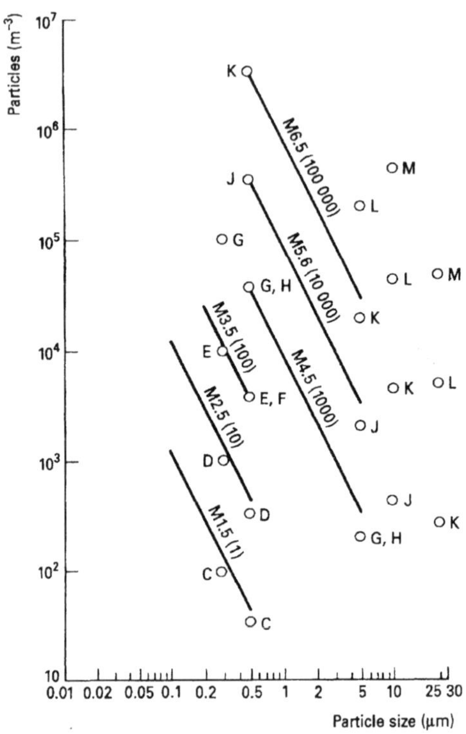

Figure 3.8 A comparison of Federal Standard 209E (1992) and BS 5295 (1989).

AIR DISTRIBUTION

(a) Turbulent. The air supply is in a conventional manner, as used for commercial air conditioning, through ceiling diffusers, without any attempt to modify the pattern of air distribution in the room. However, an absolute filter is fitted above each diffuser. Air is extracted through side-wall grilles at a low level around the room. If the source of contamination is from external sources Class M6.5 (100 000), or perhaps M5.5 (10 000), is possible in the room generally. This will not be possible if there are internal sources of contamination and clean work stations must be provided. Laminar downflow occurs at the work station: air drawn from the room by a local centrifugal fan passes through an array of absolute filters covering the whole of the ceiling of the work station and laminar downflow occurs onto the working surface. Class M3.5 (100) or better, can occur. The ceiling of the work station is usually provided with fluorescent lighting and there may be an ionising grid beneath the tubes to neutralise any electrostatic field.

(b) *Laminar downflow.* To achieve Class M3.5 (100) the whole of the ceiling in the clean room is fitted with absolute (HEPA) filters, which occupy about 85–95% of the space available, after allowing for the presence of luminaires, possibly sprinklers, and the necessary framework to support the 600 × 1200 mm filter cells (see Figure 3.9a). The air is introduced uniformly and provides virtually unidirectional (laminar) airflow down

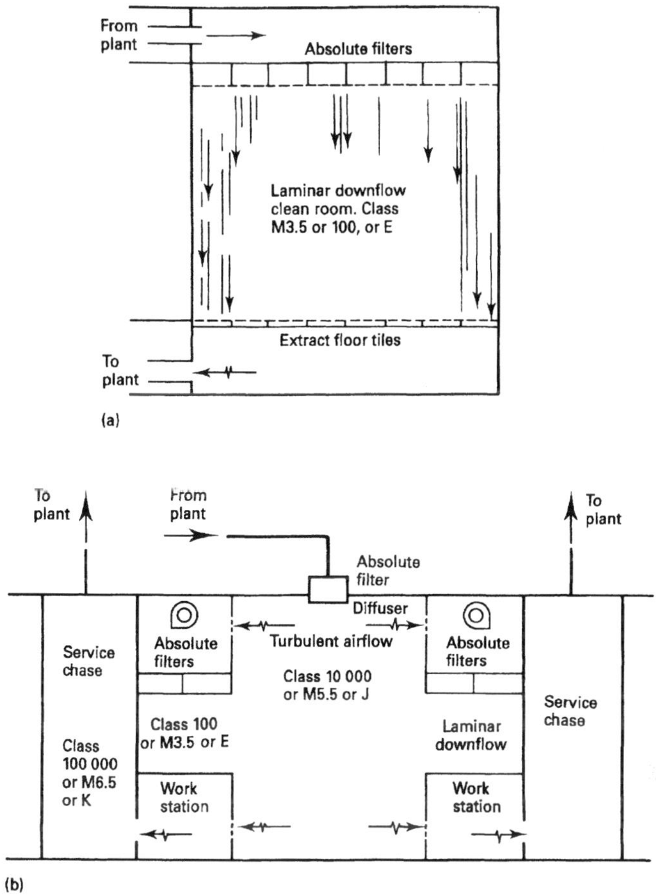

Figure 3.9 (a) Laminar downflow from a ceiling covered with absolute filters (100% gross area, 85–95% net area) to a perforated extract floor (15% to 25% open area). (b) A diagrammatic cross-section of a tunnel arrangement. The central area has turbulent air distribution but work stations have laminar downflow and are able to provide Class 100 (M3.5 or E), or even Class 10 (M2.5 or D). The clean tunnels are separated by service chases which contain the pipes gases and fluids for the processes and may also be used for the extract air, as shown.

to the floor of the room. In the best examples, the floor comprises modular gratings (with about 15–25% open area) behind which washable or disposable pre-filters are sometimes fitted. The objection to the presence of pre filters under the floor modules is that the clean-up procedure is complicated in the event of a chemical spill. It is often better to locate pre-filters elsewhere, in the extract system, where access for maintenance is easier. An alternative, less effective arrangement uses side-wall extract grilles at low level around the periphery of the room. With both techniques the intention is to achieve unidirectional airflow vertically downwards to the level of the working surfaces at about 750–950 mm above floor level. Below this level a mixture of laminar and turbulent air distribution occurs when side-wall extract is used. The presence of the equipment, the people in the room and their movement, disturb the airflow patterns and introduce a measure of turbulence which tends to lift particles from contaminated surfaces and deposit them onto the clean products being made, with adverse consequences.

An extract floor must have the following properties:

(1) The strength to stand concentrated and rolling loads without giving undesirable deflection beyond acceptable limits.
(2) An acoustic quality compatible with the design noise rating in the clean room.
(3) Safety for flame spread.
(4) It should be earthed.
(5) It must not emit particles or gases under conditions of normal wear.
(6) The floor supports must be stable.

The space beneath an extract floor in a laminar downflow clean room should not be too shallow, otherwise the extract distribution over the floor will be prejudiced and turbulence may be induced at the floor, with a chance that contaminating particles will be lifted in the clean room and spoil the class of cleanliness. Sometimes, the extract floor is suitably reinforced and the space beneath constructed as a basement where dangerous and other gases and liquids can be piped and fed upward to the process equipment in the clean rooms overhead. Plant may also be accommodated in such a basement. Floor-to-ceiling heights must be adequate.

Clean rooms may be constructed as large single spaces (2000–3000 m^2 floor area) and maintained at a desired degree of cleanliness, using the appropriate form of air distribution. Multiple work areas may then be provided within this area, as required, with better classes of cleanliness. An alternative is to use what is termed a tunnel arrangement (see Figures 3.9b and 3.10). Clean tunnels are separated by service chases, or aisles, and the working area can easily be increased by lateral expansion.

Cross-flow clean rooms have been used in the past and are cheaper but do not give the best results. Laminar airflow is possible from a clean wall (fully covered with absolute filters) across the room to a dirty, extract wall on the opposite side. Hence not only does the standard of cleanliness deteriorate across the room but, since turbulent mixing of the airstreams does not occur, a substantial horizontal temperature gradient may develop.

ABSOLUTE FILTER INSTALLATION

It is no good installing high quality, expensive, absolute filters if dirty air is allowed to pass through flanking paths between the filter cells into the clean room. Three methods are in use to deal with this possibility.

(1) Gaskets are provided between the filter cells and their supporting framework. After installation the cells are drawn up tightly, by means of adjustable screws, onto the gaskets between the underside of the cells and the supporting frames.
(2) A downstand flange, on the underside of each filter cell, rests in a channel of gel, on the upperside of the supporting framework (see Figure 3.11).
(3) The filter cells are provided with gaskets between their underside and the supporting framework but the air supply is ducted to each individual absolute filter cell. It is also arranged that there is a duct connection between the extract air system from the clean room and the ceiling void. This ensures that the space above the filter cells is at a

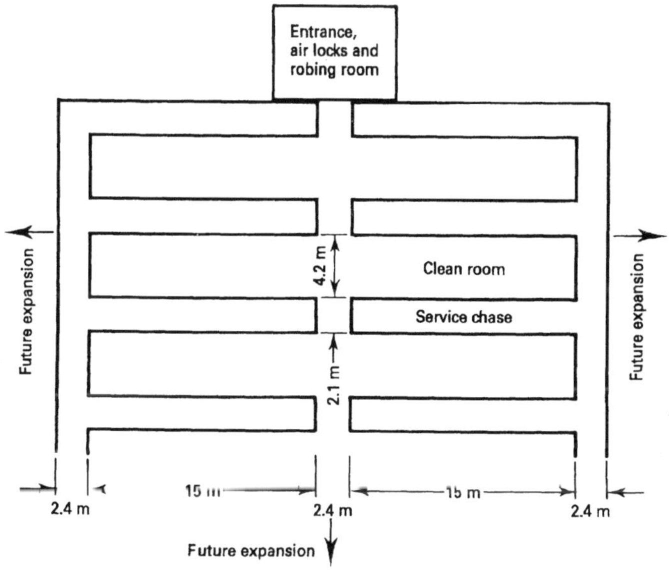

Figure 3.10 Typical plan of a tunnel clean room arrangement with approximate dimension.

negative air pressure, with respect to the clean room. Any flanking leakage between the filter cells will then be of clean air out of the room (see Figure 3.12). The method tends to be expensive.

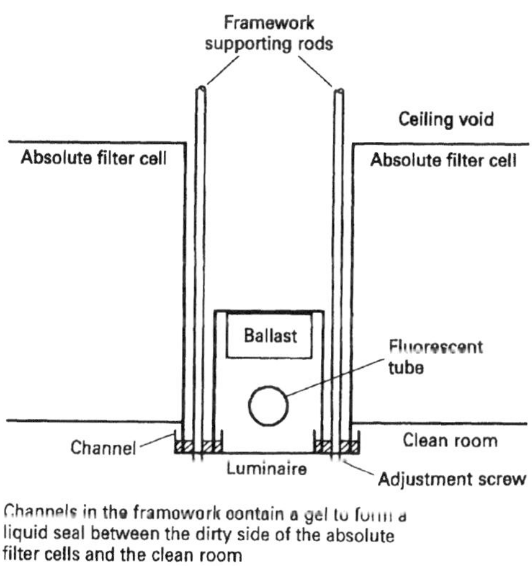

Figure 3.11 A diagram of the method of absolute filter support that provides a liquid seal to prevent flanking air leakage past cells.

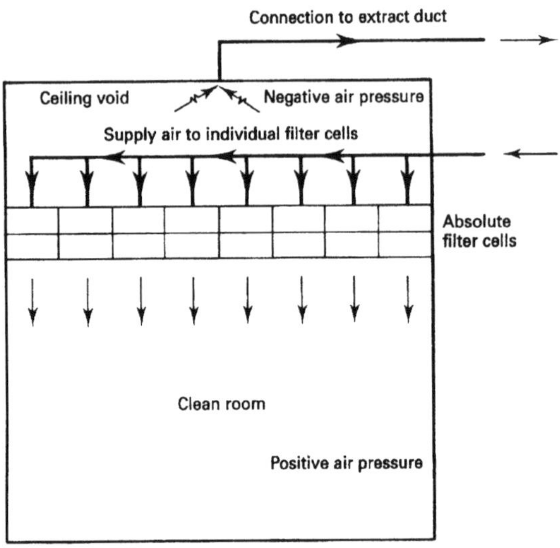

Figure 3.12 Supply air is ducted to each absolute filter cell. The ceiling void has a duct connection to the extract system and hence it is at a negative pressure with respect to the clean room. Any flanking leakage between filter cells is of clean air out of the room.

Sometimes the luminaire is not recessed into the ceiling, as in Figure 3.11, but projects into the clean room. It is then common for the diffusing enclosure of the fluorescent tube to be of tear-drop section, so as to avoid disturbing the downward-flowing streamlines of air.

PLANT ARRANGEMENTS

Figure 3.13 shows a typical schematic arrangement of the air handling plant used for a clean room. Other arrangements are possible but the important features are:

(1) The air quantity handled by the primary plant is related to the very high air change rate that results from the need to have a face velocity over the ceiling in laminar downflow clean room of 0.45 m s^{-1}.
(2) The fresh air quantity that must be provided is much greater than that related to the occupancy because it has to make good the air exhausted from the room by the process equipment.

In Figure 3.13 an absolute (HEPA) filter and a pre-filter is shown as part of the make-up air handling plant. Such an absolute filter should have an atmospheric dust spot efficiency of not less than 95%. The pre-filtering should include a conventional, commercial filter with an arrestance of about 80%, followed by a filter with an atmospheric dust spot efficiency of 30%, followed in turn, by a filter with an atmospheric dust spot efficiency of 80%. This will prolong the life of the 95% filter and the lives of the terminal filters in the ceiling of the laminar downflow clean room.

PROCESS EXHAUST SYSTEMS

To carry out the processes in the clean room various acids, solvents and gases are delivered

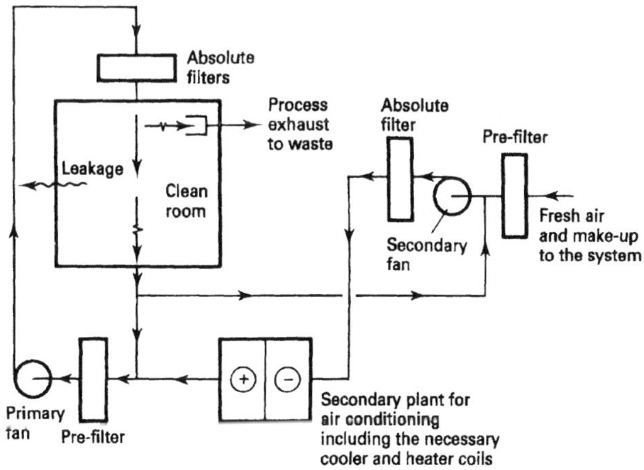

Figure 3.13 A schematic of the air handling arrangement for a clean room. The primary plant for a laminar downflow system will be handling about 540 air changes per hour if the face velocity over the filters in the ceiling is 0.45 m s^{-1} and the floor-to-ceiling height is 3 m. For smaller heights the rate can exceed 700 air changes per hour. The amount of fresh air handled by the secondary system and conditioned is governed almost entirely by the airflow rate exhausted by the process equipment in the clean room.

in piping dedicated to the nature of the gas or liquid handled. Many of these gases and liquids are corrosive, inflammable and poisonous. Some are pyrophoric, that is, spontaneously combustible in the presence of air, without the aid of ignition. Reference to the safety limits for exposure to such substances must always be made and strict adherence to the regulations for fire safety and escape followed. The exhaust airflow rates related to the processes are very considerable and can be from 5 to 50 l s^{-1} m^{-2}, referred to the floor area. The exhaust ductwork used for fumes should be generously sized to cater for possible changes in the future.

DESIGN CONDITIONS

Because the occupants must wear fully covering, protective clothing in laminar downflow clean rooms they require a dry-bulb temperature of not more than 20°C and this may commonly be subject to an acceptable tolerance of ±0.5 K although, for some processes, a tolerance as close as ±0.05 K is essential. According to ASHRAE (1995), control zones should be small enough to be able to deal with the comparatively large thermal inertia that results from the high air change rates in laminar downflow clean rooms. Relative humidities between 30% and 50% are often acceptable, with limits of between ±5.0 and ±0.5%, depending on the process. Some processes are very sensitive to the higher humidities and also to changes in humidity.

The performances of the air-conditioning system and the associated refrigeration plant are critically related to the outside design conditions chosen. The comparatively large amount of fresh air handled, to balance the process exhaust quantities, emphasises this. It follows that the outside summer design states, commonly chosen for commercial installations, are unsuitable for clean room applications in the semi-conductor industry. The outside summer design dry-bulb and wet-bulb should be higher than the usual commercial choice and reference made to the CIBSE (1982) data. In addition to this, because the performance of the system will be spoiled if the refrigeration plant cannot reject heat to outside, air-cooled condensers and cooling towers should be selected for an outside dry-bulb and an outside wet-bulb, respectively, that is at least two degrees higher than the summer design value used for the rest of the air conditioning system.

A stand-by plant is essential and provision should also be made for space to accommodate additional plant in the event that the clean room facility is extended at a future date.

Noise levels and acceptable displacements arising from vibration must be established prior to design. The building, the processes and the system of air conditioning are interrelated and must be considered as a whole in these respects (see Chapter 7). The equipment used for the processes is often sensitive to vibration. In general, displacements should not exceed one tenth of the line width.

It must not be forgotten that the plant used for abatement of the fumes and gases exhausted from the processes and discharged to waste may also be a source of objectionable noise.

FAN SELECTION AND FAN POWER

Owing to the high airchange rates in clean rooms and the big pressure drop across absolute filters, the fan powers involved become very large. This has a great effect on the size of the refrigeration plant and on energy consumption. Fan selection and operation therefore become very significant.

The most efficient fans are backward-curved centrifugals with aerofoil impeller blades, and axial flow fans. The former have a maximum total efficiency of about 89% and the latter a maximum of about 87%. The peak efficiency of a centrifugal occurs at a fairly steeply sloping part of the pressure–flow rate curve (see Jones, 1994) and this is an advantage since an increase in system resistance produces a smaller decrease in airflow rate than would otherwise be the case. The peak efficiency of an axial flow fan occurs on a part of the pressure–flow rate curve that is steeper than that of the centrifugal fan, with a correspondingly better effect.

Fan power, W_f, in kW, is defined by Equation (2.7)

$$W_f = p_{tF}\dot{v}_t/\eta \tag{2.7}$$

where p_{tF} is the fan total pressure in kPa and $\dot{v}_t$ is the volumetric airflow rate in $m^3\ s^{-1}$ at a temperature t and η is the total fan efficiency as a fraction. As an absolute filter gets dirty the pressure drop across it rises, the airflow rate reduces and the fan efficiency diminishes. Increasing the fan speed will restore the air flow rate to its design value, but at a higher fan total pressure, corresponding to the extra resistance of the dirtier filter. The total efficiency is restored to its former value, in accordance with the fan laws. The best and most efficient way of increasing the fan speed is by increasing the frequency of the electrical supply to the fan motor, through an inverter. Some other techniques, such as the use of eddy current couplings, are not efficient and do not save energy. If axial flow fans are used, the pitch angle of the impeller blades may be increased to offset the effect of the dirt collected by the absolute filters, and the airflow rate returned to its design value. Absolute filters are expensive and consequently the pressure drop across a filter is usually allowed to build up to several times its value when clean. One set of absolute filters in the ceiling of a laminar downflow clean room can last for as much as three years. At the end of this time the semi-conductor products might be overtaken by obsolescence and the whole facility and its services could be shut down and up-graded. While the installation is in use, fan speeds would be regularly increased to keep the airflow rate at its design value. Such fan speed increases are often done manually, at regular intervals of about a week, or at times dictated by operational experience.

REFRIGERATION LOAD

For the case of a Class M3.5 (100) laminar downflow clean room, the contributions of the primary fan and the fresh air load are significantly more than one would calculate for a commercial building. Comparative calculations for a typical clean room and a commercial west-facing office module yield the following:

Item	Clean room		Commercial office	
Treated floor area	100 m^2		14.4 m^2	
	kW	%	kW	%
Sensible heat gain	10.0	7.1	0.76	55.5
Latent heat gain	0.5	0.4	0.13	9.5
Fresh air load	42.5	30.2	0.30	21.9
Primary fans	76.0	54.0	0.18	13.1
Secondary fans	11.7	8.3	–	–
Total	140.7 kW	100.0%	1.37 kW	100.0%

For the commercial office the item for primary fan power covers the supply and extract fans. There are no primary and secondary fans. Note that illumination levels are high in clean rooms, and values up to 4000 lux are not uncommon. Wild emissions from processing equipment also increase the sensible heat gains.

The specific refrigeration loads are 1407 W m^{-2} for the clean room and 95 W m^{-2} for the office module. For an entire commercial building the cooling load would be less than the figure quoted for the module because diversity factors would be taken into account for people, lights and business machines. However, it is evident that refrigeration loads and fan powers are much higher for a clean room.

INSTALLATION AND COMMISSIONING

All materials for ductwork, pipework and plant must not degrade, or emit particles or gases during their operational lives. Circular section ducting is preferred. Plant and materials used for the air-conditioning system must be cleaned and sealed by the manufacturers at their works, before delivery to site. After delivery, all plant and materials must be cleaned again, sealed, and kept until required in a clean, weatherproof storage room. After erection, cleaning should be carried out once more and the system sealed until commissioning is to begin. Air must never be allowed to flow through the system without the absolute filters being in place.

The three successive phases of commissioning are:

(1) As built – the installation is complete and the building is ready but the semi-conductor processing equipment has not yet been installed.
(2) At rest – the processing equipment has been installed but not yet set to work.
(3) Operational – the processing equipment is working and in production.

System performance is demonstrated by particle counting, using actual counting for the larger sizes and light scattering for the smaller particles. Standardised methods are prescribed in BS 5295 (1989).

Continuously monitoring and recording conditions of cleanliness, temperature, humidity, air pressure, and so on, is essential after commissioning. Some semi-conductor manufacturers consider that extensive monitoring of as many variables as possible, in both the clean rooms and the various systems and piped services, leads to a significant improvement in the quality of the chips produced, with fewer faults and a higher yield. Maintenance is improved and the examination of records is a diagnostic aid in the determination of production problems.

ABATEMENT

After playing their parts in the production processes, the various liquids and gases used (many of which are dangerous) and the related exhaust fumes, must be collected and disposed of in a manner that causes no hazard to the environment. Two techniques employed are absorption and adsorption, to remove the objectionable substances, but some of the waste products cannot be dealt with in this fashion and dedicated, safe disposal by contractors appears to be adopted.

3.15 Hospitals

Hospitals in the UK have not always been fully air conditioned, except for new buildings in urban areas where ambient noise and atmospheric pollution necessitate having closed, double glazed windows (see Hunt, 1976). Air conditioning has been restricted by economic considerations and has been confined to places where it was essential, for example operating theatres. Although balanced, mechanical ventilation has been used to limit the spread of smells from service areas, for example kitchens and places providing a source of odour or infection, wards have often relied on the natural ventilation afforded by opening windows. In warmer climates, on the other hand, the need for air conditioning is more apparent and, according to ASHRAE (1995), it is claimed that its provision in wards does not merely give comfort but is actually conducive to successful therapy and promotes a more rapid recovery by patients.

In the design of air conditioning or ventilating systems it is essential to restrict the dissemination of infection between different areas and to discourage the spread of smells. Four aspects of design help to achieve these aims:

(1) Different departments should be treated by independent systems.
(2) A balance must be secured between the supply and extract air quantities to maintain a slightly negative or positive pressure in certain areas, according to the desire to limit the emission of microbial infection and/or smells.
(3) A relatively bacteria-free air quality should be achieved by adopting a proper standard of filtration efficiency.
(4) Enough fresh air must be handled to dilute odours to an acceptably low level.

HEPA filters (see Section 3.14) are effective in achieving a low bacteria count by removing virtually all particles from the airstream down to one micron (1 μm) in size. Smith and Rowe (1977) claimed that the standards of fresh air supply recommended for hospitals were based on outdated and insufficient studies and offer evidence that as much as 80% of the air supplied can be recirculated from the treated space provided, of course, that it is properly filtered. The rates of fresh air supply recommended by ASHRAE are, nevertheless, quite satisfactory. CIBSE, on the other hand, does not make recommendations directly but refers to the figures quoted in the various hospital building notes, published by the Department of Health. There is every reason to suppose that both these sources give satisfactory recommendations.

It is important that air washers, sprayed cooler coils, and the like, should not be used because they may be the focal points for the production and dissemination of microbial and other infection. When humidification is necessary it should be done by the injection of sterile, dry steam into the ducted airstream (see Section 3.15) and this steam should be generated in the vicinity of the point of use to avoid energy losses from distribution mains.

Although induction and fan coil systems have been used in hospitals, the presence of secondary cooler coils, emergency condensate trays and secondary filters or lintscreens inside the treated space creates a harbouring place for microbes and possible source of infection. Of the air–water systems available, that best suited to hospital wards and similar areas, is the chilled ceiling, with an auxiliary air supply (Section 2.12). Through-the-wall room air conditioners are not really suitable because they also form local centres of potential infection and their use should only be regarded as a temporary expedient. All-air, centralised systems offer a good solution for hospitals generally, and of the three most apt constant volume with reheat, double duct and variable air volume, the first two will be the most expensive in both capital and running cost and the third the cheapest in these respects as well as the best. One favourable feature of the variable air volume system is its ability to run at low capacity and low noise levels with low energy consumption at night-time, when cooling loads are reduced and patients are sleeping in wards.

Hospitals must run for 24 hours each day and this imposes a need for a measure of stand-by in the refrigeration machines, air-handling units, pumps and boiler plant. It is also necessary to have stand-by, diesel driven, electrical generators for at least the essential hospital services. Whenever stand-by plant is considered, it should always be chosen and installed in a way that will not allow it to deteriorate. Thus belted, running motors should be fitted as stand-by on fans, rather than motors that are static when on stand-by and will suffer bearing deformation by brinelling.

An essential feature of the design of modern hospitals is energy conservation. Measures for conserving energy must be considered with the proper tests for cost effectiveness applied. See Chapters 8 and 9.

3.16 Operating theatres

Many operating theatres have been air conditioned using conventional methods of turbulent air distribution but, as with clean rooms (Section 3.14), this proves unsatisfactory in the prevention of airborne contamination. There are three sources of bacterial contaminants: the supply of air to the theatre; the surgical team; and the emissions of bodily vapours and organisms from the patient. To deal with these and to prevent turbulent air movement and convection currents from lifting pathogens, mixing them by entrainment with the supply air and depositing them on the patient, the best steps to take are as follows:

(1) Provide special body exhaust suits for the surgical team, with extract openings at the crotch, armpits and nose, or body exhaust systems each comprising a face mask with a removable transparent visor, a hood, and a gown of low permeability material from

which air is exhausted under the hood and upward beneath the gown. Flexible hoses are attached to the backs of such body exhaust ensembles and a microphone is fitted in each face mask. Members of the surgical team are kept refreshed by the continuous flow of conditioned air and fatigue is diminished.

(2) Provide an exhaust system at the operating table itself, for the removal of the emissions and organisms from the patient, in the vicinity of the wound.

(3) Provide a laminar downflow system of air distribution over the patient, the surgical team and the operating table, at a mean speed of between 0.3–0.4 m s^{-1} which is high enough to wash away any pathogens that may be deposited. It has been found by Howarth (1972) that velocities as high as 0.6 m s^{-1} are unsatisfactory because the exposed internal tissues of the patient are over-cooled and dried out by evaporation

(4) Provide terminal HEPA filters to remove virtually all dust particles above 0.3 μm in size. It is also necessary to install local exhaust in the vicinity of the anaesthetist to carry anaesthetic vapours away from him or her.

According to Green *et al.* (1962) airborne bacteria invariably attach themselves to dust particles and very small water droplets, 80% of the particles/droplets exceeding 2 μm in size and most of bacteria conveyed being of about 2 μm in size. Fungus cells and spores are in the range of 3–20 μm and are not attached to carrier particles. Particles of size less than 0.3 μm play no significant part in conveying pathogens. It is general practice to use 100% fresh air because although it contains bacteria and other contaminants, these are seldom of the type (streptococci and staphylococci) that are major sources of wound infection, according to Schicht and Steiner (1972). Fresh air is thus very much better than recirculated air from operating theatres. It has been reported by Ma (1965) that staphylococci thrive better in dry air (humidity less than 35%) and moist air (humidity greater than 65%), than at intermediate humidities. This evidence has been verified and extended by Green (1974) to claim that there is a general decrease in the life of many airborne microorganisms when the humidity lies between 20% and 45%. Earlier work by Dunklin and Puck (1948) showed similar results and tendencies.

Typical operating theatres have plan dimensions of about 6 × 6 m with floor-to-ceiling heights of approximately 3 m. It is impractical to use fully the downflow techniques of clean rooms over the whole of the area of the theatre and, in any case, an extract-ventilated floor poses structural problems and is restrictive in theatre use. One approach by Howarth (1972) that is most effective in producing laminar downflow where it is wanted is to provide an enclosure of plan dimensions 3 × 3 m within the theatre and surround the surgical team and patient with it: the supporting services team (anaesthetist, and so on) is in the theatre, outside the enclosure, which is air conditioned by conventional air distribution which uses 100% fresh air and HEPA filters. The enclosure comprises rigid, transparent, plastic, demountable panels, suspended from the ceiling or supported by four corner posts. The side panels are of modular construction, allowing openings with various dimensions to be provided in appropriate positions for the access of X-ray machines, etc. It is usually arranged that only the part of the patient that is being operated on is within the enclosure, the rest of the body, together with the anaesthetist, is outside. The laminar downflow of air escapes under the bottom edges of the enclosure, which do not reach to the floor, and prevents the inflow of bacterial contaminants. Members of the surgical team may leave the enclosure and later return, provided they put on a freshly sterilised body exhaust suit. Wrapped, sterile instruments can be passed from the outer theatre into the enclosure. Convection currents from lights over the table tend to spoil the laminar downflow. One solution to this is supporting the lights from the ventilating ceiling within the enclosure in a special way, and another is in having the lights outside the enclosure but with their beams directed through its transparent walls onto the patient.

A modification of the laminar downflow enclosure is to use partial walls according to Whyte *et al.* (1974) that do not extend very far down from the ceiling. The aim is to simulate the characteristics of the core of a freely flowing jet over the surgical team and patient by restraining the natural expansion of the laminar downflow over a short distance. Apparently successful laminar downflow can be achieved at the table level provided that the airflow initially delivered is 20–25% more than would be handled by an enclosure extending to within 600 mm or so, of the floor. A possible disadvantage seems to be that entrainment of air from the outer portion of the theatre could contaminate sterile instruments stored outside the clean area. A third version by Allander and Abel (1948), used in Scandinavia, encloses the clean, downflow area in a peripheral air curtain.

The conditions maintained within the laminar area, in the outer part of the theatre and in the ancillary rooms that complete the operating theatre suite, are 21 ± 1°C and 50 ± 5% saturation with adjustable set points. It is worth providing humidity control in the vicinity of 50% not just because of the probable reduction in the life of bacteria, but because of the reduced risk of electrostatic discharges. At humidities less than about 40% the static charges which build up on non-conducting materials do not leak away to earth but accumulate and may give spark discharges, eventually with dangerous results if inflammable or explosive anaesthetic gases are used. It is also advisable to size the plant to be able to maintain 24°C for short-term, major surgery, if required.

The air-conditioning system should be all-air, constant volume, with multiple reheaters and sterile dry steam humidification, 100% fresh air should be drawn in through a louvred inlet located in a non-contaminated area. Intakes near ground level have the risk of introducing street pollution and the soil bacteria associated with gas gangrene. Three levels of filtration (Section 3.14) are desirable. It must be remembered that removing dust is not really synonymous with purifying the air of bacteria, even though the latter use dust as a carrier. Bacteria reproduce with great rapidity and the penetration of only a few through a filter can destroy the purity of the downstream air very quickly and extensively. Hence a prefilter, with the main object of protecting the air-conditioning plant from the grosser particles is the first line of defence. This should be followed by a high efficiency filter, say a bag filter, located on the discharge side of the fan to ensure that any air leakage thereafter is of clean air out of the system, rather than vice-versa. Finally, best quality HEPA filters should be fitted in the duct system, as close as possible to the supply air terminal outlets. These filters must be installed very carefully and great care taken that air cannot bypass them. Bacteria can live and multiply inside filters if the right conditions of humidity and temperature prevail. Dirt held by the filter can contain sufficient organic matter to nourish bacteria. High humidity assists such bacterial growth. Unfortunately, air leaving a cooler coil is frequently in a nearly saturated state and, in this respect, some degree of reheat could sometimes be worthwhile, if such a problem were encountered, provided the economic and other penalties imposed by wasteful reheat are accepted. Fortuitous reheating by adiabatic compression at the fan and by duct heat gain may often be enough to lower the humidity. There is also the risk of capillary condensation within a filter if the local humidity exceeds 70%. In spite of these potential hazards the fact remains that well-designed installations with multiple filtration and terminal HEPA filters feeding a laminar downflow air supply can successfully give average infection rates of 1.5%, or lower, to be compared with rates approaching 9% when conventional techniques of air distribution are used according to Howarth (1972).

Microorganisms thrive on wet surfaces, particularly if these are lukewarm. Cooler coils, whilst indispensible, are therefore a potential source of airborne contamination and steps must be taken to ensure rapid and effective condensate drainage from them, with face velocities no greater than 2.25 m s^{-1} to ensure no liquid droplets are carried over. Eliminator plates are never 100% efficient and fitting them should never be regarded as a licence to tolerate carryover from a cooler coil. Air washers are particularly bad and are to be condemned, as are sprayed cooler coils, because of the large amount of contaminated, recirculated water. Humidification should not be provided by water droplets at all, whether by direct aerosol injection or by spinning discs. It should only be done by the injection of dry, or slightly superheated steam, which will then be sterile and contain no water droplets. The humidifiers should be remote from the upstream side of any filter, but should always be followed by a filter because of the risk of microbial cultures forming in any patches of possible condensate in the subsequent ductwork. Ducting must be airtight and have provision made for cleaning and disinfection. Fumigation points should be provided for all plant and ducting. Circular ducts are preferred to rectangular ones and there should be no rough edges internally. Dampers are to be avoided.

A stand-by air conditioning plant should be provided for operating theatre suites. The extent and form of the stand-by is a matter of engineering judgement and economics but the aim should be to allow the operating suite to continue to function, without a loss of comfort or air purity, in spite of a failure of power from the electrical mains or a breakdown of a critical component in the air conditioning plant. This implies the need for a stand-by electrical generator, belted running motors on all fans and stand-by pumps in parallel. If water chilling plant provides refrigeration then the choice should lie between 2 × 100% or 3 × 50% plants with chilled water piped in series. If direct expansion plants are used the solution is not so easy. It is not good practice to oversize compressors because this promotes a tendency to undesirably low evaporating pressures with the likelihood of frosting on cooler coil surfaces and motor burn-outs, if hermetic or semi-hermetic compressors are used. The best solution is to provide 2 × 100% air handling and 2 × 100% associated, condensing sets. An alternative could be to use one air-handling unit fitted with a pair of 100% direct-expansion cooler coils, in series in the airstream, if the extra pressure drop can be tolerated, or in parallel in the airstream if sufficient space is available for this in the plant room and if tight shut-off dampers can be fitted to close off the coil not in use. Separate condensing sets would be piped up to the cooler coils. Hot gas valves should be installed to stabilize evaporating pressure and condenser pressure must also be regulated (Section 2.3). If cooling towers are used, then 3 × 50% or 2 × 100% units should be selected, with their ponds interconnected. When cooling towers are used, the best practices for construction, operation and maintenance, recommended by the Department of Health and CIBSE must be followed. Hygienic operation is essential for water-cooled systems of refrigeration plant. Air-cooled systems are to be preferred.

Heat gains ought to be easily established. Lamps for operating tables run off 24 V and liberate between 500 and 1000 W as a rule. An additional 1000 W would constitute the gain from other lamps and appliances, and the surgical team would comprise between four and eight people.

3.17 Constant temperature rooms

The distinction between a constant temperature room and any other, more conventional, conditioned space is that temperature gradients, whether vertical or horizontal, cannot be tolerated outside specified limits. Thus if 21 ± 1°C is specified it means that nowhere in the room can the dry-bulb be otherwise. In conventional applications it is customary to supply air at about 8 or 9°C below the room temperature to offset sensible heat gains. Obviously, this is not acceptable for a constant temperature room. The supply temperature must be at the bottom end of the specified tolerance to maintain the room at the top end under design heat gains and vice-versa for design heat losses. It is impracticable to select a cooler coil to chill air through such a small range of temperature and so the method adopted is to use two plants: a secondary one, handling a small amount of air cooled and dehumidified in the normal way to cope with the design heat gains, and a primary plant circulating a much larger amount of unconditioned air to reduce the difference between room and supply temperature to the required value. Figure 3.14 shows such an arrangement. Alternative plant combinations, with less ducting, may suggest themselves and still fulfil the object of good mixing between the primary and secondary airstreams before they enter the room; for instance, the ceiling plenum itself might be used as a mixing chamber. The outside air introduced is enough to meet the ventilation needs and, if large, may require the use of pressure relief flaps in the room to aid natural exfiltration. Since the connexions between the primary and secondary duct systems are close together (points a and b in Figure 3.14) the ducted airflow pattern is stable. If a blow-through cooler coil is used in the secondary plant the full benefit of avoiding the temperature rise by fan power can be achieved and proper coil performance obtained if the fan discharge transition piece to the coil is straight, long and gentle, in order to give smooth airflow over the face of the coil.

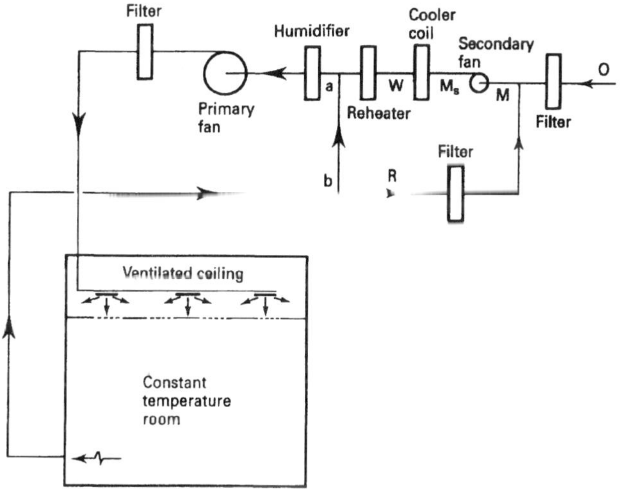

Figure 3.14 Schematic diagram of an air-conditioning system for a constant temperature room.

It is important that the walls and floor of the room are well insulated to ensure that surface temperatures seldom go outside the specified tolerance and that adequate vapour sealing is provided if there is a requirement for humidity control. Because the temperature difference between room and supply air temperatures is small, air distribution must be by a ventilated ceiling, wall or floor (see Section 3.14) since the air change rate will be so high. It is then vital that the ceiling plenum chamber be both properly lagged and vapour-sealed and that all structural holes, cracks and crevices, where pipes, ducts or conduits penetrate the enclosure, or where the building construction is poor, are positively sealed. Radiant gains from high temperature equipment within the room sometimes cause significant local temperature variations and these must then be screened. The location of the controlling thermostat is critical and a satisfactory position is often immediately behind the face of the extract grille. Unless careful attention is given to all these matters, particularly structural sealing, the installation will be a failure.

A constant temperature room is often required to be continuously conditioned and this means that some revision of the outside design state normally chosen for winter is necessary. It is suggested that in the UK a value of 5°C less than the conventional outside winter design temperature is appropriate.

EXAMPLE 3.6

Two modules of the hypothetical office block (Section 1.2) have their windows blocked in and insulated and are to be used as a constant temperature room for which the sensible heat gains are 2000 W, the latent gains are 400 W and the heat loss is 3500 W, for inside conditions of 21 ± 0.5°C dry-bulb with 50 ± 5% saturation when the outside conditions are 28°C dry-bulb, 19.5°C wet-bulb (sling) in summer and –6°C saturated in winter. Three people are to occupy the room. Determine the necessary primary and secondary airflow rates, the maximum cooler coil refrigeration duty and the maximum reheater duty. The plant is to run continuously.

Answer

First, the psychrometric considerations for the state to be maintained in the room must be established. The system will be designed so that 21.5°C is achieved under peak heat gain and 20.5°C when there is a maximum heat loss. Coupling this with 50 ± 5% means that a control rectangle must be drawn on the chart, i.e. the shaded area in Figure 3.15, if the tolerances of temperature and humidity are compatible, as they are in this case. Drawing a parallelogram is the wrong approach. If this was done one would be aiming to control the room at the point *P* (Figure 3.15) when heat gains were greatest, implying that if the sensible gains diminished there would be a temporary excursion to the left of *P* (shown by the arrow in the figure), taking the state outside the control domain specified, which is not acceptable in a constant temperature room. Note that it is not always possible to draw a rectangle, if the tolerances of temperature and humidity specified are incompatible. For example, for 21 ± 2°C with 50 ± 5%, drawing a control rectangle is impossible, as is shown in Figure 3.15 by the large dashed-line parallelogram. The solution then is to tighten the temperature tolerance until the construction of a rectangle that has a practical proportional band of dew-point temperature control is possible.

The summer design room condition in the control rectangle is at the point *R*, in the figure, at 21.5°C dry-bulb, a moisture content of 8.374 g kg^{-1} (corresponding to 20.5°C and 55%)

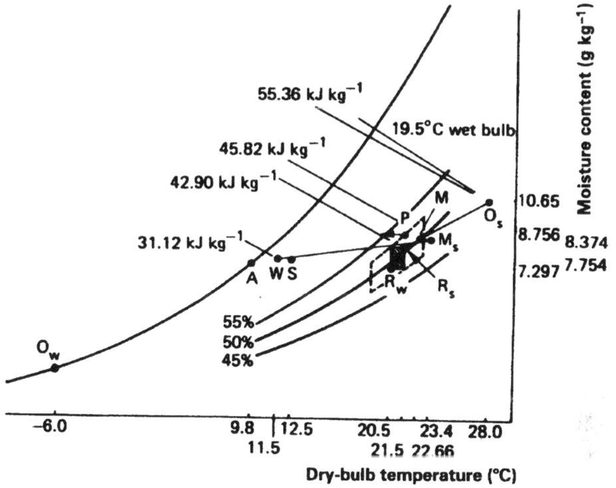

Figure 3.15 Psychrometry for Example 3.6.

and an enthalpy of 42.90 kJ kg^{-1}. In winter the control point will be at R_w, namely, 20.5°C and 7.297 g kg^{-1}, corresponding to 21.5°C and 45%.

With an occupancy of three people, taking a fresh air allowance of 1.3 l s^{-1} m^{-2} over a treated floor area of 28.8 m^2, gives 37.44 l s^{-1}, or 0.0443 kg s^{-1} at the room state. This exceeds the CIBSE recommendations of 8 l s^{-1} for each person when there is no smoking and is adopted. A separate air-conditioning plant would be needed for the constant temperature room because it must run continuously, whereas the main plant serving the rest of the office block shuts down at night. An inspection of the psychrometric chart suggests that a suitable off-coil temperature might be 11.5°C, leading to the selection of a cooler coil with four or six rows. If a blow-through cooler coil is adopted for the secondary plant there will be no penalty of temperature rise from fan power, after the cooler coil. However, the primary fan is likely to be running at a fan total pressure of, say, 0.25 kPa, as a provisional estimate, giving a rise of 0.25 K across it provided that the fan motor is not in the airstream. If we allow a further 0.75°C for duct heat gain, the secondary air would enter the conditioned room at 12.5°C, if it were unmixed with primary recirculated air and a draw-through coil used.

The humid specific heat of air at the room state is (1.012 + 1.89) × 0.008374 = 1.028 kJ kg^{-1} and so, denoting the primary and secondary mass flow rates by $\dot{m}_p$ and $\dot{m}_s$, respectively, the following equation holds for air entering the conditioned room at a temperature of 20.5°C and a sensible heat gain of 2 kW:

$$2.0 = (\dot{m}_p + \dot{m}_s) \times 1.028 \times (21.5 - 20.5)$$

and so $\dot{m}_p + \dot{m}_s = 1.946$ kg s^{-1}. At the room state the specific volume is 0.8455 m^3 kg^{-1} and so the volumetric flow rate is 1.645 m^3 s^{-1}, representing an air change rate of 1.645 × 3600/(28.8 × 2.6) = 79.1 per hour. This is clearly impossible to distribute with conventional diffusers or grilles, so a ventilated ceiling, or floor or wall, is essential, as expected.

If the primary air recirculated from the room were unmixed with secondary air, its supply temperature would be 21.5 + 0.25 = 21.75°C, ignoring any duct heat gain as trivial since ambient air is likely to be at a temperature not much different from 21°C. Therefore, 21.75 $\dot{m}_p$ + 12.5 $\dot{m}_s$ = 20.5($\dot{m}_p$ + $\dot{m}_s$), from which $\dot{m}_s$ is 13.5% and $\dot{m}_p$ 86.5% of the total supply mass flow rate. The secondary flow rate is therefore 0.135 × 1.946 = 0.263 kg s^{-1}. Of this 0.0443 kg s^{-1}, or 16.8%, is from outside and 83.2% is recirculated. The mixed state (*M*) of the secondary air before it enters the fan will be

$$t_m = (0.832 \times 21.5) + (0.168 \times 28) = 22.6°C$$

$$g_m = (0.832 \times 8.374) + (0.168 \times 10.65) = 8.756 \text{ kg}^{-1}$$

The secondary fan total pressure is likely to be about 0.75 kPa and so the mixed air will rise by 0.75°C, giving a state (M_s), onto the secondary cooler coil of t_{ms} = 23.4°C, g_{ms} = 8.756 g kg^{-1} and h_{ms} = 45.82 kJ kg^{-1}. Since only the secondary air deals with the latent heat gain in the room, the latent heat of evaporation of water can be taken as 2454 kJ kg^{-1} at 21.5°C. Therefore

$$0.400 = 0.263 \times (0.008374 - g_s) \times 2454$$

and so g_s, the moisture content of the secondary air after the cooler coil is 7.754 g kg^{-1}. Identifying state *S*, on a psychrometric chart as 12.5°C and 7.754 g kg^{-1}, 1 K can be set back for primary fan power (0.25 K) and secondary duct gain (0.75 K) to state *W* (11.5°C and 7.754 g kg^{-1}). Joining *W* to M_s (23.4°C and 8.756 g kg^{-1}) by a straight line and projecting it to cut the saturation curve at *A* (9.8°C) allows the appropriate cooler coil contact factor to be determined as (23.4 – 11.5)/(23.4 – 9.8) = 0.88, suggesting a four-row or six-row cooler coil, depending on the face velocity. Incidentally, if a blow-through coil is used, it is imperative that the duct connexion from the fan discharge onto its face is such as to ensure smooth airflow over the coil.

From a chart or interpolating in tables, the enthalpy leaving the cooler coil is 31.12 kJ kg^{-1}. The cooling load is thus 0.263 × (45.82 – 31.12) = 3.87 kW of refrigeration. This should be checked as independently as possible. Therefore, noting that the secondary volumetric airflow rate, measured at the room temperature is 0.263 kg s^{-1} × 0.8455 m^3 kg^{-1} × 0.8455 m^3 kg^{-1} = 0.222 m^3 s^{-1}, the following can be calculated using Equation (2.3) where necessary.

		kW
Primary fan power:	(0.25 × 1.645 × 358)/(273 + 21.5) =	0.500
Primary duct gain:	=	–
Secondary fan power:	(0.75 × 0.222 × 358)/(273 + 21.5) =	0.202
Secondary duct gain:	(0.75 × 0.222 × 358)/(273 + 21.5) =	0.202
Fresh air load:	0.0443 × (55.36 – 42.90) =	0.552
Sensible heat gain:	=	2.000
Latent heat gain:	=	0.400
Total cooling load:		3.856 kW

which is in good agreement with the other result.

Proportional plus integral plus derivative control should be used with a good quality hot gas valve and a hot gas header to control the temperature of the air leaving the direct-expansion cooler coil. The reheater should be controlled in a similar mode from a return air thermostat. For a plant of this small duty it would not be economically viable to vary the proportions of fresh and recirculated air to exploit the natural cooling capacity of the cold outside air in winter and so minimise running costs. Instead, minimum, fixed fresh air should be used with the refrigeration plant running throughout the year. This makes it more than ever necessary to stabilise condensing pressure. Humidification is best done with a dry steam injector, under proportional plus integral control, from a dry steam generator, probably energised electrically and located close to the point of use.

3.18 Computer rooms

A characteristic feature of the sensible heat gains is that the energy liberated by the computer units is so large as to dwarf the significance of the other components. Latent gains are also quite small and the slope of the room ratio line is therefore very flat. The size of the computer installation influences the design approach and, although it is hard to generalise, there is evidence that direct-expansion systems are cheaper and hence more appropriate when total cooling loads are less than about 175 kW of refrigeration, but chilled water may be a better proposition for duties exceeding this. There must be some overlap, of course, and the availability of water chillers with capacities as small as 15 kW of refrigeration makes the distinction between the use of the two forms of cooling somewhat blurred. The use of self-contained, incremental air-conditioning units, actually located in the conditioned space, adds a further note of uncertainty about the dividing line between chilled water and direct-expansion, in computer air conditioning.

(1) *Direct-expansion systems of conventional type*. With the smaller installations, although the sensible heat gains from the computer may still amount to a large proportion of the total, it is often possible to achieve conditioning with a fairly conventional system design: direct-expansion air cooling with an air, or water, cooled condensing set, either as a self-contained entity for the smallest schemes, or as a split system for larger duties. Ducting is generally needed for air distribution to the grilles, or diffusers that can be used if the air change rate is less than about 20 per hour, as is likely with small duties. No separate extract fan is required and the system runs with about 5% fresh air as a fixed amount throughout the year. Since occupancy is small, little fresh air is necessary for ventilation but a small surplus of supply over extract is desirable to achieve a slight, but unspecified, positive pressure in the computer room. As a general rule it is misleading to quote this as a percentage of the air supply rate, for example, 5%, because exfiltration, and so pressurisation, depends on the tightness of the building structure. With high air change rates it may be impossible for 5% of the air supplied to exfiltrate and the room can then be regarded as part of the duct system with the airflow rate into and out of it coming to equilibrium at a value of less than the 100% design figure. The method of controlling temperature and humidity will depend on the tolerances acceptable to the computer manufacturers. These can be quite varied among the different makes, some accepting wide swings in temperature and humidity, as long as condensation does not occur. It is common to provide conditions that are comfortable for the occupants if the computer tolerances are broader than the comfort zone. Cycling the compressor motors or unloading cylinders, in sequence with a heater battery, from a return air thermostat may be an acceptable form of control in many cases, provided that the motor does not cycle more often than a safe, maximum number of times per hour – as few as, say, six for

small hermetic machines. If the compressor motor continues to run, loading and unloading cylinders can generally be more frequent than the maximum number of motor starts, without harm. There is always the risk of frosting on the air cooler coil, with the ultimate consequence of motor burnout for hermetic compressors and, as discussed in Section 2.3, this is exacerbated by the use of larger fresh air quantities and may be made worse if humidification is not needed for the environment of the particular make of computer. This is because the wet-bulb temperature onto the cooler coil falls as winter operating conditions approach with a consequent lowering of the evaporating temperature. A check should always be made on the risk of such a situation developing and steps taken to pre-empt it by the use of a hot gas valve which is not always possible with small, packaged units by switching off the refrigeration plant when it is not needed, or perhaps by minimising the use of fresh air. An approach sometimes adopted successfully is to provide a small, separate, ventilation plant that handles filtered, minimum fresh air, tempered in winter, and delivers it directly to the room, thus allowing the air-conditioning unit to deal with only recirculated air and achieving a fairly constant state onto its cooler coil. If this is done the sensible and latent components of the fresh air introduced must be added to the sensible and latent heat gains calculated for the room, to be dealt with by the air-conditioning unit. However the system is controlled it is essential to stabilise the condensing pressure, either by means of a liquid level back-up control or by modulating the condenser fan speed through a solid-state controller or an inverter, responding directly to condenser pressure. If a water-cooled condenser is used pressure should be controlled directly from the condenser by regulating the water flow rate through it. With larger installations, exceeding about 70 kW of refrigeration, an economic argument in terms of owning and operating cost, or present value, over the life of the plant, may possibly be made for the use of variable proportions of fresh and recirculated air quantities. In such a case the plant might run with minimum fresh air until the outside air temperature is cool enough to permit the refrigeration plant to be switched off. In this way the wet-bulb temperature onto the cooler coil and the cooling load is kept up until the switch-off temperature and the risk of frosting, compressor cycling and motor burn-out much reduced.

Controlling temperature by sequencing the capacity of the refrigeration plant and the heater battery, with a high limit humidity override to bring back part of the refrigeration capacity when needed for dehumidification (the heater battery continuing to control temperature by temporarily cancelling part of the cooling) may not be satisfactory in all cases. It can then be necessary to run the refrigeration plant to give a nominally constant temperature after the cooler coil, its evaporating temperature being regulated by a hot gas valve and header, and to control room temperature by reheating. Humidity control is achieved in its most stable form with a sprayed cooler coil, fitted with a good quality hot gas valve that can turn down to perhaps 5% of full load. The large thermal inertia of the mass of cold spray water in the recirculation tank under the direct-expansion cooler coil adds a high degree of steadiness to the control of air temperature off the coil. Variations in the on-coil state or in the evaporating temperature are damped. The big disadvantages of a sprayed coil are its capital cost, the maintenance problems it presents and the hygienic risk. Scaling and corrosion seem to combine to discourage its use in many instances. The alternative is to use steam injection. Adequate control over humidity is possible and, with a proper choice of humidifying equipment, scaling and corrosion problems are minimised. With steam injection it is important to inject the vapour into an airstream that can accept it: if the state of the air is near saturation initially, little further moisture can be added and the injected steam will immediately form condensate.

(2) *Self-contained incremental units within the computer room.* A natural development of small direct-expansion systems has been to package them with an external appearance and finish similar to that of the computer cabinets and to locate them in the computer room itself. Such packages commonly comprise: one or more hermetic compressors, each with its own direct-expansion cooler coil and condenser, offering a nominal capacity in the range 10–60 kW of refrigeration for a state of 22°C dry-bulb, 15.3°C wet-bulb onto their evaporators, throwaway air filters, steam pan humidifier with a disposable inner surface to reduce the problems of scaling, forward curved centrifugal fans, variable speed drives, drip-proof driving motors, reheaters using conventional media or hot gas, liquid receivers sized for low load operation and an automatic control system for temperature and humidity. Temperature control is effected from a thermostat positioned in the recirculated air stream within the cabinet which cycles one or more compressors in sequence with the heater battery. Modulating control over the reheat with the compressors running continuously can also be achieved. The risks of cycling the compressors too frequently should have been considered by the manufacturers. The injection of water vapour from the steam pan humidifier is often into some of the recirculated air which is arranged to bypass the cooler coil, thus avoiding the risks of condensation mentioned earlier. Humidity may be controlled by sequencing the steam pan humidifier with one of the compressors, overriding temperature control when necessary to achieve dehumidification.

Such packages come in various forms but manufacturers commonly offer four versions:

(a) an air conditioning unit with a remote air-cooled condenser. There are then the usual limitations on the lengths and vertical distances of the hot gas and liquid lines
(b) an air-conditioning unit with an in-built water-cooled condenser, fed from a remote cooling tower
(c) an air-conditioning unit with an in-built glycol-cooled condenser fed from a remote, forced draught heat exchanger, after the style of a motor car radiator. There are no limitations on the pipe runs and water treatment problems are avoided
(d) an air-handling unit fed with chilled water from a remote chilling plant.

Of these four, the glycol-cooled version is probably most popular and has certainly achieved effective results in reliability and control. One or more incremental packages, depending on the load, are located in the computer room and they may extract air from the room, at the top, condition it and discharge it into a floor void for subsequent delivery to the room through floor grilles or perforated panels, strategically positioned. Alternatively, the cabinets extract air from the room through grilles in their sides, condition it and discharge it upwards, through ducting into a ventilated ceiling. The downflow system is the more popular. Although a separate ventilating plant can be used to provide the fresh air needed, the incremental cabinets then doing all the load, both sensible and latent, it is not uncommon for the auxiliary ventilation system to be modified to cool and dehumidify the fresh air in summer, and vice-versa in winter. The plant then copes with the latent loads, transmission gains and losses, and perhaps also the gains from people and lights, leaving the incremental cabinets to do the sensible cooling associated with the computer load only. In this way the room and the various ancillary areas, such as tape stores, and so on, can be kept conditioned through 24 hours, even when the computer plant is off. For the incremental cabinets to do sensible cooling their evaporating temperatures must be properly controlled at values high enough to avoid condensate forming on the cooler coils.

(3) *Chilled water systems*. With large installations, better flexibility, stand-by, control and lower running costs may be achieved by using chilled water from multiple, air-or-water-cooled refrigeration plants. Two pumped chilled water circuits are necessary: a primary circuit delivering lower temperature water to the coils in the air-handling (ventilating) unit dealing with the latent loads and a secondary circuit feeding higher temperature chilled water to those cooler coils in the plant dealing with the computer loads and doing sensible cooling only. In this way, reductions in heat emission from the computers can be matched by reductions in the cooling capacity of the sensible cooling coils, instead of using wasteful reheat to cancel overcooling.

(4) *Air distribution*. There are three possibilities of air distribution. If the air change rate is less than about 20 per hour, conventional side-wall grilles or ceiling diffusers may be used. If this rate is exceeded, a ventilated ceiling or floor is the only answer and in either case it cannot be too strongly stressed that the plenum chamber or void must be sealed and then properly lagged and vapour sealed. Failure to ensure this will cause the installation to be also a failure. The ventilated ceiling will usually consist of a mixture of live and dead tiles, unless the air change rate is very high, i.e. over about 200, when the whole of the ceiling will be live. The live tiles should be positioned over the items of computer equipment producing the heat gains but, since they are removable, it is always possible to reposition them during commissioning to get the best results. The argument in favour of a ventilated floor is that a floor void is needed in any case to accommodate cables feeding the computer cabinets and possibly other services. The raised floor usually consists of 600 mm square tiles, made of plywood, faced beneath with a fire-retardant material, or die-cast steel. A variety of upper surface finishes is available, from vinyl to carpet. The minimum clear floor depth possible is about 75 mm, up to a standard maximum of about 500 mm. Clear depths greater than this are usually possible in non-standard forms, probably with cross bracing (stringers) underneath to give stability. Less than 100 mm clear space, after allowing for cables, is useless for supply ventilation. The same considerations apply for the throw and distribution of air in a floor void as in a ventilated ceiling (see Section 5.10). The tiles rest on supporting pedestals, very often without stringers, that allow the easy laying in of cables. Complete interchangeability, using suction cups for the simple removal of the tiles is usual, to accommodate rearrangement of computer cabinets which are fed with cables from underneath through specifically cut holes. It is possible to have steel or anodised aluminium grilles in the tiles for the supply of conditioned air, or the tiles may be perforated. Supply air grilles are often located around the perimeter of the room and should be sized for face velocities of about 0.5 m s^{-1}. Perforated tiles commonly have open areas of 15–25%, giving velocities through the holes of about 2.5 m s^{-1} for face velocities over the floor of 0.45 m s^{-1}. Such air velocities are generally acceptable to people, with shod feed but bare legs, standing near the tiles or even on them. Placing tiles containing grilles near the peripheral walls tends to keep them out of the main avenues of foot traffic but perforated tiles are frequently positioned adjacent to the air inlet grilles in the lower parts of the computer cabinet side panels. Vertical temperature gradients of as much as 5 K over a floor-to-ceiling height of 3 m have been measured when supplying cooling air through perforated floor tiles into an unoccupied room, at the velocities mentioned Such gradients usually disappear when the room is in normal use or can be corrected by increasing the velocity of air flow through the holes in the tiles. The upper limit of such an increase must be human comfort, which can only really be established on a subjective basis. Manufacturer's recommendations for sizing perforated tiles and grilles should be followed.

As with ventilated ceilings, sealing, lagging and vapour sealing the floor void is vital.

3.19 Combined heat and power

Conventional power stations generate electrical power in a wasteful manner, only about 35% of the energy in the fuel used is converted into electricity and the remaining 65% is wasted as heat loss at the power station and electrical distribution losses in the grid. On the other hand, if the waste heat is reclaimed and the electrical power distributed over comparatively short distances in the locality, the overall efficiency can increase to 85%. In the UK such an installation is termed a combined heat and power (CHP) system (see CIBSE, 1994) and when the electricity generated provides for all the power needs on the site it is sometimes called a total energy system. In USA the term co-generation is adopted, instead of combined heat and power. Figure 3.16 shows a simplified diagram of a reciprocating engine driving an alternator, with heat recovery from the engine cooling jacket and the exhaust.

CHP systems not only reclaim much of the wasted heat and minimise distribution losses but also give lower emissions of CO_2. According to the Encrgy Efficiency Office (1994) the emission of CO_2 from a gas-driven reciprocating engine connected to an alternator, is only about 43% of the emission by a conventional power station.

The prime movers for CHP systems are most commonly reciprocating internal combustion engines, running at between 360–1200 rpm. They may be spark-ignition, with maximum electrical generating capacities up to about 13 MW, or compression ignition, with

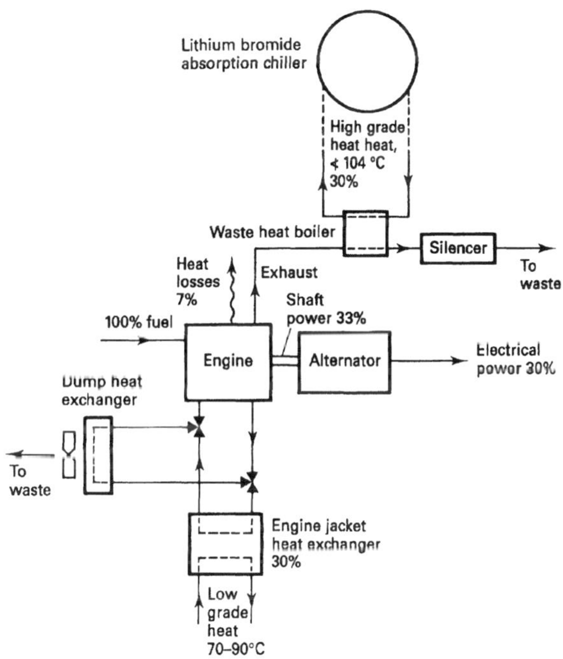

Figure 3.16 A simplified diagram of CHP. The heat exchanger for the oil cooler is not shown.

capacities from 20 kW to about 20 MW. Natural gas or diesel oil is used as a fuel. Dual fuel engines are also possible, offering alternative energy sources. Selection is generally arranged to provide about 20 000 operating hours between minor overhauls and about 50 000 hours between major maintenance shut-downs. (It is to be noted that reciprocating engines intended for emergency stand-by operation are not suitable for operation in CHP systems. They run at much higher speeds and are not intended for prolonged periods of use.) Gas turbines are available, with electrical generating capacities from 40 kW to 130 MW, running at much higher speeds, between 3500 and 50 000 rpm. Various sorts of steam turbine can also be used. Slower running machines, operating at temperatures less than the maximum allowable, will have longer lives and be more reliable.

The typical efficiency of an alternator peaks at just over 90% of its rated load. At reduced duties the efficiency falls off but could still be as much as 80% at 25% load. A steady-state overload up to 25% can be accommodated by industrial generators but the engine or turbine driving the generator must also be able to cope with the same overload.

The electrical output is likely to be 11 kV, 3 phase, 50 Hz and this must be balanced by the requirements of the electrical systems supplied, comprising three phase and single phase motors, lighting, small power, and so on. At times when the site does not want all the electrical power available from the alternator the surplus power can be sold to the National Grid or, possibly, to other users.

Exhaust temperatures from engines are between 250 and 650°C, depending on the type of engine, the fuel used and the load. This allows the use of a waste heat boiler to produce high grade heat for operating an absorption refrigeration machine to produce chilled water for air conditioning purposes. Absorption machines should be two-stage, giving a coefficient of performance of about 1.2, instead of 0.65 with a single stage machine. A lithium bromide absorption machine requires dry steam at a pressure not less than 115 kPa (104°C) and desirably rather more. High and medium temperature hot water has been used in the past but has generally proved unsatisfactory: thermal expansions and contractions, associated with the difference between the flow and return water temperatures at the generator, have caused difficulties. The best source of energy for the generator in an absorption machine is dry steam.

If suitably pressurised, the flow temperatures from both the oil cooling heat exchanger and the heat exchanger used for cooling the engine itself can be high enough for an absorption machine but the oil temperature should not be too high, otherwise engine performance suffers and its life is shortened. Lower cooling water-flow temperatures are preferred (80–90°C) and the recovered heat is then used for conventional heating purposes.

When the requirement for heat on the site or in the building dealt with by the CHP installation is small, it is necessary to reject the surplus heat, through a dump heat exchanger, to outside, if the generation of electrical power is to continue.

Packaged CHP plants are available, with electrical outputs in the range 20–370 kW.

The conventional way of sizing CHP systems is to determine a base load for heating. The CHP plant is sized to meet this load and will then run for a substantial period of time, not less than 4500 hours per year being desirable to achieve a simple pay back period of 3 or 4 years, according to CIBSE (1994). An alternative approach, proposed by Griffiths (1995), argues that capital and running costs are the most important factors when sizing a CHP system. The economies of scale give significant reductions in the cost per kW of electrical power for larger machines and, further, operating efficiency is better for larger machines. It is considered that fuel costs are likely to be five times the capital cost of the system over its life and hence, even a small improvement in efficiency will give worthwhile economic benefit. Maintenance costs also reduce as machine size increases. The answer to the problem of sizing is claimed to be multiple calculations to establish the size of CHP plant that will give the best financial case. This requires computer simulation.

Some measure of stand-by plant is essential for a system that is expected to run for long periods of time. For an ideal case, in addition to the duty machines (engine-alternator combinations) there should be one machine standing by and ready for immediate use, a second machine undergoing routine maintenance and, possibly, a third machine broken down and undergoing repair. Practical economics may suggest otherwise for a particular case.

An aspect of CHP that must not be ignored is the noise produced. The plant room must be provided with adequate acoustic treatment and the exhausts from the engines must be fitted with silencers. Vibration must also be dealt with. These considerations require that the plant room building itself must be suitable: proper treatment for noise and vibration is intimately related to building construction. See Chapter 7.

In principle, the factors that affect economic viability are as follows:

(1) The minimum electrical power requirement for the proposed installation should be 1 MW. Smaller minimum powers may also give an acceptable solution but not such a good economic case.
(2) The load factor should be 40% or more.
(3) The buying price of the fuel should be not more than one-fifth of the cost of buying electricity in the normal way, during the period of the daytime tariff.
(4) The requirement of thermal energy (for heating and absorption refrigeration) should be at least equal to the demand for electrical power and lighting. Furthermore, the daily profiles for heat and electricity consumption should be reasonably balanced. There must be an economic demand for the heat reclaimed during the time that the electrical power is being generated.

It follows that applications such as hospitals, hotels, or swimming pools, needing heat over 24 hours, must be worth considering. Similarly, cases where there is a residential component or where there is the possibility of exporting heat to a nearby residential site, are also attractive.

The fact that electricity may be sold to the National Grid or, possibly, to other users, must not be overlooked when weighing up the economics. The minimum desirable electrical output of 1 MW, quoted above, might then be reduced, but it must be borne in mind that the price received for exported electricity is likely to be a good deal less than the cost of buying it from the local electricity distribution company.

Exercises

1 Estimate the cooling load for the hypothetical hotel considered in Examples 3.1 and 3.2 at 13.00 h sun time in July, making the same assumptions for diversity factors. Use Equation (1.2) and psychrometric tables to establish the outside enthalpy at 13.00 h sun time. (*Answer* 454 kW of refrigeration)

2 For the hypothetical office block (Section 1.2) plot maximum cooling load against percentage glazing for a given orientation, and maximum cooling load against orientation for a given percentage of glazing. Make use of Table A.10 and assume a lighting load of 30 W m^{-2} and a population density of 9 m^2 per person. Repeat the exercise for a lighting load of 20 W m^{-2}.

3 Repeat the calculations for Example 3.3 but assume the illumination in the supermarket is 1500 lux, the open refrigerated display cabinets absorb 40 000 W of total heat and the population density is 4 m^2 per person.

Notation

Symbols	Description	Unit
A_D	Du Bois surface area of a human body	m^2
A_p	Area of a pool surface	m^2
d	Particle size	μm
g_r	Room air moisture content	$kg\ kg^{-1}$ or $g\ kg^{-1}$
g_s	Supply air moisture content	$kg\ kg^{-1}$ or $g\ kg^{-1}$
h	Height of a human body	m
M	Numerical part of a metric clean class name	–
m	Mass of a human body	kg
n_h	Number of persons arriving at a pool per hour	h^{-1}
n_p	Number of persons bathing per unit area of pool surface	m^{-2}
p	Number of particles per cubic metre	m^{-3}
p_s	Vapour pressure of an air–water vapour mixture	mbar or Pa
p_{tF}	Fan total pressure	kPa
p_w	Saturation vapour pressure of water	mbar or Pa
Q_e	Latent heat gain to a pool hall	kW
Q_{fa}	Fresh air load	kW
Q_l	Latent heat gain	kW
Q_{lc}	Latent cooling capacity	kW
Q_{ra}	Recirculated fan power load	kW
Q_{ref}	Refrigeration load	kW
Q_s	Sensible heat gain	kW
Q_{sf}	Supply fan power and duct gain	kW
Q_{tfp}	Power of fan coil unit driving motors	kW
q_l	Latent heat gain by evaporation from a pool surface	$W\ m^{-2}$
t	Air dry-bulb temperature	°C
t_r	Room air dry-bulb temperature	°C
t_s	Supply air dry-bulb temperature	°C
v	Air velocity over a pool surface	$m\ s^{-1}$
$\dot{v}_t$	Volumetric airflow rate at a temperature t	$m^3\ s^{-1}$ or $l\ s^{-1}$
W_f	Fan power	kW
W_o	Evaporation rate for an occupied pool	$kg\ s^{-1}$
W_u	Evaporation rate for an unoccupied pool	$kg\ s^{-1}$
$\Delta\theta$	Average length of stay of bathers in a pool	h
ε	Normalised activity factor for an occupied pool	–
η	Fan total efficiency as a fraction	–

References

Allander, C. and Abel, E., Unterschung eines Enblassystems fur Operationasraume. *Decema Monographie*, **69**, 297–306, 1948.

ASHRAE Handbook, *Heating, Ventilating and Air Conditioning Applications*, SI edition, 1995.

Biasin, K. and Krumme, W., Evaporation in an indoor swimming pool. *Electrowarme International*, **32,** A3, 115–29, 1974.

Braham, G. D., Wise use of energy in swimming pool design. *The Heating and Ventilating Engineer*, December, 6–8, 1975.

BS 476, Part 7, Fire tests on building materials and structures, Part 7: Method for classification of the surface spread of flame of products, 1993.

BS 476, Part 8, Fire tests on building materials and structures, Part 8: Test methods and criteria for fire resistance of elements in building construction, 1972.

BS 5295, Environmental cleanliness in enclosed spaces. Part 0: General introduction, terms and definitions for clean rooms and clean air devices, Part 1: Specification for clean rooms and clean air devices, Part 2: Method for specifying the design, construction and commissioning of clean rooms and clean air devices, Part 3: Guide to operational procedures and disciplines applicable to clean rooms and clean air devices, Part 4: Specification for monitoring clean rooms and clean air devices to prove compliance with BS 5295 Part 1. 1989.

Butcher, E. C., Fire progressive spread and growth. BRE information paper, October, 1987.

CIBSE Energy Efficiency Guide, Section 3.11, Combined heat and power, 1994.

CIBSE Guide, Section A2, Weather and solar data, 1982.

CIBSE Guide, Section B2, Ventilation and air conditioning requirements, CIBSE, 1992.

CIBSE, Information technology in buildings, Applications Manual AM7, CIBSE, 1992.

Doe, L., Gura, G. H. and Martin, P. L., Building services for swimming pools. *JIHVE*, **35**, 261–86, 1967.

Doone, R. E., Lighting and the integrated environment in multiple stores. *The Steam and Heating Engineer*, February, 1971.

Du Bois, D. and Du Bois, E. E., A formula to estimate approximate surface areas if height and weight are known. *Archives of Internal Medicine*, **17,** 864, 1916.

Dunklin, E. W. and Puck, T., The lethal effect of relative humidity on airborne bacteria. *J Experimental Medicine*, **87**, 87–101, 1948.

Federal Standard 209E, Airborne particulate cleanliness classes in clean room and clean zones. September, 1992.

Good Practice Guide 115, An environmental guide to small-scale combined heat and power. Energy Efficiency Office, Department of the Environment, 1994.

Good Practice Guide 116, Environmental aspects of large-scale combined heat and power. Energy Efficiency Office, Department of the Environment, 1994.

Green, G. H., The effect of indoor relative humidity on absenteeism and colds in schools, *ASHRAE Trans, Part II*, 131–41, 1974.

Green, V. M., Vesley, D., Bond, R. G. and Michaelson, G. S., Microbiological contamination of hospital air. I. Quantitative studies. *Applied Microbiology*, **10**, 561–6, 1962

Griffiths, R., Baseload sizing – just say 'no'. *Energy World*, March, 11–14, 1995.

Hinkley, P. L., Some notes on the control of smoke in enclosed chopping centres. Fire Research Note No. 875, May 1971, Fire Research Station, Borehamwood, Herts, 1971.

Howarth, F. H., The prevention of airborne infection during surgery. *Proceedings of the International Symposium for Contamination Control*, Swiss Federal Institute of Technology, Zurich, *Journal of the Society of Environmental Engineers*, **55**, 31–3, 1972.

Hunt, E. L., Building engineering services at the new Charing Cross Hospital (Fulham). *The Building Services Engineer*, **44**, 41–9, 1976.

Jamieson, H. C., Presidential address. *JIHVE*, **27**, 245, 1959.

Jones, W. P., *Air Conditioning Engineering*, 4th edn, Fig 15.20, p. 431, Edward Arnold, 1994.
Kay, P., Doing the pools. *Building Services*, May, 45–6, 1994.
Knight J. and Jones W. P. (1995) *Newnes Building Services Pocket Book*, Newnes.
Ma ,W. Y. L., Air conditioning design for hospital operating rooms. *JIHVE*, **33**, 165–79, 1965.
Martin, P. L., A study of occupancy in chain stores. *JIHVE*, **33,** 99–102, 1970.
Morgan, H. P., Smoke control methods in enclosed shopping complexes of one or more storeys, a design summary. *BRE* **34** (K2.3), 1976.
Murphy, V., Department store air conditioning. *Building Systems Design*. February, 41–3, 1970.
Pilkington, Glass and thermal safety, Pilkington Flat Glass Ltd, 1980.
Reeker, J., Water evaporation in indoor swimming pools, the results of practical tests. *Klima and Kalte Ingenieur*, **6,** 1, 29–33, 1978.
Schicht, H. H. and Steiner, W., The contribution of air conditioning to asepsis in the operating theater. *Sulzer Technical Review*, **4**, 1972.
Smith, R. M. and Rae, A., Odour and ventilation in hospital wards. *The Building Services Engineer*, **44**, 265–71, 1977.
Stinson, R. G., Shopping centres HVAC. *Building Systems Design*, February, 36–8, 1970.
Thiesenhusen, H., Studies concerning the rate of water evaporation as a function of water temperature, atmospheric humidity and wind speed. *Gesundheits Ingenieur*, **53**, 8, 113–19, 1932.
Thornley, D. L., Auditoria – a review of present-day HVAC practice. *JIHVE*, **37**, 170–83, 1969.
Valerio, E. L., Air conditioning bowling centres. *Air Conditioning, Heating and Ventilation*, March, 75–81, 1959.
Whyte, W., Shaw, B. H. and Bailey, P. V., An assessment of partial walls for a laminar downflow system. *International Symposium for Contamination Control*, London, 24 September 1974, *Journal of the Society of Environmental Engineers*, **66**, September, 1974.

4

Water Distribution

4.1 Pipe sizing

The basis of pipe-sizing is the Fanning equation

$$\Delta H = (4flv^2)/(2gd) \tag{4.1}$$

in which ΔH is the head lost in metres of fluid flowing, f is a friction factor defined by Equation (4.2), l is the length of pipe, v is the mean velocity of flow, g is the acceleration arising from the force of gravity and d is the internal pipe diameter. Generally, waterflow in a pipe is turbulent and so Colebrook and White's formula applies

$$1/\sqrt{f} = -4 \log [(k_2/3.7d) + (1.255/(\mathrm{Re})\sqrt{f})] \tag{4.2}$$

in which k_s is the absolute roughness of the pipe wall (in metres) and (Re) is the Reynolds number. Since pressure and head are related by

$$p = \rho g H \tag{4.3}$$

Equations (4.1) and (4.2) can be modified, for the case of turbulent flow, to yield an expression for the mass flow rate, M, in terms of the pressure drop per metre, Δp_l

$$M = -4(N_1 \Delta p_l d^5)^{0.5} \log [(k_s/3.7d) + (N_2 d/(N_1 \Delta p_l d^5)^{0.5})] \tag{4.4}$$

in which, using the notation adopted by the CIBSE guide, $N_1 = \pi^2 \rho/32 = 0.30842\rho$ and $N_2 = 1.255\ \pi\mu/4 = 0.98567\mu$, μ and ρ being the absolute viscosity and density, respectively, of the fluid flowing.

Solving Equation (4.4) involves tedious computation and the CIBSE (1977) guide publishes tables for the flow of low temperature hot water at 75°C in clean, medium grade, black steel pipes to BS1387: 1967. However, chilled water flowing in pipelines is likely to be at about 7.5°C and cooling water at about 30°C. With lower water temperatures the kinematic viscosity increases and hence the Reynolds number reduces. Equation (4.2) shows that the friction factor increases as the Reynolds number falls, for a given pipe roughness. It follows that the rates of pressure drop for pipes carrying chilled water exceed those quoted in the

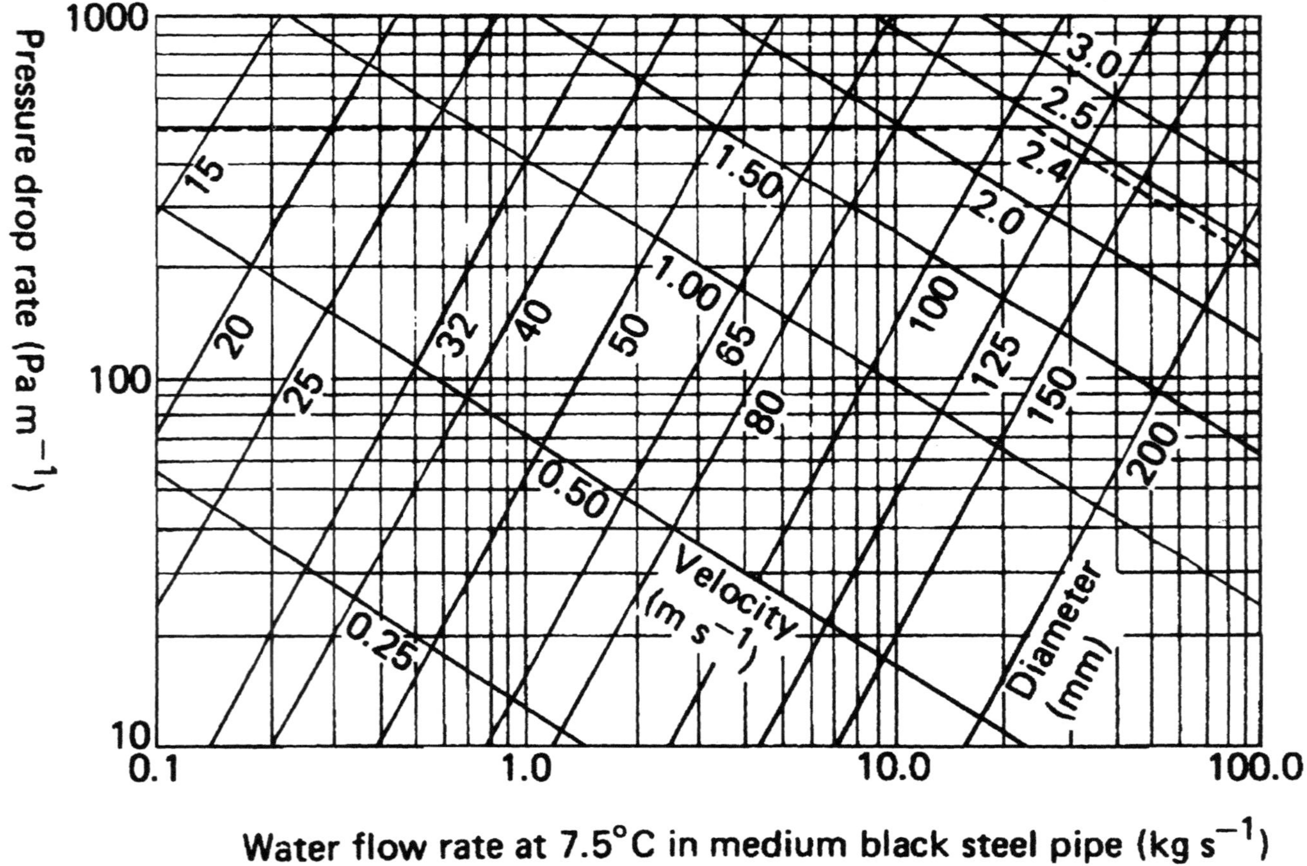

Figure 4.1 Pipe sizing chart for water at 7.5°C in clean, medium-grade, black steel pipes (reproduced by kind permission of Haden Young Ltd).

CIBSE tables for the flow of LTHW. For equal mass flow at a velocity of 2.4 m s^{-1} the pressure drop rate at 7.5°C is about 30% greater than at 75°C over the range of pipe sizes from 25 to 200 mm.

Figure 4.1 is derived from Equations (4.1)–(4.4) and may be used to size pipes carrying chilled water at 7.5°C. It is suggested that the same figure also be used for sizing pipes carrying condenser cooling water at about 30°C. For such open circuits the dirtier pipework and greater surface roughness will take up the small margin in pressure loss.

The loss through most fittings is expressed as a factor, k multiplying the velocity pressure. By means of Equation (4.3) substitution for ΔH in Equation (4.1) gives pressure loss in N m^{-2}

$$\delta p = (2\rho f l v^2/d) \tag{4.5}$$

The velocity pressure is $1/2\rho v^2$ and if k is unity the pressure lost through the fitting, δp would be numerically equal to this

$$\delta p = 1/2\rho v^2 = (2\rho f l v^2)/d$$

This can be expressed as the equivalent length of pipe, l_e, that has a pressure loss numerically equal to one velocity pressure. The determination of f by Equation (4.2) is tedious and l_e is usually determined from the tables used for pipe sizing in the CIBSE guide (1977). Tests have established values of k for various fittings and a representative selection is published. The loss through a fitting can consequently be expressed as so many metres of equivalent straight pipe.

EXAMPLE 4.1

Determine the pressure drop past a 90° malleable, cast-iron elbow of 25 mm diameter for a flow rate of 0.5 kg s^{-1} at 75°C.

Answer

From the CIBSE guide (1977) Table C4.12 for medium-grade, black steel pipe, the pressure drop rate for a flow of 0.5 kg s^{-1} in a 25 mm pipe is 349 Pa m^{-1}, by interpolation, and the equivalent length is 1.1. From CIBSE, Table C4.36, $k = 0.8$, so the length of straight pipe with the same pressure loss as the fitting is $0.8 \times 1.1 = 0.88$ m and the actual loss is $0.88 \times 349 = 307$ Pa.

Notable among the fittings absent from the CIBSE table is the strainer. Figure 4.2 shows the resistance to flow for brass and bronze, *y*-type strainers, based on data published by Spirax Sarco. We see that for a 200 mesh, 25 mm strainer with 0.5 kg s^{-1} flowing, the loss is about 3 kPa, which is a significant figure. As the strainer gets fouled its resistance increases towards infinity, ultimately stopping the flow. Figure 4.3 shows a strainer fitted with a drain cock and a hose for convenient maintenance.

As a general principle, pipe should be sized using velocities in the range of 1–2 m s^{-1}, but because larger pipe, fittings and valves are much more expensive than the smaller sizes (see Figure 4.4), it is desirable to keep the water velocity high for the larger pipes, up to a maximum of about 2.4 m s^{-1}, set by the likelihood of excessive erosion, especially at elbows. On the other hand, smaller pipes carrying water at this speed have high pressure drops, leading to pumps with uneconomically large heads. The dashed line in Figure 4.1 shows that

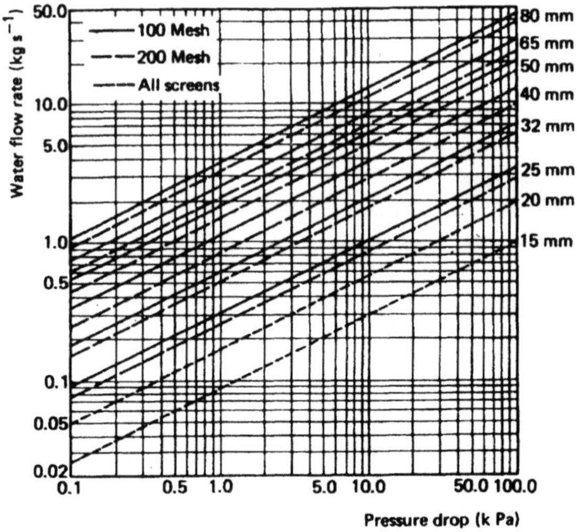

Figure 4.2 Typical pressure drop data for brass or bronze Y-type strainers (after Spirax Sarco).

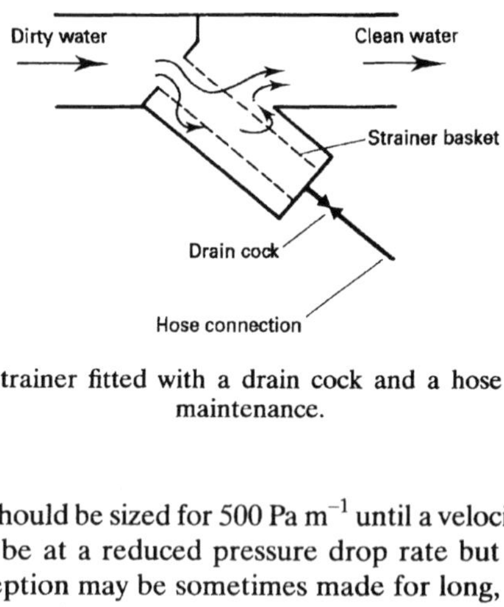

Figure 4.3 Diagram of a strainer fitted with a drain cock and a hose connection for convenient maintenance.

it is suggested that pipes should be sized for 500 Pa m^{-1} until a velocity of 2.4 m s^{-1} is reached. After this, sizing should be at a reduced pressure drop rate but retaining 2.4 m s^{-1} as a limiting velocity. An exception may be sometimes made for long, large diameter, straight, heavy grade, steel mains, carrying non-corrosive water, where the limiting main velocity may be as much as 4.0 m s^{-1}.

EXAMPLE 4.2

Size pipes to carry 1, 10, 30 and 50 kg s^{-1} of chilled water at 7.5°C, quoting the pressure drop rates and velocities.

Answer

Using Figure 4.1 the following can be determined

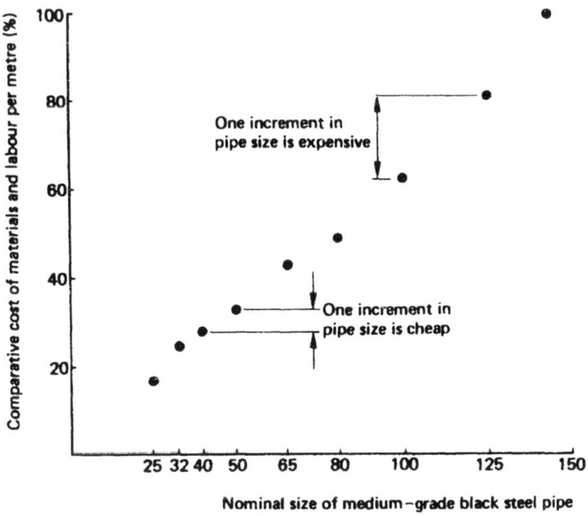

Figure 4.4 Comparative costs of medium-grade, black steel piping.

1 kg s^{-1}, 32 mm diameter, 380 Pa m^{-1}, 0.95 m s^{-1}
10 kg s^{-1}, 80 mm diameter, 500 Pa m^{-1}, 1.95 m s^{-1}
30 kg s^{-1}, 125 mm diameter, 360 Pa m^{-1}, 2.3 m s^{-1}
50 kg s^{-1}, 200 mm diameter, 87 Pa m^{-1}, 1.45 m s^{-1}

If Table C4.12 had been used for water at 75°C, the following would have been obtained

1 kg s^{-1}, 32 mm diameter, 321 Pa m^{-1}
10 kg s^{-1}, 80 mm diameter, 444 Pa m^{-1}
30 kg s^{-1}, 125 mm diameter, 329 Pa m^{-1}

There is no entry for 200 mm pipe, but the pressure drop rates are less than for water at 7.5°C. The lines in Figure 4.1 for velocity and diameter should really be very shallow curves but are drawn straight, because of the small size of the diagram, without much loss of accuracy.

4.2 The design of piping circuits

Most hydraulic circuits used in building services have an open water surface in them somewhere, but it is common practice to speak of a closed circuit when water is pumped in a continuous loop fitted with a connexion either to a feed and expansion tank (Figure 4.5a) or to a pressurisation unit. An open system, on the other hand, involves water flowing under gravity from one level to another (Figure 4.5b) or its being pumped from a lower to a higher level (Figure 4.5c).

Nine important principles must be followed when designing pipe circuits.

(1) Centrifugal pumps must always be primed.

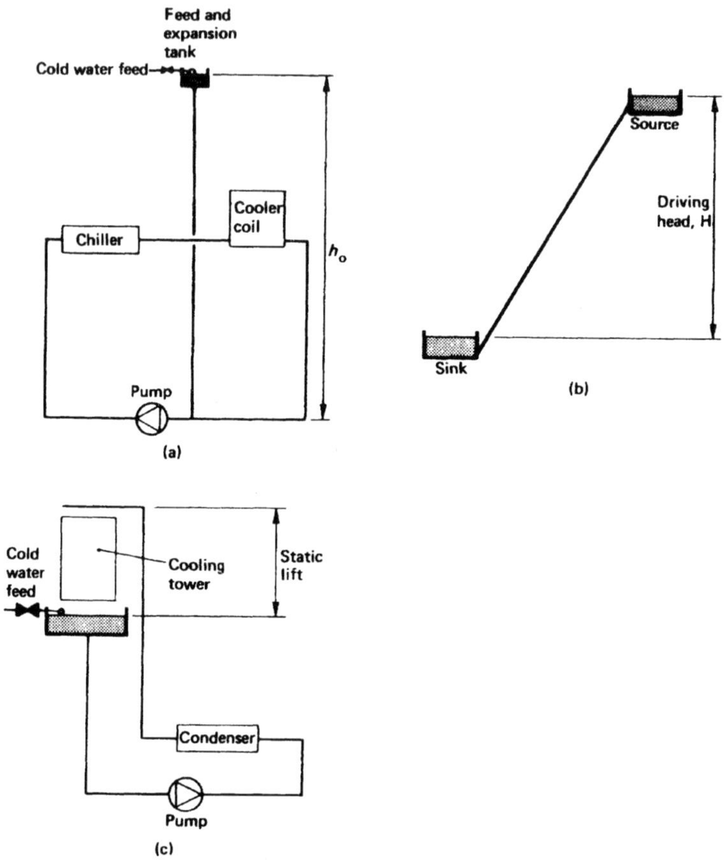

Figure 4.5 (a) Closed pipe circuit connected to a feed and expansion tank. (b) Open pipe circuit. Water flows under gravity from one level to another. (c) Open pipe circuit. Water is pumped from a lower level to a higher level.

(2) Cavitation must be avoided.
(3) Air must not be drawn into the system.
(4) All air must be vented from the system.
(5) In open circuits, as much of the pipework as possible must be below the static water level.
(6) All control valves must be below the static water level in open circuits.
(7) Attention must be paid to position head.
(8) Open vents, used when there is the risk of warmed water boiling, must not be located so as to discharge continuously into the feed and expansion tank.
(9) The connexion from the feed and expansion tank must be made at a place such that the above principles are followed.

Chilled and cooling water circuits are invariably two-pipe, although in certain situations a one-pipe system is a possibility. Figure 4.6(a) shows a conventional two-pipe layout in

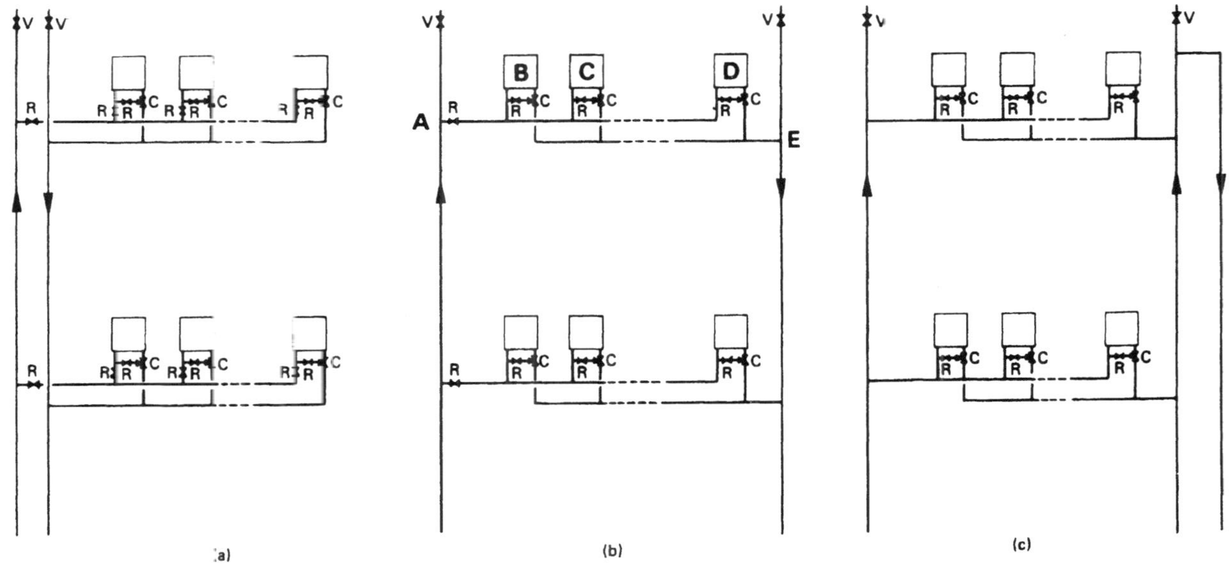

Key: V = Air vent ;
R = Regulating valve ;
C = Automatic control valve

No isolating valves are shown

Figure 4.6 Typical piping arrangements: (a) Two-pipe. (b) Two-pipe ladder with reversed return branches. (c) Complete reversed return.

which regulating valves (R) are fitted to balance both branches and units. A two-pipe ladder with reversed-return branches is shown in Figure 4.6(b), and this is very convenient for feeding fan-coil units, the same pressure drop occurring for any horizontal branch, say **AE**, no matter which unit, **B**, **C**, or **D**, the water flows through. Fitting regulating valves at each unit is often unnecessary but regulation must still be provided for each horizontal branch. A complete reversed-return system is shown in Figure 4.6(c). Theoretically, no regulating valves are needed at all but an extra riser is required, which makes the arrangement expensive, and so less popular.

Instead of the open feed and expansion tank shown in Figure 4.5(a), a pressurisation unit is sometimes connected to the pump suction. For example, it may be impossible, because of restraints imposed by the building construction, to obtain enough head from such a tank to ensure that the entire system of piping and plant is adequately pressurised. A hydro-mechanical pressurising method is then needed. Whichever method is used it must be remembered that the tank or unit must also provide enough volume to accommodate the thermal expansion and contraction of the water in the system, as it responds to temperature changes in the course of its operation. The minimum working pressure of the water in the system must always exceed the saturation vapour pressure corresponding to its temperature, otherwise water will flash to steam. A safety margin of 12 K, beneath the boiling point related to the pressure of the water, is necessary. Thus if the lowest possible absolute pressure of the water in the system is 90.94 kPa, for which the saturation temperature is 97°C, then the highest allowable temperature in the system is 85°C. The expansion vessel in the unit may be pressurised by a pair of intermittently running pumps (one duty, one stand-by), or by a pair of continuously running pumps (one duty, one stand-by). A diaphragm is used to separate the air and water, in order to prevent oxygenation of the water and consequent increased risk of corrosion.

It is quite common for the temperature of the water supplied to a heat exchanger to be different from that produced by the heating or chilling plant. This is particularly so when a refrigeration plant produces chilled water for several circuits, some of which are required to provide sensible cooling and so need a controlled flow temperature higher than the flow temperature from the chiller. Similarly, different operating temperatures may be needed for systems handling hot water. To cope with these situations primary and secondary piping circuits are needed. Figure 4.7 illustrates three common piping arrangements.

In Figure 4.7(a) the secondary circuit provides chilled water for fan coils units, chilled ceilings/beams or perimeter induction units, that must give sensible cooling only. The primary air cooler coil provides dehumidified air controlled by temperature sensor C1, which regulates a motorised three-port mixing valve, R1. Secondary chilled water is supplied at a temperature controlled by C2, which regulates the flow through the motorised three-port mixing valve, R2. Individual secondary units are controlled by motorised two-port throttling valves (R3, R4, etc.). As the valves throttle under conditions of reduced cooling duty, the pressure drop across the valves increases, reducing the effectiveness of their proportional control and causing noise. Hence it is desirable to fit a motorised three-port mixing valve (Rn) at the index unit. A pair of pressure sensors, P1a and P1b, is located in a suitable position across the flow and return lines, often between two-thirds and three-quarters of the distance from pump discharge to the index terminal (to sense a pressure difference of not more than 50 kPa). These are used to relieve the pressure, either by reducing the speed of the secondary pump or by partly closing a motorised two-port valve located at pump discharge (shown as Rd in the figure) as the differential pressure increases. As an alternative, the relief valve is sometimes located in a by-pass (not shown in the figure) between the pump discharge and the nearby return line; this opens as the pressure difference rises. The three-port valve at the index unit serves to keep the secondary chilled water line alive during conditions of light load. The pressure drop in the primary circuit between A and B must be small, so that any variations in the secondary flow rate drawn from the primary circuit at A, do not affect the primary flow rate through the chiller, which will freeze if the flow rate drops. It is vital that the flow rate drawn from the primary circuit to feed the secondary circuit does not exceed the rate pumped in the primary circuit. If, as is often the case, the secondary pump duty is more than the primary duty, then steps must be taken to rectify the situation. A by-pass between the points W and X in the primary circuit would allow the primary duty to be increased, but this will have consequential effects on the selection of the chiller and the sizing of the primary circuit, with increased capital costs. A better solution might be to

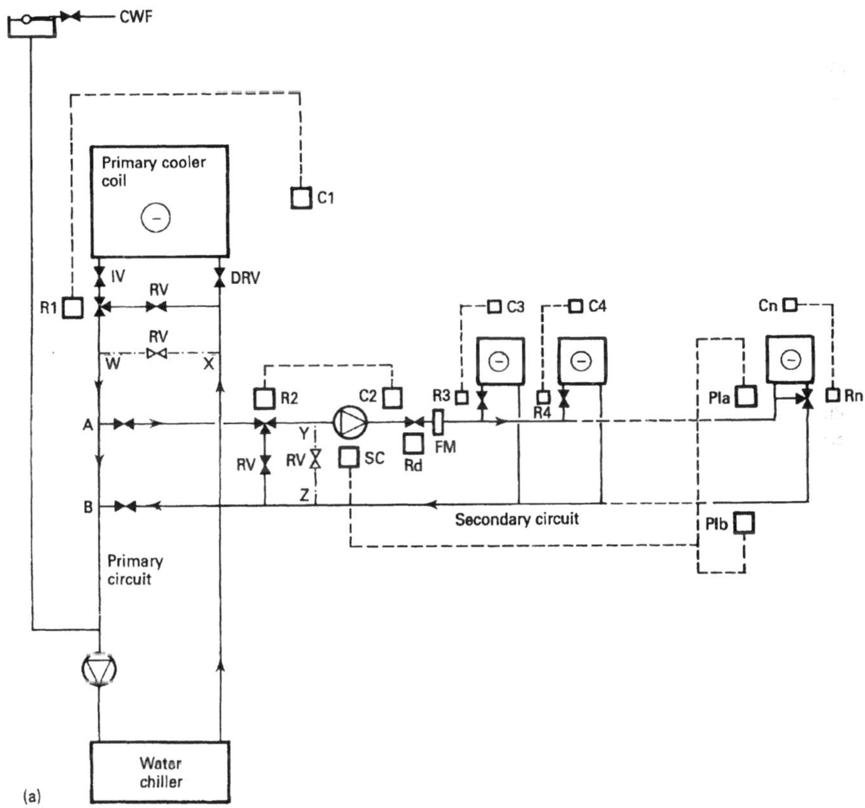

Figure 4.7 (a) Secondary circuit used for sensible cooling.
Notation: IV: Isolating Valve
DRV: Double Regulating Valve
RV: Regulating Valve
R: Control Valve Regulator
C: Temperature Sensor
P: Pressure Sensor
SC: Speed Controller
FM: Flow rate Measurement

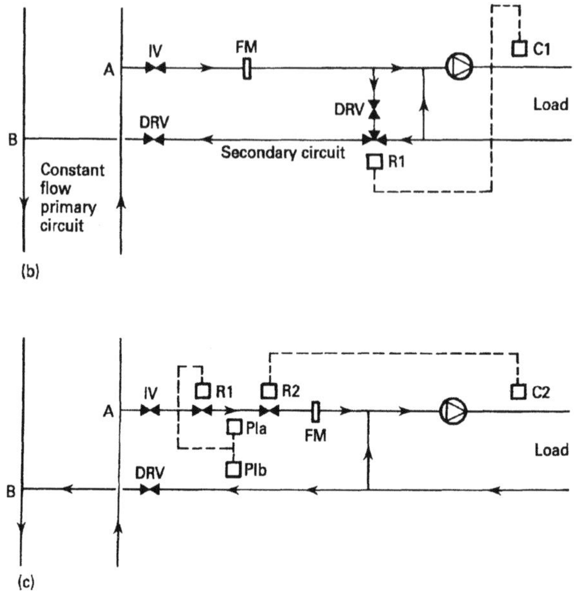

Figure 4.7 (b) Injection circuit with constant primary flow. (c) Injection circuit with variable primary flow.

fit a by-pass in the secondary circuit, between the points Y and Z. The motorised valve R2, must be sized so that the port fed from the primary circuit is fully open at design load in order to give good control. The presence of the by-pass between Y and Z will increase the temperature of the secondary water fed to the sensible cooling units, reducing their capacity and increasing capital cost if larger units have then to be selected. Another possibility is to reconsider generally the selection of the units doing sensible cooling, so as to avoid the use of a by-pass. Whatever solution is adopted it must be based on a proper consideration of primary and secondary water-flow rates and temperatures.

Figure 4.7(b) shows an injection circuit where the secondary connections are across the primary flow and return. Since the flow in the primary circuit is constant, the pressure difference across the flow and return mains can be used to promote flow through the secondary circuit. A temperature sensor, C1, maintains a constant secondary flow temperature through the action of the motorised, three-port, mixing valve, R1.

In Figure 4.7(c) another case of an injection circuit is shown where the flow in the primary circuit is variable. In this case it is necessary to introduce a pressure regulating valve, R1, that maintains a constant pressure difference across the secondary flow and return connections by means of the differential pressure sensors P1a and P1b. A temperature sensor, C2, can then maintain a constant secondary flow temperature through the action of the two-port motorised valve, R2. Figure 4.8 shows some piping arrangements used for water chillers.

In Figure 4.8(a) a pair of equal capacity chillers is piped in series. The advantage is that one of the two chillers runs at full duty, and hence maximum efficiency, down to 50% load. The order of the leading and lagging machines would be varied so as to equalise wear and prolong system life. The full chilled-water flow rate passes through each chiller and this gives a high pump head with potentially high energy consumption for the primary pump. The

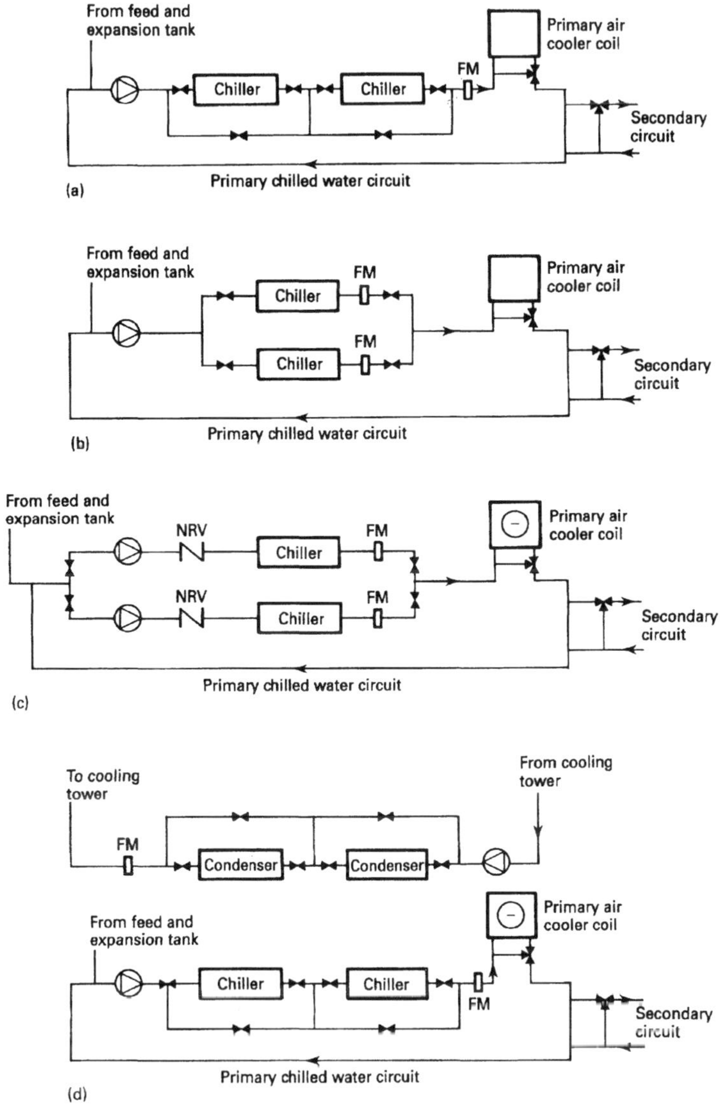

FM: Flow measuring device
NRV: Non-return valve

Figure 4.8 Some piping arrangements for water chillers. (a) Two chillers in series. (b) Two chillers in parallel with a common pump. (c) Two chillers in parallel with individual pumps. (d) A pair of chillers and condensers connected in series-counterflow.

water temperature drop through the pair of chillers is larger than would normally be the case for a single chiller or for two chillers in parallel and this has an effect on the selection of the cooler coils. A stand-by pump (not shown) would always be provided.

Figure 4.8(b) illustrates two chillers in parallel with a common pump (stand-by pump not shown). The primary pump head is smaller than with chillers in series because the water flow is in parallel. This is a popular arrangement for chillers using reciprocating compressors with capacity control by unloading pairs of cylinders, as the return chilled water temperature falls with a reduction in system load. It is usual to alternate the unloading of cylinder pairs between the compressors so that their cooling duties remain similar and do not become greatly out of phase. The stepped variation in chilled water flow temperature is greater than would be the case if a pair of reciprocating chillers were piped in series, as in Figure 4.8(a). If more than two chillers were to be in parallel this method would not be used, the following arrangement being preferred.

In Figure 4.8(c) the chillers are in parallel but each has its own pump and the stand-by pump (not shown) would be common to the pair. Pump duties and powers are smaller than with arrangement (b) and system energy consumption can be reduced as the refrigeration load falls, by off-loading pumps. This piping method offers more operational flexibility than does the use of a single pump.

A method sometimes adopted for centrifugal chillers is shown in Figure 4.8(d). Condenser cooling water and chilled water flow in opposite directions. Thus, the chilled water flow temperature might be 5°C, on the right-hand end of the diagram next to a cooling water flow temperature of 26°C, to give a difference of 21 K. At the left-hand end of the figure the cooling water return temperature could be 33°C, next to a chilled water return temperature of 12°C, implying a constant pressure difference between the condenser and the evaporator. It is claimed that this arrangement gives an efficient regime of operation.

Chillers using cylinder unloading as a method of capacity variation must never be controlled from chilled water flow temperature. Attempting this will cause the compressor to cycle on–off very frequently when operating on its last step of cooling capacity, with the ultimate certainty of motor burn-out. Chilled water only takes a second or two to flow through the chiller. It follows that system stability requires that there must be sufficient mass of water in the system to reduce to an acceptable level the possible frequency of machine starting. Examining the variables involved (see Jones, 1994) leads to the following equation

$$m = (215\, q_r)/(f\Delta t) \tag{4.6}$$

where

q_r	Average cooling capacity of the last step of refrigeration	kW
m	Mass of chilled water in the system	kg
Δt	Allowable variation in the chilled water return temperature	K

The mass of chilled water in the system, m, is the sum of the water contents of all the relevant piping (see Table 4.1), water chillers, cooler coils, storage vessels, etc. An approximate alternative to Equation (4.6) for estimating the necessary storage, in litres, is given by $30V/n$, where V is the chilled water pump duty in l s^{-1} and n is the permitted number of starts per hour. This approximation is usually an over-estimate of the storage needed.

The allowable number of starts per hour are:

Hermetic centrifugal compressors:	2
Hermetic reciprocating compressors:	4–6
Open reciprocating compressors:	8–10

Table 4.1 Water content of piping at 10°C

Nominal diameter (mm)	Internal diameter medium grade black steel (BS 1387) (mm)	Mass of water per metre of pipe run (kg m^{-1})	Internal diameter heavy grade black steel (BS 1387) (mm)	Mass of water per metre of pipe run (kg m^{-1})
10	12.4	0.121	11.3	0.100
15	16.1	0.203	14.9	0.174
20	21.6	0.366	20.4	0.327
25	27.3	0.585	25.7	0.519
32	36.0	1.018	34.4	0.929
40	41.9	1.378	40.3	1.275
50	53.0	2.206	51.3	2.066
65	68.7	3.706	67.0	3.525
80	80.7	5.113	79.1	4.913
90	93.15	6.813	91.55	6.581
100	105.1	8.673	103.5	8.411
125	129.95	13.259	128.85	13.036
150	155.4	18.961	154.3	18.694

4.3 Centrifugal pumps

When turning in the normal way, the backward-curved vanes on the impeller rotate the water and impart kinetic energy to it, directing the liquid outwards from the suction eye, over the vanes to the casing, where it flows away at high pressure. The kinetic energy of the water leaving the impeller, corresponding to its velocity head, is converted into potential energy, corresponding to its static head, either by a volute casing or by a diffuser casing. The former is the more usual design for pumps used in building services and with it most of the energy conversion takes place as the volute finally expands to the pump outlet.

The total pump head, H, or pressure, p_t is analogous to fan total pressure and is defined by

$$H = (H_d + v_d^2/2g) - (H_s + v_s^2/2g) \tag{4.7}$$

where H_d and H_s are the static heads at pump discharge and pump suction, respectively, v_d and v_s being the corresponding velocities of water flow. Since $v^2/2g$ is velocity head, Equation (4.7) states that the total pump head is the difference between the total heads at pump discharge and suction, in accordance with Bernoulli's theorem. Total pump head is the total energy per unit weight of water flowing and, in the same way, the total pump pressure (see Equation (4.3)) is the total energy per unit volume of water flow. It follows that the pump power imparted to the water by the impeller is defined by

$$W_p = \dot{m}gH = \dot{V}p_t \tag{4.8}$$

in which W_p is the pump power, $\dot{m}$ is the mass flow rate of water and $\dot{V}$ is the volumetric flow rate. It follows also, from Equation (4.3), that the total pump pressure can be expressed as

$$p_t = (p_{sd} + 0.5\rho v_d^2) - (p_{ss} + 0.5\rho v_d^2) \tag{4.9}$$

where p_{sd} and p_{ss} are the static pressures at discharge and suction, respectively.

The static pressure datum for a system is established by the connexion from the feed and expansion tank, if the circuit is closed, as in Figure 4.5a. In the case of an open circuit, such as in Figure 4.5c, the pond of the cooling tower is the feed and expansion tank and the static water level in it establishes the system datum pressure. If the feed and expansion tank has a water level at a height h_o above the centre-line of the impeller it imposes a static head of h_o, or a static pressure of $p_o = \rho g h_o$, on a closed circuit at the point of connexion, whether the pump runs or not. If the connexion is made at pump suction, $p_o = p_{ss}$ and this is the lowest static pressure in the system. When the connexion is at the pump discharge, $p_o = p_{sd}$ and this is the highest static pressure in the system. It is generally desirable to make the connexion at the pump suction in chilled water circuits so that there is no risk of cavitation (see Section 4.8) if a large pressure drop should occur across an item of plant elsewhere in the system.

EXAMPLE 4.3

(a) A centrifugal pump delivers 7.6 l s^{-1} through a closed pipe circuit and has suction and discharge connexions of 80 and 65 mm, respectively. The water level in the feed and expansion tank is 15 m above the centre-line of the impeller. Pressure gauges are fitted at suction and discharge and the connexion from the feed and expansion tank is at virtually the same position as the suction gauge. When the pump runs normally the discharge gauge indicates 250 kPa. Taking the density of water to be 1000 kg m^{-3}, determine the total pump pressure and power.
(b) Calculate the static pressure at pump suction if the connexion from the feed and expansion tank is at pump discharge.

Answer

(a) $V = 7.6 \text{ l s}^{-1} = 0.0076 \text{ m}^3 \text{ s}^{-1}$

The internal diameters of 80 mm and 65 mm medium grade steel tube are 0.0807 and 0.0687 m, respectively. Hence, the area of the suction inlet is 0.005115 m^2 and that of the outlet is 0.003707 m^2. Therefore, $v_s = 1.486 \text{ m s}^{-1}$ and $v_d = 2.045 \text{ m s}^{-1}$; $h_o = 15$ m, therefore $p_o = (1000 \times 9.81 \times 15)/1000 = 147.15$ kPa. Hence, $p_{ss} = 147.15$ kPa and $p_{sd} = 250$ kPa, and from Equation (4.9)

$$p_t = (250 + [(0.5 \times 1000 \times 2.045^2)/1000]) - (147.15 + [(0.5 \times 1000 \times 1.486^2)/1000])$$
$$= 252.09 - 148.25 = 103.84 \text{ kPa}$$

From Equation (4.8)

$$W_p = 0.0076 \times 103.84 = 0.789 \text{ kW}$$

(b) The same increases of static and total pressure occur across the pump but now $p_o = p_{sd} = 147.15$ kPa. Therefore

$$p_{ss} = 147.15 - (250 - 147.15) = 44.3 \text{ kPa}$$

It is to be noted that the influence of velocity pressure on the total pump pressure is small. It is often ignored and the difference in the readings of the gauges at discharge and suction taken as the pump pressure. For the above example, the total pump pressure would then have been taken as 250 – 147.15 = 102.85 kPa. It is also to be noted that if the gauges are at different levels different readings are obtained and must be accounted for.

EXAMPLE 4.4

The static pressure indicated by a gauge on the inlet side of a chiller is 100 kPa. The gauge on the outlet is mounted 1 m higher than that at the inlet and indicates 50 kPa. What is the drop of static pressure across the chiller?

Answer

If the gauge at the outlet were lowered by 1 m, to bring it to the same level as the gauge at the inlet, the increase in position head imposed by the water level in the feed and expansion tank would be 1 m, corresponding to 9.81 kPa. Hence the pressure drop across the chiller is really 100 – 59.81 = 40.19 kPa.

The mechanical efficiency, η, of a pump is the ratio of the rate of energy input to the water (Equation (4.8)) to the power applied to the impeller shaft, W_s. Thus

$$\eta = 100\ W_p/W_s = 100\ Vp_t/W_s \tag{4.10}$$

Since $p_t = \rho g h$, it follows that W_p is proportional to the density of the fluid and this may be significant when chilled brines or glycols are handled.

The difference $W_s - W_p$ arises from bearing losses, skin friction and turbulence within the pump, losses accompanying expansions and contractions and the energy losses incurred by the formation of eddies, all of which constitute losses of pump pressure between the places where this is measured at suction and discharge. There are also losses of pump capacity because of leakage between the impeller and its casing through the clearance spaces, and because of seepage through the gland.

Theoretical considerations of the relative velocities at entry to and exit from the impeller vanes show that a linear relationship exists for the static pressure developed and the volumetric flow rate. Figure 4.9 shows how the losses mentioned above combine to give a real curve, like an inverted parabola, very similar to the characteristic curve build-up of a centrifugal refrigeration compressor. Although the actual shape of the curve is established by test results for a constant pump speed (n) and impeller diameter (d), its position on the p–V coordinate system, the pressure developed and the power absorbed are related by six pump affinity laws.

$$V_2 = V_1\ (n_2/n_1) \tag{4.11}$$
$$p_{t2} = p_{t1}\ (n_2/n_1)^2 \tag{4.12}$$
$$W_{p2} = W_{p1}\ (n_2/n_1)^3 \tag{4.13}$$
$$V_2 = V_1\ (d_2/d_1) \tag{4.14}$$
$$p_{t2} = p_{t1}\ (d_2/d_1)^2 \tag{4.15}$$
$$W_{p2} = W_{p1}\ (d_2/d_1)^3 \tag{4.16}$$

The third law is based on the assumption that the efficiency–volume curve is a fixed shape, for a given pump and impeller, independent of speed. It follows that the power absorbed at the pump impeller can also be plotted against the volumetric flow rate (Figure 4.10).

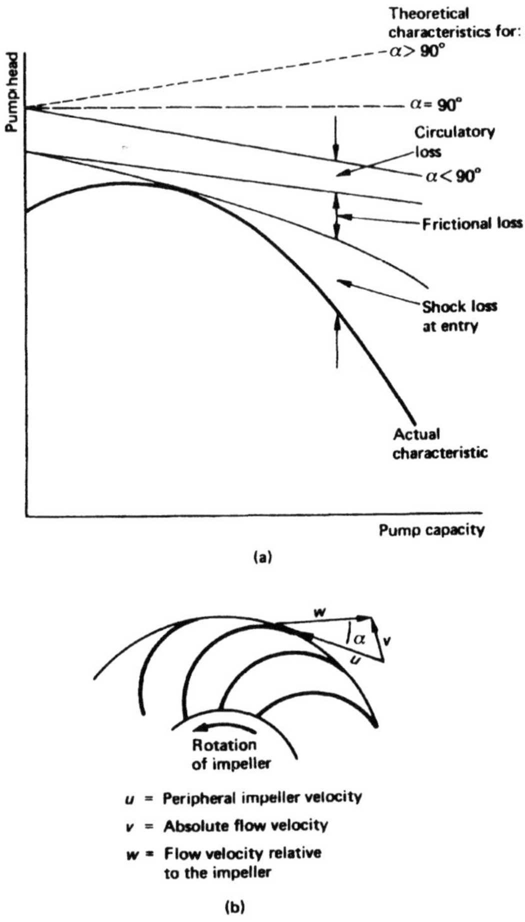

Figure 4.9 Build-up of a pump characteristic. (a) Head-capacity characteristic. (b) Discharge velocity triangle.

As with fans in duct systems, the actual duty of a pump-piping combination can only be determined from the intersection of the pump and system characteristics, on the same p–V coordinates. A straightforward interpretation of Equation (4.5) assumes that the pipe friction factor, f, is a constant and that the pressure loss is proportional to the square of the velocity of water flow and hence to the square of the volumetric flow rate. This square law assumption fits the first two pump laws (Equations (4.11) and (4.12)).

EXAMPLE 4.5

A piping circuit (Figure 4.11) carries 1 kg s^{-1} of chilled water and is of 32 mm nominal bore, medium-grade steel with a pressure drop rate of 380 Pa m^{-1} (Fig. 4.1) and an equivalent length of 1.6 m. Calculate the pressure loss in the system, assuming that pipe lengths without

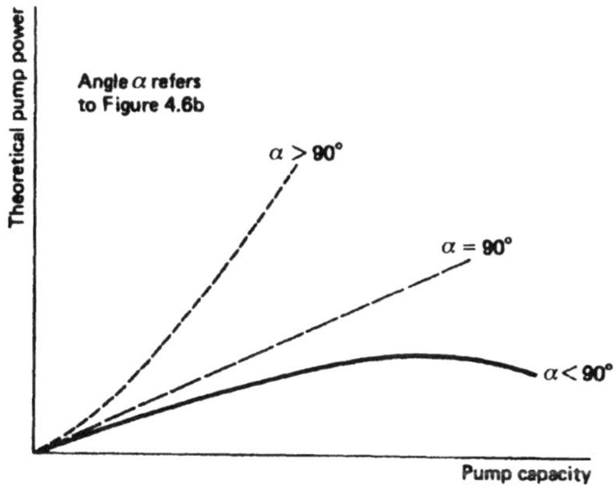

Figure 4.10 Power-capacity of a pump. See Figure 4.9 for the meaning of α.

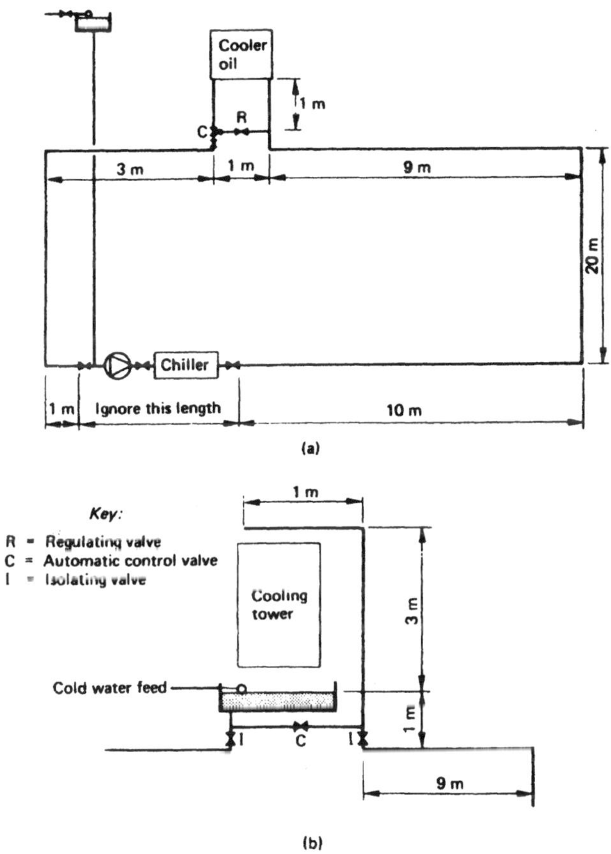

Figure 4.11 Closed and open piping circuits for Examples 4.5 and 4.6. (a) Closed chilled water circuit. (b) Open cooling water circuit.

dimensions are negligible. The authority of the control valve is 0.3 and the losses of head through the cooler coil and the chiller are 1 and 4 m, respectively. Gate valves are used for isolating purposes and malleable cast-iron 90° elbows are fitted where shown.

Answer

From Table C4.36 in the CIBSE (1977) guide, $k = 0.7$ for an elbow and 0.2 for a gate valve. With a divergent tee having equal diameters on all branches, $k = 0.2$ for straight-through flow. Adding dimensions given in Figure 4.11(a) the straight pipe length is 10 + 20 + 9 + 1 + 1 + 3 + 20 + 1 = 65 m. The total equivalent length of pipe can now be calculated:

Straight pipe:	= 65
Elbows: $6 \times 0.7 \times 1.6$	= 6.72
Tee (branch to the coil): 0.2×1.6	= 0.32
Tee (connexion from the F & E tank): 0.2×1.6	= 0.32
Gate valves: $5 \times 0.2 \times 1.6$	= 1.6
Total equivalent length of straight pipe:	= 73.96 m

The authority of the control valve refers to the part of the circuit where variable flow occurs, namely, through the coil and local pipe connexions, the loss through the bypass being made equal to this by adjusting the regulating valve, R, during commissioning. Hence, the pressure loss through the control valve is:

$$[0.3/(1 - 0.3)][(2 \times 380)/1000 \text{ (for straight pipe)}$$
$$+ (0.32 \times 380)/1000 \text{ (for the tee)}$$
$$+ 9.81 \text{ (for the cooler coil, Equation (4.3))}] = 4.58 \text{ kPa}$$

The total pressure loss through the system may now be calculated:

Equivalent straight pipe: $(73.96 \times 380)/1000$	= 28.10
Chiller: $(1000 \times 9.81 \times 4)/1000$	= 39.24
Cooler coil: $(1000 \times 9.81 \times 1)/1000$	= 9.81
Control valve:	= 4.58
Total pressure loss:	= 81.73 kPa

Assuming a coefficient of performance of 4, the condenser associated with the chiller, above, would probably need a cooling water flow rate of 1.25 kg s^{-1}. The rate of pressure drop for this flow in a 32 mm pipe exceeds the suggested limit of 500 Pa m^{-1} although not by much. Because of the fouling that inevitably occurs in open cooling water systems it might be wise to choose one pipe size larger, i.e. 40 mm. From Figure 4.1 it can be determined that this size has a loss of about 250 Pa m^{-1}, with a velocity of about 0.91 m s^{-1}, using an internal diameter of 41.9 mm from CIBSE Table C4.4. From CIBSE Table C4.12, the equivalent length is 1.9 m.

EXAMPLE 4.6

Using the modification to Figure 4.11(a) shown by Figure 4.11(b), where an open cooling tower has been substituted for the cooler coil and assuming that a condenser with the same pressure drop replaces the chiller, calculate the total system loss for a flow rate of 1.25 kg s^{-1} in 40 mm nominal bore tube. There is no feed and expansion tank because the tower pond provides for this.

Answer

Adding the modified piping dimensions in Figures 4.11(a) and (b), gives 10 + 20 + 9 + 1 + 3 + 1 + 3 + 20 + 1 = 68 m, beginning at the outlet from the condenser. The total equivalent length of pipe can now be determined:

Straight pipe:	= 68
Elbows: $7 \times 0.7 \times 1.9$	= 9.31
Tees (past the tower bypass): $2 \times 0.2 \times 1.9$	= 0.76
Gate valves: $5 \times 0.2 \times 1.9$	= 1.9
Contraction leaving the cooling tower pond: 0.5×1.9	= 0.95
Total equivalent length of straight pipe:	= 80.92 m

Note that for the butterfly valve in the bypass to exercise effective control no water must flow over the tower when it is fully open. It follows that the loss of head across the butterfly valve, fully open, is 3 m, i.e. the static lift. However, in this case the bypass does not form part of the index circuit. The total system loss is, therefore

Equivalent straight pipe: $(80.92 \times 250)/1000$	= 20.23
Condenser: $(1000 \times 9.81 \times 4)/1000$	= 39.24
Pressure loss corresponding to the static lift: $(1000 \times 9.81 \times 3)/1000$	= 29.43
Total pressure loss:	= 88.9 kPa

When the pump stops, all the water in the system above the static water level drains back under gravity into the pond of the cooling tower.

EXAMPLE 4.7

For the case in Example 4.6 determine the minimum necessary distance between the working water level and the overflow from the pond, assuming its plan area is 0.5 m^2.

Answer

From Figure 4.11(b), there is 3 m of 40 mm nominal bore tube above the water level. Taking the internal diameter of the pipe as 41.9 mm its water content is calculated as 0.0042 m^3. Hence, the bottom of the overflow must be at least 9 mm above the static water level, if wastage of water is to be avoided. This may not seem much but, if there is very little piping above the water level there will be very little drain back. Unfortunately, most cooling towers have shallow ponds that cannot accommodate much water and hence it is essential that the piping above the static water level is kept to a minimum, if wastage of water is to be avoided when the pump shuts down.

4.4 The interaction of pump and system characteristics

As with fan and duct systems, the only way of establishing the actual duty of a pump installed in a piping system is to plot the characteristics of both, using common pressure–volume coordinates. The characteristic curve for a pump having a given impeller and rotational speed is obtainable from the manufacturers and the system curve is determined by assuming a square law.

EXAMPLE 4.8

(a) Assuming that 1 l s^{-1} is numerically equal to 1 kg s^{-1}, plot the system pressure–volume characteristic curve and determine the actual duty achieved if a pump with the following performance for a speed of 24.17 rev s^{-1} is fitted in the closed circuit used in Example 4.5.

l s^{-1}	0.2	0.4	0.6	0.8	1.0	1.2	1.4	1.6
kPa	82	83	83.5	84	83.7	82.2	80.7	76.8

(b) Plot the system characteristic curve for the open circuit used in Example 4.6 and determine the duty obtained if the same pump is used.

Answer

(a) Assuming a square law relationship between pressure loss and water flow rate, the system has the following characteristic:

l s^{-1}	0.2	0.4	0.6	0.8	1.0	1.1	1.2	1.3
kPa	3.3	13.1	29.4	52.3	81.73	98.9	117.7	138.1

This is plotted in Figure 4.12, with the data for the pump. An intersection occurs at point A and it can be seen that the duty obtained is about that wanted, 1.0 l s^{-1} at 81.8 kPa.

(b) With the open system, enough pressure must be developed to lift the water by 3 m before any flow occurs at all, although thereafter a parabolic system law may be assumed, as before. Example 4.6 shows that the total system loss is 88.9 kPa but that of this, 29.43 kPa is ascribed to the static lift. The loss through the rest of the system for a flow rate of 1.25 l s^{-1} is therefore 59.47 kPa and it is this figure that must be used with the square law assumption. The following may be calculated:

l s^{-1}	0	0.2	0.4	0.6	0.8	1.0	1.25	1.4
Loss in pipe, etc. (kPa)	0	1.5	6.1	13.7	24.4	38.1	59.47	74.6
Static lift (kPa)	29.43	29.43	29.43	29.43	29.43	29.43	29.43	29.43
Total system loss (kPa)	29.43	30.93	35.53	43.13	53.83	67.53	88.9	104.03

Plotting this on Figure 4.12 shows that the intersection with the pump curve now occurs at point B and the duty is only about 1.18 l s^{-1}. This is not enough.

If the pump were belt-driven, it might be possible to speed it up, by changing the pulleys, in accordance with the pump law given by Equation (4.11)

Speed required $= 24.17 \times 1.25/1.18 = 25.6$ rev s^{-1}

The pump would then have the characteristic shown by the dashed line in Figure 4.12, the intersection would be at point C and the desired duty of 1.25 l s^{-1} obtained.

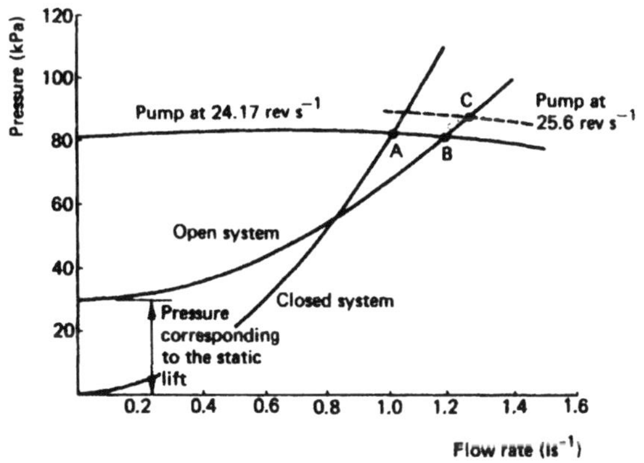

Figure 4.12 Closed and open pump circuits for Example 4.8.

If the pump were driven directly from its motor, changing speed might not be a practical proposition. The solution then is to use a larger impeller, within the same volute. Equation (4.14) shows that the flow rate is directly proportional to the impeller diameter, at constant pump speed and fluid density, and so this could be used to size a new impeller. If it is not possible to do this, because of the size of the pump casing, a larger pump must be chosen, its impeller diameter being selected to give the duty wanted.

4.5 Variable flow systems

When a constant water flow rate can be ensured through boilers and chillers, by means of primary and secondary circuits, two-port throttling valves may be adopted to regulate the capacities of heater batteries and cooler coils in the secondary system. In fact, with larger installations it is highly desirable that variable flow be used so that full advantage may be taken of load diversity, the sizes of the mains and pumps being reduced and running costs minimised. Although such two-port valves will be sized for a pressure drop when fully open that gives an authority of between 0.2 and 0.4, pressures throughout the system will change as the valves modulate independently to match load changes and, at low flow rates, the pressure drop across the few valves remaining open will be exceedingly large.

EXAMPLE 4.9

Figure 4.13 shows a simplified secondary pipe circuit feeding five cooler coils from a primary main in which the pressure is constant at 1 kPa, there being no significant drop between X and Y, nor between Y and Z. Pipes for which a dimension is not shown are of negligible length and the loss through fittings is ignored. The design water flow rate is $0.2\,l\,s^{-1}$ through each cooler coil, with a corresponding pressure drop of 0.5 kPa. An automatic, motorised, modulating valve (C1, C2, etc.) is fitted in the flow line to each coil. (a) Calculate the total pump pressure and the pressures at pump suction and discharge, assuming that the index run is through control valve C5, which is sized for an authority of 0.3.

(b) If all control valves, except C5, are fully closed, calculate the pressure that this remaining open valve must absorb if it is to pass its design flow rate. For simplicity it is assumed that the pump used has a flat, horizontal characteristic.

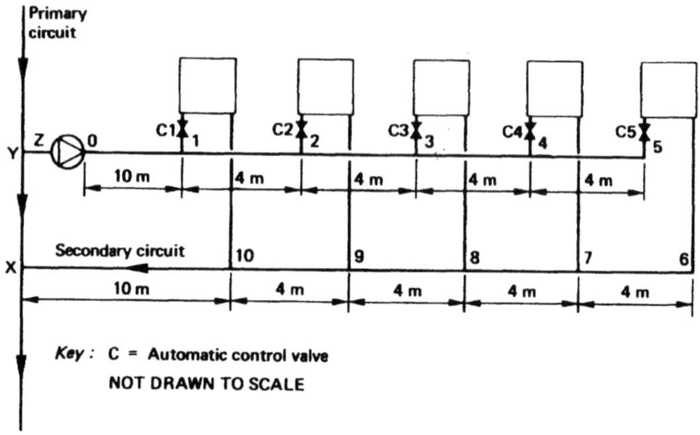

Figure 4.13 Piping circuits for Examples 4.9 and 4.10.

Answer

Item	Flow rate ($l\ s^{-1}$)	Size (mm)	Pressure drop rate ($k\ Pa\ m^{-1}$)	Length (m)	Pressure drop (kPa)
0–1	1.0	32	0.380	10	3.80
1–2	0.8	32	0.250	4	1.00
2–3	0.6	32	0.150	4	0.60
3–4	0.4	25	0.270	4	1.08
4–5	0.2	20	0.250	4	1.00
0–5					7.48
6–X					7.48
Cooler coil					0.50
Subtotal					15.46

(a) For an authority of 0.3

0.3 = (valve loss)/(valve loss + loss in rest of system)

Hence, valve loss = $(0.3 \times 15.46)/(1.0–0.3) = 6.63$ kPa, and the total system loss = 6.63 + 15.46 = 22.09 kPa.

If a pump is selected to deliver 1 $l\ s^{-1}$ with a total pressure of 22.09 kPa then, when installed where shown in Figure 4.12, the pressure at pump suction, Z, will be 1.0 kPa and that at pump discharge, 0, will be 23.09 kPa.

(b) When only 0.2 s^{-1} is flowing through the circuit to the index unit, the pressure loss through the piping can be assessed by assuming a square law as a reasonable approximation for Equations (4.2) and (4.5), without reference to Figure 4.1. For example, if the pressure loss in section 0–1 is 0.380 kPa m^{-1} when 1.01 s^{-1} is flowing then it is 0.380 $\times (0.2/1/0)^2 = 0.0152$ kPa m^{-1} when 0.2 $l\ s^{-1}$ is flowing. Hence the presures at various points, numbered throughout the system, may be determined

Item	Flow rate ($l\ s^{-1}$)	Pressure drop rate ($kPa\ m^{-1}$)	Length (m)	Pressure drop (kPa)	Pressure at a point in the system (kPa)
0					23.090
0–1	0.2	0.0152	10	0.1552	
1					22.938
1–2	0.2	0.0156	4	0.062	
2					22.876
2–3	0.2	0.0167	4	0.067	
3					22.809
3–4	0.2	0.0675	4	0.270	
4					22.539
4–5	0.2	0.2500	4	1.000	
5					21.539

Flow conditions X–7 are similar to those in 0–4, giving a pressure drop of 23.09 – 22.539 = 0.551 kPa. Hence, since the pressure at X is 1 kPa, it must be 1.551 kPa at 7. The loss from 7–6 is 1 kPa, the same as for 4–5, and the loss through the coil is 0.5 kPa, for 0.2 $l\ s^{-1}$ flowing. The pressure on the coil side of the index valve, C5, is therefore 1.551 + 1 + 0.5 = 3.051 kPa. So the index valve must absorb 21.539 – 3.051 = 18.488 kPa and will be almost closed if it is to pass only 0.2 $l\ s^{-1}$. Apart from the stress imposed on the valve the amount of stem movement available for full control of the water flow is very much reduced and control will tend to degenerate from proportional to two-position, probably with noticeable water noise and mechanical juddering. (In fact, most control valves can absorb up to about 50 kPa.)

Figure 4.14 shows how the pressure to be absorbed by the index valve will increase from 6.63 kPa under design full load conditions to 18.488 kPa at the partial duty considered above.

It is worth considering here the meaning of the system characteristic curve. Any piece of pipe, or any fitting or item of plant, has a system characteristic that has nothing to do with the pump which may be connected to it. The system characteristic shows the relation between the flow rate through the pipe, or other item, and the pressure loss, in accord with the equations mentioned, especially Equations (4.4) and (4.5). However, for simplicity, it is common to adopt a square law, pressure loss proportional to flow rate squared, without introducing much error, on the assumption that turbulent flow is occurring. As soon as any change in the system is made, for example by partially closing a control valve, it becomes an entirely different system and has quite a different characteristic. Thus the curve (a) in Figure 4.14 depicts the flow–pressure loss relation for the system as designed and balanced, by regulating valves at each cooler coil in addition to the control valves shown in Figure 4.13, to pass a total of 1.0 $l\ s^{-1}$, 0.2 $l\ s^{-1}$ flowing through each coil. If the index valve is removed and replaced by a short piece of frictionless pipe, its new system characteristic is shown by curve (b). Curve (c) shows the system when all the valves are shut, except at the index coil, which continues to pass 0.2 $l\ s^{-1}$, under its own, local, thermostatic control. Similarly, curve (d) is for the system passing 0.2 $l\ s^{-1}$ to the index coil only, but assuming that the index control valve has been removed. To produce these curves the starting points for the application of a square law were: (a) 1.0 $l\ s^{-1}$, at 22.09 kPa, (b) 1.0 $l\ s^{-1}$ at 15.46 kPa, (c) 0.2 $l\ s^{-1}$ at 22.09 kPa and (d) 0.2 $l\ s^{-1}$ at 3.602 kPa (= 22.09 – 18.488).

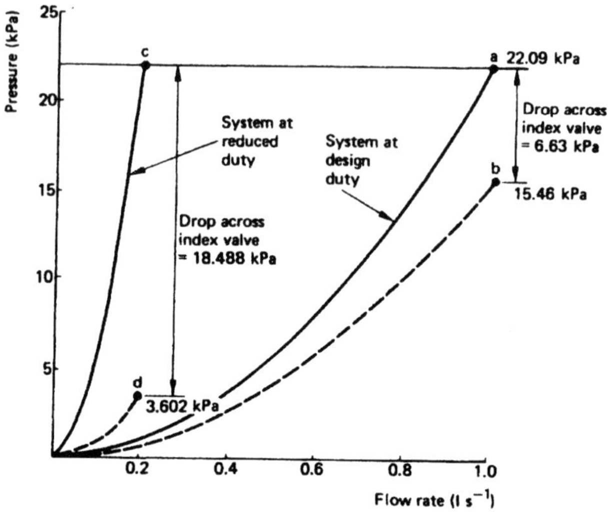

Figure 4.14 System curves with variable flow.

EXAMPLE 4.10

Plot pressure against position in the system for curves (a) and (c) in Figure 4.14 and also for the case when all control valves are partly closed under thermostatic control to pass $0.1\ \mathrm{l\ s^{-1}}$ each.

Answer

Pressure through the system when it is passing $0.2\ \mathrm{l\ s^{-1}}$ has been tabulated in answering Example 4.9(b) and this can be plotted to show the pressure distribution in the system described by curve (c). The results from example 4.9(a) can be used to establish pressure throughout the other two systems.

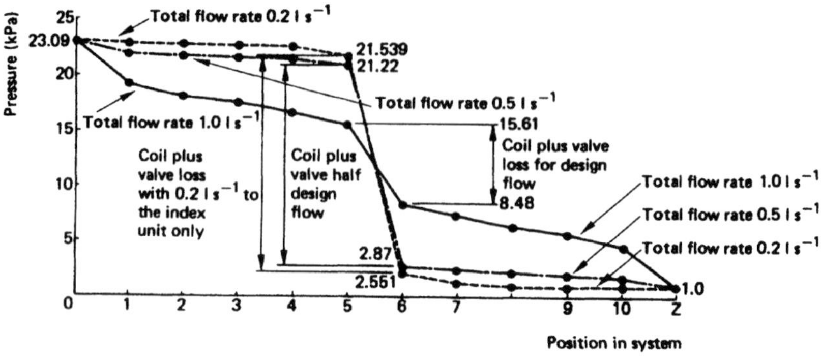

Figure 4.15 System pressure distribution with variable flow.

Item	Flow rate ($l\ s^{-1}$)	Pressure drop (kPa)	Pressure (kPa)	Flow rate ($l\ s^{-1}$)	Pressure drop (kPa)	Pressure (kPa)
0			23.09			23.09
0–1	1.0	3.8		0.5	0.95	
1			19.29			22.14
1–2	0.8	1.0		0.4	0.25	
2			18.29			21.89
2–3	0.6	0.6		0.3	0.15	
3			17.69			21.74
3–4	0.4	1.08		0.2	0.27	
4			16.61			21.47
4–5	0.2	1.00		0.1	0.25	
5			15.61			21.22
Z			1.00			1.00
Z–10	1.0	3.8		0.5	0.95	
10			4.8			1.95
10–9	0.8	1.0		0.4	0.25	
9			5.8			2.20
9–8	0.6	0.6		0.3	0.15	
8			6.4			2.35
8–7	0.4	1.08		0.2	0.27	
7			7.48			2.62
7–6	0.2	1.00		0.1	0.25	
6			8.48			2.87

These results are plotted in Figure 4.15.

Pressures in a variable water flow system vary considerably as the flow rate changes. It is to be noted that, in Example 4.9, the pressure at pump discharge is constant because a flat, straight-line, pump characteristic was assumed. In a real case there would be some variation as the point of intersection with the system curve shifted up or down the curved pump characteristic.

Figure 4.15 shows that, whilst the pressure absorbed across the index valve is 6.63 kPa (= 15.61 – 8.48 – 0.5) for design duty, it rises to 18.488 kPa (= 21.539 – 2.551 – 0.5) as the flow falls to $0.2\ l\ s^{-1}$. To ease this situation there are three possibilities: reducing pump speed, bypassing water across the coil circuit, throttling flow at pump discharge. The first solution is the best and is increasingly used, with inverters, where the penalty of the extra capital cost of the speed control device is more than balanced by the saving in running cost over the system life (see Section 8.4). As a more expensive alternative to motor speed variation, a variable speed hydraulic coupling is possible. According to Etheridge (1976) such variable couplings have a maximum drive efficiency of about 95% and this reduces proportionately as the transmitted speed falls. Using this property with the third pump law (Equation (4.13)), it can be verified that the power absorbed by the drive is proportional to the square of the speed ratio and so

$$W_{p2} = (W_{p1}/0.95) \times (n_2/n_1)^2 \quad (4.17)$$

Pressure changes in the variable flow piping circuit are used to regulate the flow control devices mentioned. The best approach seems to be to choose a pair of places, one in the flow and one in the return, about two-thirds of the way from the pump towards the index unit, and measure the differential pressure. For the circuit shown in Figures 4.13, 4.14 and 4.16, one pressure-sensing probe might be located at position 3 in the flow and the other at 8, in the return. It is good sense to provide places for pressure-sensing probes in about three different pairs of locations, between two-thirds and three-quarters of the way to the index unit, to allow some repositioning on site during commissioning, in the event that the first choice is not the best. Under design flow conditions the sensor at 3 would measure 17.69 kPa and that at 8 a value of 6.4 kPa, giving a differential of 11.29 kPa. If nothing were done, the pressure at 3 would rise to 22.809 kPa and that at 8 would fall to 1.281 kPa, providing a differential of 21.518 kPa. Supposing that a reasonable proportional band for the pressure sensors were 5 kPa, then the differential could rise to 11.29 + 5 = 16.29 kPa. This variation in pressure difference is needed to tell the pump to reduce speed or to open a bypass valve across the coil circuit, i.e. between section 0–1 in the flow and X–10 in the return). Most automatic control valves can cope with pressure drops across them of up to 50 kPa.

If a throttling valve is used in the main at pump discharge, not as much power is saved at reduced flow as with pump speed reduction. Firstly, the overall design pump pressure must be greater so as to give the main throttling valve a reasonable authority and, secondly, the product of the pressure drop across this valve and its volumetric flow rate (Equation (4.8)) represents a loss of power.

4.6 Pump types

The centrifugal pump with a volute casing is the type most commonly used for building services and, among the many styles available, one method of classification is according to the drive arrangement:

(1) Integral canned rotor – the impeller is fixed to the shaft of the rotor, which revolves in a rotor can filled with water. The stator and the electrical connexions are outside the rotor can and, since the pump shaft does not emerge from the casing, there is no need for a shaft seal to limit water leakage. Such pumps are common in domestic heating installations and, if they are light enough in weight, may be fixed directly in the pipeline without additional support. Otherwise, they are floor-mounted. Such pumps must never be used in chilled water lines because condensation then occurs on the electrical connexions and causes continual difficulties.
(2) Direct-coupled – the motor shaft is coaxial with and attached to the impeller shaft through a flexible coupling, both motor and pump being mounted on a common base-plate. Difficulties in shaft alignment are sometimes experienced and a shaft seal is necessary to limit leakage from the pump casing.
(3) Close-coupled – the motor and pump are separate but the impeller is mounted on an extension of the motor shaft. There is consequently no possibility of misalignment but motor noise can be transmitted directly into the piping system. Although the pump has no bearings, as those of the motor serve for both, a shaft seal is needed.
(4) Belt-driven – vee-belts and pulleys provide the drive connexion between the pump and its motor, the whole assembly being fixed on a common base-plate. A shaft seal is necessary.

Modifying the pump performance is done by changing the impeller in cases (2) and (3), by altering the pulleys in case (4) and by electrical methods for case (1).

That part of the pump where the impeller shaft passes through the casing is formed into what is called a stuffing box, containing either a packed gland or a mechanical seal. The former contains replaceable, asbestos string or the like, lubricated with graphite grease or something similar. A screwed top to the gland may be tightened to control seepage but some leakage is desirable to give shaft lubrication and cooling at the gland. The packing is replaced when worn. If there is a negative pressure at the pump suction air can be drawn into the system and to prevent this occurring a lantern ring may be fitted to divert a small amount of water from the pump discharge to the gland. A mechanical seal comprises a spring-loaded labrynth of carbon–ceramic or carbon–stainless steel faces, through which there is virtually no leakage but which may offer some difficulty in replacement should this be necessary. Both types of seal can be used for water temperatures up to 100°C but beyond this special arrangements for materials and cooling are needed. For mechanical seals in particular, the working pressure and temperature are critical to performance and cooling water from an external source must be supplied when they operate above 100°C. It is essential that the piping system is properly flushed out before a pump with a mechanical seal is run and all seals must be provided with flushing connexions to keep the seal interfaces free from contamination. Mechanical seals should only be fitted to pumps that are of suitable design, namely, those with short, stiff shafts having adequate bearings. Mechanical seals must not be regarded as bearings. Seal faces will be damaged if the pump runs without water, even as a momentary dry run, when the electrician tests the direction of rotation, is sufficient to cause serious damage and render the seal useless.

Other aspects of classification are according to construction: horizontal or vertical impeller shafts, split casings which facilitate impeller changes, cleaning and maintenance, single or double suction impellers, single or multi-stage, self-priming, etc.

4.7 Margins and pump duty

Few systems suffer because the pump is slightly oversized since it is nearly always possible to reduce capacity by changing impellers or speeds, but many are in difficulties if the system resistance has been underestimated. If a serious underestimation has been made and no margin added, it may not always be possible to get the desired duty by altering the impeller or the speed and an entirely new and larger pump may be needed. If any margin is added to the flow rate, to cover a design uncertainty or for any other reason, then a corresponding allowance must be made to the system resistance, in accordance with Equation (4.5). Using the binomial approximation this means that a small increase in the flow rate must be accompanied by a doubled increase in the system resistance. Furthermore, there is the likelihood that the system will not be installed exactly as it was designed; almost certainly the actual installation will be more complicated and hence will offer a greater resistance to flow. It is, therefore, suggested that 5% be added to the design flow rate, that 10% be added to the calculated resistance, and that a further resistance addition, as estimated for the particular difficulty of the job in question, or 5%, whichever is the greater, be made to cover installation variations.

In general, a pump should be chosen so that the design flow rate occurs at a steeply falling part of the pressure–volume characteristic and at a near peak on the efficiency–volume curve. The virtue of this is that changes in system resistance will give small variations in the flow rate, which is most desirable with open systems where continuous fouling occurs.

Table 4.2 Some efficiencies for pumps used in building services

Flowrate ($l\ s^{-1}$)	2.2	2.8	3.3	3.9	4.4	5.0	5.6	6.1	6.7
Efficiency	59%	64%	68%	70%	71%	69%	66%	61%	53%

Reliable data on pump efficiencies are difficult to find from published pump catalogues. However, the information given in Table 4.2, based on one manufacturer's product, seems to be reasonable.

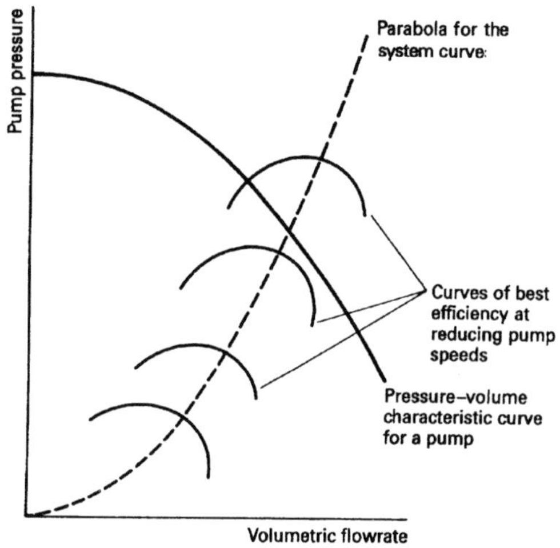

Figure 4.16 Maximum pump efficiency loops move down a parabolic path as pump speed is reduced.

The pump affinity laws (Equations (4.11)–(4.16)) are similar to the fan laws and a property shared by pumps and fans is that the total efficiency remains constant, for a given pump/fan and system, as speed is changed. Since the system law relating pressure loss and flow rate is assumed to be a square law, it is evident that an efficiency loop on the pressure–volume characteristic for a pump moves down a parabolic path as pump speed is reduced. Figure 4.16 shows this.

4.8 Dissolved gases and cavitation

The solubility of a gas in water is given by Henry's law

$$c = Hp \tag{4.18}$$

where c is the concentration of the gas in water at a given temperature, H is a constant and p is the partial pressure of the gas. Hence the mass absorbed is directly proportional to the partial pressure of the gas and, since its volume is also proportional to this, it follows that water at a given temperature will absorb a constant volume of a given gas. Table 4.3 gives the volumetric solubilities of some common gases in water at various temperatures.

It must be emphasised that the relevant pressure in Equation (4.18) is the partial pressure of the gas. Consequently when the partial pressure of the air above water at 100°C is 101.325 kPa, the total pressure over the water will be 202.65 kPa because water at 100°C in equilibrium with its gaseous phase also exerts a vapour pressure of 101.325 kPa. Only under such circumstances will 0.012 volumes of air be dissolved in one volume of water at 100°C. The corollary is that, since water at 100°C and under a total pressure of 101.325 kPa exerts this as its saturated vapour pressure, it can contain no dissolved air at all.

EXAMPLE 4.11

Determine the volume and mass of air that could be dissolved in 1 m^3 of water at 20°C if the ambient atmosphere is at 20°C dry-bulb, 50% saturation and 101.325 total pressure.

Answer

Table 4.3 shows that 0.02 volumes are dissolved. The partial pressure of the air in the ambient atmosphere is 101.325 – 1.182 = 100.143 kPa (from CIBSE tables of psychrometric data). From the same source, taking the density of dry air at 101.325 kPa and 20°C as 1/0.8301 = 1.2047 kg m^{-3} the mass of dissolved air can be deduced as $(0.02 \times 1.2047 \times 100.143)/101.325$ = 0.0238 kg m^{-3}.

When water flows through a piping system air will tend to come out of solution as the pressure drops and as the temperature rises. The air pocket so formed restrict the area available for the flow of water and so increase the pressure drop rate, the net result being that less water is delivered. The release of air in this way is not sudden and it appears (Crocker and King, 1967) that the amount coming out of solution depends on the time the water flow in the pipe is subjected to the conditions that favour its production.

The restricting effects of air pockets are reduced as the velocity of flow increases. Higher velocities help to break up bubbles into smaller sizes and to scour out the pockets themselves. When water velocities are less than 0.2 m s^{-1} air bubbles tend to rise naturally out of water, to high points in the piping system, where there may be air bottles for collection and venting.

Table 4.3 Volumetric solubilities of gases at different temperatures according to Walker *et al.* (1967) and Crocker *et al.* (1967)

	Volume of gas absorbed per unit volume of water			
Gas	0°C	20°C	50°C	100°C
Air	0.032	0.02	0.0125	0.012
Nitrogen	0.026	0.017		0.0105
Oxygen	0.053	0.034	0.021	0.0185
Carbon dioxide	1.87	0.96	0.5	0.26
Hydrogen	0.023	0.02		0.018
Ammonia	1250.0	70.0		0
Chlorine	5.0	2.5		0
Hydrogen disulphide	5.0	2.8		0.87

During a process of air venting the pump should be off, from time to time, to allow bubbles to migrate to such air collection points. Although, in general, 0.3 m s^{-1} is taken as the lowest water velocity for a piping system to be self-purging, air is carried along with water, according to Miller (1978), when velocities exceed 0.6 m s^{-1} for pipes up to 50 mm. It is recommended by Denny *et al.* (1957) that above this the minimum velocity to make the system self-purging should be that corresponding to a pressure drop rate of 75 Pa m^{-1}. Thus for 150 mm pipe at 75 Pa m^{-1}, the minimum velocity should be about 1 m s^{-1}. Much higher minimum velocities are required to convey air along downward sloping piping, and as much as 3 m s^{-1} is suggested by Miller (1978) as necessary although such high speeds are usually to be avoided because of the risks of erosion of the pipe walls.

A distinction must be drawn between the amount of air dissolved in water and the amount that can come out of solution under changes of pressure and temperature, the latter being not easy to predict. It has been suggested by Crocker and King (1967) that as little as 10% of the quantity theoretically possible actually comes out of solution.

EXAMPLE 4.12

Cooling water at a rate of 1.25 l s^{-1} enters a condenser at 27°C and leaves at 32°C. If the initial pressure of the water is 99 kPa and the drop through the condenser is 39 kPa, determine the mass of dissolved air that could, theoretically, come out of solution.

Answer

Interpolating in Table 4.3 at temperatures of 27°C and 32°C, 0.0182 and 0.017 volumes of air, respectively, could be dissolved in one volume of water. If the air came out of solution at entry to the condenser, its pressure would be 99 kPa and, since dry air at 27°C and 101.325 kPa has a density of 1/0.85 = 1.176 kg m^{-3} (from CIBSE tables of psychrometric data), its density at 99 kPa would be (1.176 × 99)/101.325 = 1.149 kg m^{-3}. So the mass of air in solution at entry to the condenser could be 0.0182 × 1.149 = 0.0209 kg m^{-3}. At exit from the condenser, the air, if it came out of solution, would have a temperature of 32°C and a pressure of 99 – 39 = 60 kPa. The density of dry air at 32°C and 101.325 kPa is 1/0.8642 = 1.157 kg m^{-3}. Hence at 32°C and 60 kPa its density is (1.157 × 60)/101.325 = 0.685 kg m^{-3}. So the mass of air that could be dissolved in the water leaving the condenser is 0.017 × 0.685 = 0.116 kg m^{-3}. Hence, the theoretical mass of air that might be released as the water flows through the condenser is (0.0209 – 0.0116) × 0.00125 = 0.000011625 kg s^{-1}. If this air were liberated at 60 kPa, the pressure prevailing at the condenser outlet, it would correspond to a volumetric flow rate of (1000 × 0.000011625)/0.685 = 0.0170 l s^{-1}, i.e. the percentage of air in the water, by volume, would be (0.017 × 100)/1.25 = 1.4%. However, only one-tenth of this might actually be present.

With an open system using a cooling tower or air washer the air is thoroughly aerated, but in closed systems the water is not necessarily saturated with air. There are, however, other sources of free air bubbles, such as: improper venting during filling; air leakage into the system through pump glands, valve glands or poorly made joints, if these are in parts of the piping where the pressure is subatmospheric; the formation of a vortex at the outlet from a tank, particularly from the pond of a cooling tower.

Any permanent feature of a tank, not necessarily close to the outlet, may impart a rotational component to the water flowing past it; this input of energy is continuous and so the rotational movement is self-sustaining and causes the formation of a whirlpool in the water, above the outlet from the tank. Transient disturbances of the water surface produce eddies that are not self-sustaining and play no part in establishing the vortex at the outlet. The funnel of air at the centre of the whirlpool extends into the outlet pipe, its tail thinning and eventually breaking, to form bubbles that are entrained by the water and conveyed through the system. As much as 10% of the outlet flow can be air (Denny *et al.*, 1957) in extreme cases.

Discouraging the formation of a vortex at an outlet seems to be possible according to Denny *et al.* (1957), by taking one or more of the following steps: minimising the rotational flow leading to the outlet; using a larger area of outlet; increasing the depth of the water; locating the outlet near a vertical tank wall; intercepting the tail of the whirlpool by placing a baffle beneath the surface, near the outlet (Figure 4.17).

None of these steps necessarily prevents swirl being established in the outlet pipe. This can best be prevented, according to Denny *et al.* (1957), by inserting cruciform guide vanes in the pipe, some two or three diameters downstream from the outlet. The length of the vanes should be at least one outlet diameter. Although swirl can be removed in this way, a good deal of turbulence remains in the water and the presence of guide vanes alone cannot prevent a vortex forming in the tank.

To see some of these seemingly insignificantly small percentages in perspective, consider a pipe of radius R and length $2R$, containing n bubbles, each of radius r. The percentage by volume occupied by the bubbles is $200\ n\ r^3/3R^3$. For the case of a 50 mm tube, containing 1% of free air in the form of 5 mm diameter bubbles, $n = 3 \times 25^3/(200 \times 2.5^3) = 15$, but if the bubbles are only 0.5 mm in diameter, there are 15 000 of them.

There is very little published information concerning the effect of free air on the performance of centrifugal pumps. However, Stepanoff (1967) shows the results of some tests done by Siebrecht (1930), a version of these being shown in Figure 4.18. It seems that pump pressure, efficiency and power are all diminished by the presence of free air in the impeller. Taking a flow rate of 50 $\mathrm{l\ s^{-1}}$ as an example, Figure 4.18 shows that the presence of 2% free air reduces the pump pressure developed by about 3% from 189 to 183 kPa. Further along the curve, where it is falling more steeply at 70 $\mathrm{l\ s^{-1}}$, the effect is more pronounced, and the presence of 2% free air causes a drop of 11% from about 167 kPa to about 148 kPa. According to one view by Prandtl (1953) the presence of 5% by volume of free air corresponds to a 20% drop in pump speed. Any free air in a system is undesirable.

Although the term is sometimes used also to denote the formation of bubbles of gas, cavitation is really quite a different phenomenon and occurs when the pressure in a fluid falls to a value less than the saturation vapour pressure, p_w, corresponding to the local fluid

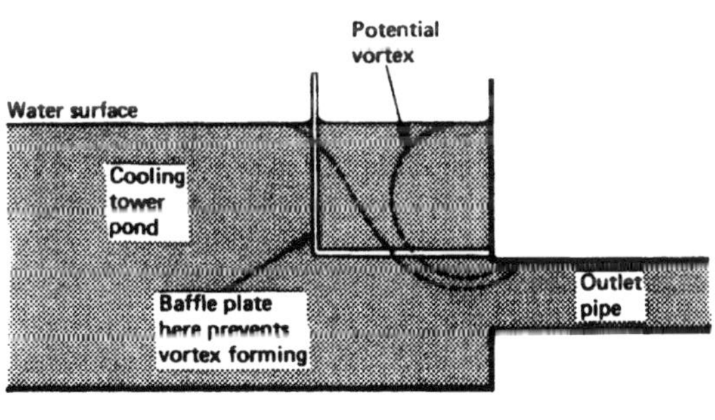

Figure 4.17 Prevention of a vortex formation by interception of its potential tail with a suitably positioned baffle plate near the outlet.

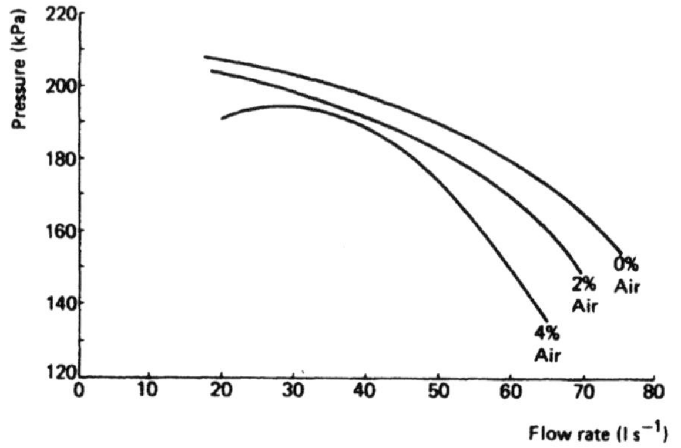

Figure 4.18 An approximate interpretation of the results by Siebrecht (1930), showing the effect of free air admitted into the pump suction.

temperature. If water flowing along a pipe at an initial pressure p_1 suffers a frictional pressure loss to a lower pressure, p_2, then if p_2 is less than p_w some of the water will flash to steam, producing bubbles of vapour that are carried along with the water. If these bubbles subsequently enter a region of higher pressure, say in the passages of the pump impeller, they suddenly collapse with considerable noise and set up pressure waves in the water that cause mechanical damage to the impeller and pipe wall, termed cavitation erosion. Prandtl (1953) postulated that cavitation was favoured by the presence of gas bubbles and dissolved air. More recently, Pearsal (1972) considers that the presence of submicroscopic gas bubbles might be the nucleii needed to give cavitation at pressures exceeding the vapour pressure. Apparently, prepressurised, de-aerated, pure water may be subjected to considerable tension of up to 300 atmospheres, without cavitation occurring, in the absence of nucleii. It is certain that cavitation is responsible for much severe damage to pump impellers, appearing as deep pitting on the blades. Moreover, it is accompanied by a pronounced fall in pump performance, affecting both head and efficiency, and this forms the basis of determining what is known as the net positive suction head (NPSH) required by the pump manufacturers. NPSH is defined in BS 5316 (1976) as the total inlet head plus the head corresponding to atmospheric pressure, minus the head corresponding to the vapour pressure, the total inlet head being the sum of static, position and velocity heads at the inlet of the pump.

As water flows from the suction flange into the pump it suffers a significant fall of pressure before the impeller starts to impart energy and so raise its pressure again. This loss of energy defines the lowest absolute pressure that is possible at the suction flange if cavitation is to be avoided. The only pressures that can be easily measured are those indicated by the gauges at the suction and discharge flanges. It is not possible, unfortunately, to measure the lowest pressure within the impeller and this can only be inferred from other observations. Several methods may be adopted to measure the NPSH required but that currently most popular is to reduce the head at the suction flange until a 3% reduction in the pump head is observed. It is clear from this that cavitation has already commenced some time earlier and it follows that, even if the NPSH required by the manufacturer is just barely provided, there will still be cavitation, and possible damage from cavitation erosion, occurring within the pump. The inception of cavitation and its extent depends upon how far the operating pump efficiency is from its maximum efficiency, reinforcing the dictum that pumps should always be chosen to work at their points of peak efficiency. Grist (1974) recommends that the NPSH available should exceed the NPSH required, measured on a 3% basis by the manufactures and termed NPSH (3%), by a factor, f

$$NPSH_{available} = f\,NPSH(3\%) \tag{4.19}$$

as given in Table 4.4.

Table 4.4 Recommended factors for applying to the MPSH required

Percentage of the flow rate at peak efficiency	30–49	50–79	80–110	111–125
f	9	6	3	12

Thus the NPSH available should desirably be at least three times that required by the manufacturers, even when the pump is working at the flow rate that gives maximum effici-

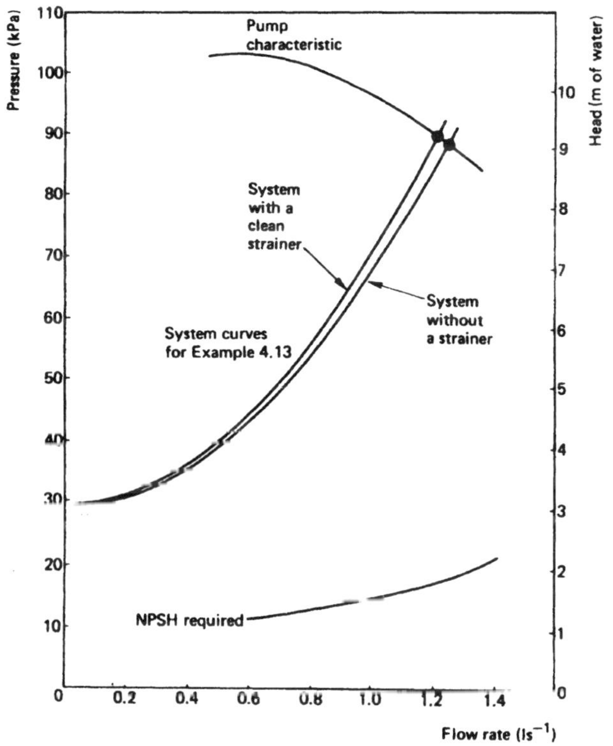

Figure 4.19 Pressure–volume and head–volume characteristic curves for a typical centrifugal pump, with the corresponding NPSH required.

ency. The value of the NPSH required rises rapidly as the flow rate increases (Figure 4.19). NPSH available is defined by

$$\mathrm{NPSH}_{\mathrm{available}} = H_{\mathrm{at}} + H_{\mathrm{z}} - H_{\mathrm{w}} \tag{4.20}$$

if the feed and expansion connexion is at the pump suction and by

$$\mathrm{NPSH}_{\mathrm{available}} = H_{\mathrm{at}} + H_{\mathrm{z}} - \Delta H_{\mathrm{p}} - H_{\mathrm{w}} \tag{4.21}$$

if the connexion is upstream of the suction branch, where H_{at} is the head corresponding to the atmospheric pressure acting on the open surface in metres of the fluid handled; H_{z} is the position head represented by the vertical distance of the open surface above the centre-line of the pump suction branch; ΔH_{p} is the friction and entry loss between the feed and expansion connexion and the pump suction branch in metres of the fluid handled: and H_{w} is the head corresponding to the vapour pressure of the fluid.

EXAMPLE 4.13

Using the appropriate information from Example 4.6, but with the assumption that a 200 mesh, 32 mm strainer is fitted in the suction line between the pond and the pump, determine the flow rate and the likelihood of cavitation occurring with water at 27°C. Take the atmospheric pressure as 101.325 kPa and the density of water as 1000 kg m^{-3}. Assume that the centre-line of the pump suction branch is 1 m below the surface of the water in the pond. The pump used has the characteristic performance shown in Figure 4.19.

Answer
First the loss in the suction line and the new overall system loss must be established. Referring to Example 4.6 and Figures 4.11(a) and (b) shows that, on the suction side of the pump, for a flow rate of 1.25 kg s^{-1}

Length of straight pipe:	=	25.0
Elbows: 3 × 0.7 × 1.9	=	3.99
Tee (after the cooling tower): = 0.2 × 1.9	=	0.38
Gate valves: 2 × 0.2 × 1.9	=	0.76
Contraction leaving the pond: 0.5 × 1.9	=	0.95
Original subtotal:	=	31.08 m
Original pressure loss: (31.08 × 250)/1000	=	7.77
Loss through a clean strainer (from Figure 4.2):	=	5.0
Revised suction loss:	=	12.77
Original total system loss:	=	88.9
Loss past a clean strainer:	=	5.0
Revised total system loss:	=	93.9 kPa

The revised system characteristic performance can now be established, assuming a square law and noting that the system loss without the static lift in 93.9 – 29.43 = 64.47 kPa

Flow rate (kg s^{-1})	0	0.2	0.4	0.6	0.8	1.0	1.2	1.25
Loss in pipe, etc. (kPa)	0	1.65	6.60	14.85	26.41	41.26	59.42	64.47
Static lift (kPa)	29.43	29.43	29.43	29.43	29.43	29.43	29.43	29.43
Total system loss (kPa)	29.43	31.08	36.03	44.28	55.84	70.69	88.85	93.90

In Example 4.8 the system characteristic performance without a strainer was calculated, so both system curves may be plotted (Figure 4.19) and it can be seen that, whereas the duty was 1.25 kg s^{-1} without a strainer it is a little less at 1.21 kg s^{-1} when a clean strainer is fitted and for this a NPSH of about 1.7 m H_2O is required.

Referring to CIBSE tables, $p_w = 3.564$ kPa at 27°C and, calculating the revised loss in the suction line as $12.77 \times (1.21/1.25)^2 = 11.97$ kPa, it can now be determined that

$$H_{at} = (101.325 \times 1000)/(1000 \times 9.81) = 10.33 \text{ m}$$

$$H_z = 1.00 \text{ m}$$

$$\Delta H_p = (11.97 \times 1000)/(1000 \times 9.81) = 1.22 \text{ m}$$

$$H_w = (3.564 \times 1000)/(1000 \times 9.81) = 0.36 \text{ m}$$

Hence, from Equation (4.20)

$$\text{NPSH}_{\text{available}} = 10.33 + 1.0 - 1.22 - 0.36 = 9.75 \text{ m } H_2O$$

The critical value of the head lost in the suction line, ΔH_p, is the head corresponding to the atmospheric pressure, plus the position head, minus the saturation vapour pressure corresponding to the temperature of the water, minus the NPSH required, namely 10.33 + 1.0 – 0.36 – 1.7 = 9.27 m H_2O. This represents a pressure loss of (1000 × 9.27 × 9.81)/1000 = 90.94 kPa.

When the strainer gets dirty this sort of pressure drop is a possibility. It is generally a bad principle to put a strainer, or any other significant resistance, on the suction side of a pump, because of the risk of cavitation. For this reason condensers and chillers, which invariably do have a significantly large pressure drop, always have their pump arranged to discharge into them and never to suck from them.

4.9 Temperature rise across pumps and heat gain to pipes

The rate of energy input to the water flowing through a pump, given by Equation (4.8), causes a small temperature rise. If c is the specific heat capacity of water, then

$$\dot{m}c\Delta t = \dot{m}gH = \dot{V}p_t = \dot{m}p_t/\rho$$

whence $\Delta t = gH/c$ or $p_t/\rho c$. If 60% is taken as the typical pump efficiency and it is assumed that half of the wasted 40% enters the water, the remaining 20% being dissipated to the surroundings, a fractional efficiency of 0.8 can be used to give the actual rate of energy input to the water. Inserting valves of 4178 J kg^{-1} K^{-1} for c, 9.81 m s^{-2} for g and 1000 kg m^{-3} for ρ, approximate expressions are obtained as follows:

$$\Delta t = 0.003 \text{ K m}^{-1} \text{ of pump head} \quad (4.22)$$

$$\Delta t = 0.0003 \text{ K kPa}^{-1} \text{ of pump pressure} \quad (4.23)$$

EXAMPLE 4.14

A pump develops a pressure of 94 kPa, corresponding to 9.61 m of water. Determine the temperature rise across the pump.

Answer

From Equations (4.22) and (4.23)

$$\Delta t = 0.003 \times 9.61 = 0.029 \text{ K}$$

$$\Delta t = 0.0003 \times 94 = 0.028 \text{ K}$$

Heat gain to insulated piping carrying chilled water has received little attention and data is scanty, one reason probably being that the loads imposed by the heat gain are only a very small proportion of the total cooling load. Table 4.5 gives some approximate heat gains. The figures refer to a temperature difference between the surface of the pipe beneath the lagging and an ambient temperature of about 30°C, but it is common to assume, without much error, that the difference between the chilled water temperature and the ambient air temperature may be used. The figures are not likely to be more accurate than ±10% and they do not include the gains to fittings, valves, and so on, which might account for an additional 10%, or even more, in plant rooms.

EXAMPLE 4.15

Determine the temperature rise accruing from pump power and heat gain to the chilled water pipes used in the circuit for Example 4.5. Take the thickness of the insulation to be 40 mm, the chilled water flow temperature to be 6.5°C, the return temperature to be 12°C and the ambient air temperature to be 30°C.

Answer

The total pump pressure calculated in Example 4.5 was 81.73 kPa and 65 m of 32 mm pipe was involved. The temperature rise across the pump is given by Equation (4.23)

$$\Delta t = 0.0003 \times 81.73 = 0.025 \text{ K}$$

From Table 4.4, the heat gain is 0.274 W m^{-1} K^{-1}. If the mean chilled water temperature is 9.25°C the heat gain to the piping is 0.274 × 9.25 × 65 = 164.7 W. With a flow rate of 1 kg s^{-1}, this corresponds to a rise in the temperature of the water of 164.7/(4718 × 1) = 0.035 K.

To bring these answers into perspective they should be looked at as percentages of the chilled water temperature rise of 5.5°, corresponding to the total cooling load:

Table 4.5 Heat gains to insulated piping

Nominal pipe size (mm)	15	25	32	40	50	80	100	150	200	250	300
Nominal thickness of lagging (mm)	40	40	40	40	40	40	45	45	50	50	50
Heat gain per metre of pipe run (W m^{-1} K^{-1})	0.203	0.238	0.274	0.309	0.354	0.491	0.525	0.696	0.797	0.968	1.121

Note: The thermal conductivity of the lagging is assumed to be about 0.043 W m^{-1} K^{-1}.

Δt for the pump power = 0.025 × 100/5.5 = 0.45%

Δt for pipe gain = 0.035 × 100/5.5 = 0.64%

Total t = 1.09%

Therefore, any margin to cover pump power and heat gain to the pipes is of the order of 1%.

Exercises

1 A closed piping circuit has a resistance of 50 kPa. If the water level in the feed and expansion tank is 40 m above the centre-line of the pump suction, to which the cold feed is connected, determine the heads and pressures that would be indicated by gauges positioned at (a) 0.5 m below pump suction and (b) 1.5 m above pump discharge.
(*Answer* (a) 40.5 m, 397 kPa, (b) 43.6 m, 428 kPa)

2 A closed water piping system comprises 130 m of straight pipe of 50 mm nominal bore, 16 elbows (k = 0.7), two tees (k = 0.2) and 6 gate valves (k = 0.2). Given that f = 0.0048, calculate the equivalent length of straight pipe that will absorb one velocity pressure and determine the total equivalent length of straight pipe.
(*Answer* 2.6 m; 163.2 m)

3 If the system in 2 handles 3 kg s^{-1} and includes a chiller having a pressure drop of 40 kPa, determine the system pressure drop using Figure 4.1 for the friction loss in pipe work.
(*Answer* 107 kPa)

Notation

Symbols	Description	Unit
c	Concentration of gas in water	–
	or specific heat capacity	J kg^{-1} K^{-1}
d	Internal pipe diameter	m or mm
f	Friction factor or factor that multiplies NPSH	–
g	Acceleration arising from gravity	m s^{-1}
	or the specific force due to gravity	N kg^{-1}
H	Head or total pump head, or Henry's constant	m
H_{at}	Head of fluid handled corresponding to atmospheric pressure	m
H_d	Static head at pump discharge	m
H_s	Static head at pump suction	m
H_w	Static head of fluid handled corresponding to its vapour pressure at a particular temperature	m
H_z	Position head corresponding to a distance z	m
ΔH	Head lost	m
ΔH_p	Head of fluid lost, corresponding to the pressure loss in a suction line	m
h	Static head	m

h_o	Static head imposed on an open surface of fluid	m
k	Friction factor for pipe fittings	–
k_s	Absolute roughness of a pipe wall	m
l	Length of a pipe	m
l_e	Equivalent length of pipe that has a pressure loss numerically equal to one velocity pressure of the fluid flowing	m
M	Mass flow rate	kg s^{-1}
NPSH	Net positive suction head	m
R	Pipe radius	m or mm
(Re)	Reynolds number	–
V	Volumetric flow rate	$m^3\ s^{-1}$ or l s^{-1}
W_p	Pump power	W or kW
W_s	Power supplied to the impeller shaft of a pump	W or kW
$\dot{m}$	Mass flow rate	kg s^{-1}
n	Number of bubbles or rotational speed of a pump impeller	– rev s^{-1}
p	Pressure or partial pressure of a gas	N m^{-2}, or Pa, or kPa
p_{at}	Atmospheric pressure	N m^{-2}, or Pa, or kPa
p_g	Gauge pressure	N m^{-2}, or Pa, or kPa
p_o	Static pressure corresponding to h_o	N m^{-2}, or Pa, or kPa
p_{sd}	Static pressure at pump discharge	N m^{-2}, or Pa, or kPa
p_{ss}	Static pressure at pump suction	N m^{-2}, or Pa, or kPa
p_t	Total pump pressure	N m^{-2}, or Pa, or kPa
p_v	Vapour pressure	N m^{-2}, or Pa, or kPa
p_w	Vapour pressure of water	N m^{-2}, or Pa, or kPa
r	Radius of a bubble	m or mm
Δp	Specific pressure loss per unit length of pipe	Pa m^{-1}
Δt	Temperature rise through a pump or by heat gain to a pipe	K
δp	Pressure loss	N m^{-2}, or Pa, or kPa
η	Total efficiency of a pump	%
μ	Absolute viscosity of a fluid	N s m^{-2} or kg $m^{-1}\ s^{-1}$
v	Mean velocity of water flow	m s^{-1}
v_d	Mean velocity of water flow at pump discharge	m s^{-1}
v_s	Mean velocity of water flow at pump suction	m s^{-1}
ρ	Density of a fluid	kg m^{-3}

References

CIBSE Guide, Section C4, Flow of fluids in pipes and ducts, 1977.
Jones, W. P., *Air Conditioning Engineering*, 4th Edition, Edward Arnold, 1994.
Etheridge, R., Variable speed pumping, *The Building Services Engineer*, 44, A45, September, 1976.
Walker, W. H., Lewis, W. K., McAdams, W. H. and Gilliland, E. R. *Principles of Chemical Engineering*, McGraw-Hill, 1967.
Crocker, S. and King, R. C., *Piping Handbook*, McGraw-Hill, 1967.
Miller, D. S., *Internal Flow Systems*, Volume 5. BHRA Fluid Engineering Series, 1978.
ASHRAE, *Handbook of Fundamentals*, 1993.
Denny, D. F. and Young, G. A. J. *The Prevention of Vortices and Swirl at Intakes*. Publication SP 583, BHRA, VIIth Congress of the IAHR, Lisbon, July 1957.
Stepanoff, A. J., *Centrifugal and Axial Flow Pumps*, 2nd edn, John Wiley, New York, 1967.
Siebrecht, W. Untersuchungen uber Regelung von Kreiselpumpen. *Z ver deut Ing*, 74, p. 87, 1930.
Prandtl, L., *Essentials of Fluid Dynamics*, Blackie, London, 1953.
Pearsal, I. S., *Cavitation*, M & B Monograph, ME/10, Mills & Boon, London, 1972.
BS 5316, Part 1, Acceptance Tests for Centrifugal, Mixed Flow and Axial Pumps; Class C Tests, 1976.
Grist E. 1974: *Net Positive Suction Head Requirements for Avoidance of Unacceptable Cavitation Erosion in Centrifugal Pumps. Cavitation*, a conference arranged by the Fluid Machinery Group of the I Mech E, Heriot-Watt University, Edinburgh, pp. 153–62, September 1974.

5

Air Distribution

5.1 The free isothermal jet

The behaviour of an air jet that is discharged into a large room at the same temperature has been well studied by Farquharson (1952), Frean and Billington (1955), Parkinson and Billington (1957), Tuve (1953) and Koestel *et al.* (1952) and is best described in terms of four sections of its length:

(1) For a short distance, up to about four equivalent diameters from the plane of the outlet, its centre-line velocity is constant.
(2) Over the succeeding four diameters of length a transition to turbulent flow occurs and in this zone the centre-line velocity, v_x, is inversely proportional to $\sqrt{x}$, where x is the distance from the plane of the outlet.
(3) After this, turbulent flow is fully established and the influence of inertial forces predominates. For the next 25–100 equivalent diameters of jet length v_x is proportional to $1/x$. This is the zone of most interest.
(4) Finally, at the end of the turbulent zone, v_x diminishes rapidly to less than 0.25 m s^{-1} and the pattern of air movement is unpredictable, viscous forces being dominant and quite small influences enough to produce random and transient changes.

Within the third zone the following empirical equation may be used to determine v_x, for most practical purposes:

$$v_x = K'Q/[x\sqrt{(A_g c_d R_{fa})}] \tag{5.1}$$

in which K' is a constant of proportionality (see Table 5.1), Q is the volumetric airflow rate, A_g is the gross area of the outlet, c_d is its coefficient of discharge which varies from 0.6 for sharp-edged orifices to 1.0 for circular openings with well-rounded edges, and R_{fa} is the ratio of free to gross area for the outlet.

In Table 5.1, R is the aspect ratio of the opening (breath/height) and v_o is the effective velocity of the jet over the section of the vena-contracta when issuing from a sharp-edged orifice, or the average velocity of discharge through an open-ended duct, as given by

$$v_o = v_c/(c_d R_{fa}) \tag{5.2}$$

where v_c is the nominal mean velocity through the area of the orifice or the open-ended duct.

Table 5.1 Values of the proportionality constant for Equation (5.1)

Type of opening	Values of K'	
	v_o = 4 to 8 m s^{-1}	v_o = 2 to 4 m s^{-1}
Round or square openings	7.0	5.7
Rectangular, free slot ($R < 40$)	6.0	4.9
Grilles ($R_{fa} > 0.4$)	5.7	4.7

The typical profile over the cross-section of a free isothermal jet in the third zone is in the form of a bell-shaped curve for velocity against distance through the jet section. Although the maximum velocity, v_x, lies along the centre-line of the jet the mean velocity over its section is only about 20–30% of this. The throw of a jet, for comfort conditioning, is usually defined as the value of x for which v_x equals 0.25 m s^{-1} and thus, on average over the cross-section of the jet, the mean residual velocity is only about 0.05–0.075 m s^{-1}. One criterion adopted for the throw is that v_x should reach 0.25 m s^{-1} when $x = L$, L being the distance from the outlet to the opposite wall, or to the end of the throw of a jet coming from another supply opening, opposite. Sometimes the distance down the opposite wall, or vertically downward from the point of meeting with the oncoming jet, to the occupied zone (1.8 m from floor level) is included as part of the throw.

An isothermal jet in free space expands naturally with an included angle of between 20–24°, air being entrained around its periphery, momentum flow (defined as the product of the air mass flow rate and the mean air velocity) conserved and its mean velocity consequently reducing. If the jet is at the same temperature as the room air generally, no buoyancy forces are present and only the natural expansion of the jet can produce air movement in the room, the tendency being that room air wafts slowly towards the supply opening and the periphery of the jet. The concept of drop is really only relevant when the jet is colder than the room air and, therefore, falls in a trajectory towards the occupied zone, under gravitational influences. The intention is then to deliver the jet from its outlet at a velocity and temperature that will decay to 0.25 m s^{-1} and room temperature by the time the occupied zone is reached, but such non-isothermal behaviour has not yet been fully investigated.

Most research has been done on the behaviour of free jets entering large rooms but, in the practical case, air is discharged from openings so as to flow along adjoining ceilings or up nearby walls, taking account of the Coanda effect, which is that the frictional loss between the jet and the surface with which it is in contact creates a pressure difference across the section of the jet that tends to press it to the surface, countering any downward buoyancy forces and delaying the drop into the occupied zone. The proximity of the ceiling to a side-wall grille or a diffuser is thus a significant feature of the selection of proprietary distribution terminals (see Figure 5.1).

EXAMPLE 5.1

A 300 mm diameter nozzle with $R_{fa} = 1.0$ and $c_d = 1.0$ delivers a horizontal jet of air into a very large room. Assuming that K' (Equation (5.1)) is 7.3, determine the centre-line velocity at a distance of 9 m from the nozzle if the initial air velocity is 5 m s^{-1}.

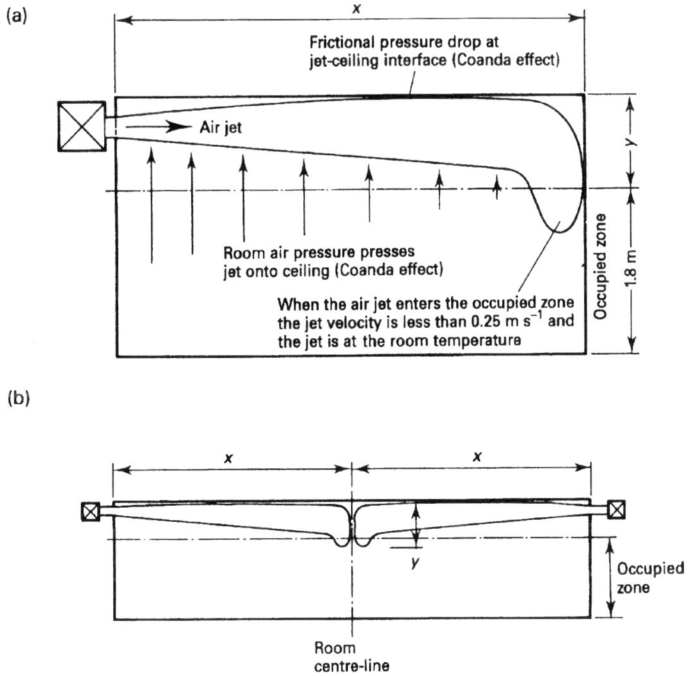

Figure 5.1 (a) The throw of a jet of air from a side-wall grille and the Coanda effect. (b) The throw of a pair of opposing side-wall grilles in a wide room.

Answer

$$Q = A_g v_o = (0.3^2\pi/4) \times 5 = 0.0707 \times 5 = 0.3535 \text{ m}^3 \text{ s}^{-1}$$

From Equation (5.1) $v_x = (7.3 \times 0.3535)/[9\sqrt{0.0707}] = 1.1 \text{ m s}^{-1}$.

5.2 The free non-isothermal jet

The behaviour of air jets issuing from outlets at temperatures other than that of the ambient air is complicated. Koestel (1955) has proposed that the drop, y, of a jet flowing horizontally from an opening of diameter D_o, at a mean initial velocity v_o, may be calculated for a horizontal distance, x, from the outlet by the equation

$$\pm y/D_o = 0.065(x/D_o)^3[(t_r - t_o)/(273 + t_r)][(gD_o)/(v_o^2)] \tag{5.3}$$

where t_r and t_o are the room and initial jet temperatures, respectively. The coefficient of proportionality for the jet, $K = K'/1.13$, has been taken as 6.5 in this equation, corresponding to $K' = 7.3$.

EXAMPLE 5.2

For the case of Example 5.1 determine the drop at $x = 9$ m, if $t_r = 22$°C and $t_o = 12$°C. Take g as 9.81 m s^{-2}.

Answer
From Equation (5.3)

$$y = 0.3 \times 0.065(9/0.3)^3[(22 - 12)/(273 + 22)][(9.81 \times 0.3)/5^2] = 2.1 \text{ m}$$

Koestel's equation makes certain simplifying assumptions, notable among which is that the slope of the jet trajectory should not be too great. This would seem to limit its practical use but, nevertheless, reasonable agreement with experimental results has been reported.

Koestel (1954) has also studied the maximum downward throw, x_{max}, of warm air discharged from a nozzle and he proposes that

$$x_{max}/D_o = \sqrt{[3.4((273 + t_o)/(t_o - t_r))(v_o^2/gD_o)]} - 2.85 \qquad (5.4)$$

EXAMPLE 5.3

Determine the maximum downward throw of a jet at an initial temperature of 40°C into a room at 20°C, for the conditions of Example 5.1.

Answer
From Equation (5.4)

$$x_{max} = 0.3[\sqrt{(3.4((273 + 40)/(40 - 20)(5^2/9.81 \times 0.3))} - 2.85] = 3.5 \text{ m}$$

An approximate relationship between jet temperature and velocity, for the horizontal, non-isothermal case, is given by

$$(t_o - t_r)/v_o = (t_x - t_r)/v_x \qquad (5.5)$$

where t_x is the mean temperature of the jet at a distance x from the outlet. Adjustable nozzles, of circular cross-section, are often used to supply air to large enclosed spaces such as exhibition halls or arenas. The long throw possible is used to deliver the ventilating air over long distances to the places where it is needed, for example from ducts at the perimeter walls or balconies to the centre of the space.

5.3 Side-wall grilles

Laboratory studies by Müllejans (1966) suggest that the path of a horizontal, non-isothermal jet is best described in terms of the Archimedean number, (Ar), defined for a room by

$$(\text{Ar}) = g(t_w - t_o)D_h/(T_r u_r^2) \qquad (5.6)$$

in which t_w is the temperature of the heated surfaces in the room, T_r is the absolute room temperature, u_r is a fictitious mean air velocity through the cross-section of the room and D_h is its mean hydraulic diameter, defined by $D_h = BH/2(B + H)$, B and H being the breadth and height of the room, respectively.

Further laboratory studies by Jackman (1970) have attempted to use the Archimedean number to establish a design procedure for the selection of side-wall grilles. The method proposed, however, usually yields a supply air quantity that requires a larger temperature difference, room-to-supply, for offsetting the sensible heat gains than would be determined by the practical considerations of psychrometry. The method is not recommended and it is better to adopt conventional, commercial procedures for selecting the necessary supply air temperature and to use Equation (2.3) to calculate the supply air quantity, following this by a grille selection from a manufacturer's catalogue. There is no agreed definition of throw but good practice suggests that any equation used to describe the throw should include an indication of the terminal velocity on which it is based, thus

$$T_{0.25} = x + y \tag{5.7}$$

where $T_{0.25}$ is the throw to a terminal velocity of 0.25 m s^{-1}, x is the horizontal component of the throw and y is the vertical component (see Figure 5.1a). When air supply outlets blow towards each other, as in wide rooms, the horizontal and vertical components are interpreted as shown in Figure 5.1(b).

An alternative definition of throw is to a terminal velocity of 0.5 m s^{-1} at a position that is 75% of the horizontal distance from the supply outlet to the opposite wall (or boundary of the zone). Following the principle of Equation (5.7) the throw would then be described by

$$T_{0.50} = 0.75x \tag{5.8}$$

It is considered by Koestel and Young (1951) that the jet velocity will decay to a lower value, consistent with comfortable conditions, when it enters the occupied zone. A velocity of 0.5 m s^{-1} was chosen because Rydberg (1962) considered it was difficult to measure lower values with accuracy.

The cheapest way of supplying air to a conditioned space is very often by side-wall grilles rather than by ceiling diffusers. For the distribution to be effective the pressure drop along the duct in which the grilles are fitted should be small compared with the drop across the grille and its associated dampers and vanes, in order to assist balancing and sizing the duct by static regain helps. It is very important to ensure smooth airflow from the duct normally into the face of the grille if the anticipated throw, drop, spread and noise level are to be achieved. Figure 5.2(a) shows that the effects of static pressure (p_s) and velocity pressure (p_v) give a resultant direction of airflow that is angled to the face of the grille. The throw anticipated from the catalogue selection data will not be achieved and more noise than expected will be generated. In Figure 5.2(b) it is seen how the use of adjustable turning vanes can direct the correct proportion of the airstream normally onto the face of the grille, to achieve the desired throw, in accordance with catalogue data. The presence of the turning vanes generates some noise, as also does the partial closure of any balancing dampers. Figure 5.2(c) shows how the presence of a spigot may help to improve the airflow onto the face of the grille and give a better direction to the leaving airstream. The minimum length of spigot should not be less than 1.5 times the width of the grille face but the manufacturer's advice should be sought.

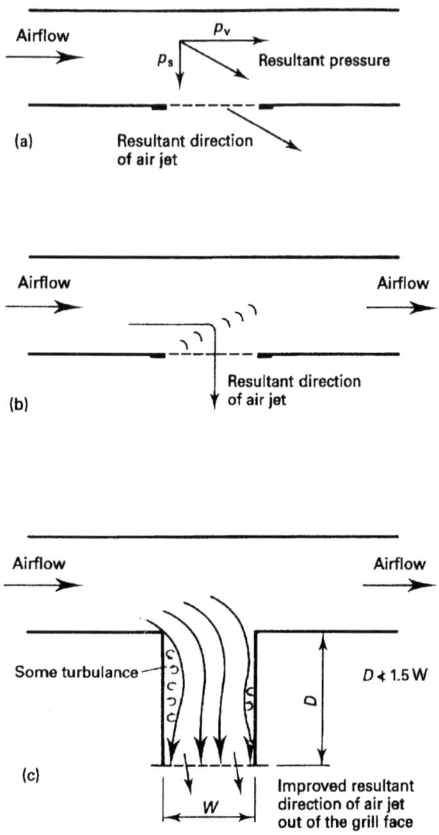

Figure 5.2 (a) Angled airflow from a grille face. (b) The use of adjustable turning vanes to obtain smooth airflow out of the face of a grille. (c) The use of an extended spigot to help smooth airflow into a grille face.

Values of throw to a terminal velocity of 0.25 m s^{-1} for a typical grille located near a flat ceiling and fitted with opposed-blade dampers and two sets of direction-control vanes are given in Table 5.2 for sizes from a commercial range. Figure 5.3 shows a plot of the throw against the volumetric flow rate for the full range of sizes available.
The following conclusions can be drawn:

(1) For a given grille size, corresponding to a part of the curve, the throw is approximately proportional to the volumetric flow rate.
(2) The throw is more sensitive to changes in the flow rate at the lower volumes.
(3) A shorter throw can be achieved for a given total flow rate by using several small grilles rather than by a few large ones.

Furthermore, if the throw for grilles with their vertical vanes set at the 45° position (90° included angle) is plotted, as shown by the dashed line in Figure 5.3, the throw is seen to be about half that obtained for the 0° vane setting, with vertical vanes parallel.

Table 5.2 Values of throw to a terminal velocity of 0.25 m s^{-1} for grille sizes from a commercial range

			Available sizes of grilles with nominal dimensions in mm																			
						Size A 406 × 102 305 × 127 254 × 152				Size B 457 × 102 356 × 127 305 × 152 203 × 203				Size C 508 × 102 406 × 127 336 × 152 254 × 203				Size D 610 × 102 457 × 152 406 × 152				
	Velocity through the free area (m s^{-1})	NR (dB)	Pressure drop across opposed-blade dampers and grille (Pa)			Flow rate (l s^{-1})	Throw to a terminal velocity of 0.25 m s^{-1} (m)			Flow rate (l s^{-1})	Throw to a terminal velocity of 0.25 m s^{-1} (m)			Flow rate (l s^{-1})	Throw to a terminal velocity of 0.25 m s^{-1} (m)			Flow rate (l s^{-1})	Throw to a terminal velocity of 0.25 m s^{-1} (m)			
Vertical vane setting (°)			0	22.5	45		0	22.5	45		0	22.5	45		0	22.5	45		0	22.5	45	
	1.5	21	2	3	4	47	5.5	4.3	2.7	54	5.8	4.6	3.0	66	6.7	5.5	3.4	73	7.0	5.5	3.4	
	2.0	21	4	5	7	64	6.4	5.2	3.4	73	7.0	5.5	3.4	87	7.6	6.1	3.7	99	7.9	6.4	4.0	
	2.5	21	7	8	12	80	7.3	5.8	3.7	82	7.6	6.1	4.0	109	8.2	6.7	4.3	123	8.8	7.0	4.6	
	3.0	21	9	11	16	97	7.9	6.4	4.0	111	8.5	6.7	4.3	130	9.1	7.3	4.6	146	9.8	7.9	4.9	
	3.5	26	13	14	22	113	8.5	6.7	4.3	130	9.1	7.3	4.6	151	9.8	7.9	4.9	172	10.7	8.5	5.2	
	4.0	30	17	19	29	127	9.1	7.3	4.6	146	9.8	7.9	4.9	175	10.7	8.5	5.2	196	11.3	9.1	5.5	
	5.0	36	27	30	45	160	10.1	7.9	5.2	184	11.0	8.8	5.5	217	11.9	9.4	6.1	245	12.5	10.1	6.4	
	6.0	42	39	44	66	193	11.3	9.1	5.5	222	12.2	9.8	6.1	260	13.1	10.4	6.4	295	13.7	11.0	7.0	

Note: All grilles are assumed to be fitted with opposed-blade dampers and two sets of manually adjustable direction-control vanes. Noise ratings are for fully open dampers with smooth, uniform airflow at right angles to the plane of the grille. Room effect is taken as zero (see Section 7.8). Vane settings of 22.5° and 45° increase the NR value by 1 and 7 dB, respectively.

In practice, it is very likely that higher NR values will be experienced because airflow is seldom smooth into a grille face and because most grilles have partially closed dampers in order to achieve the correct airflow rate.

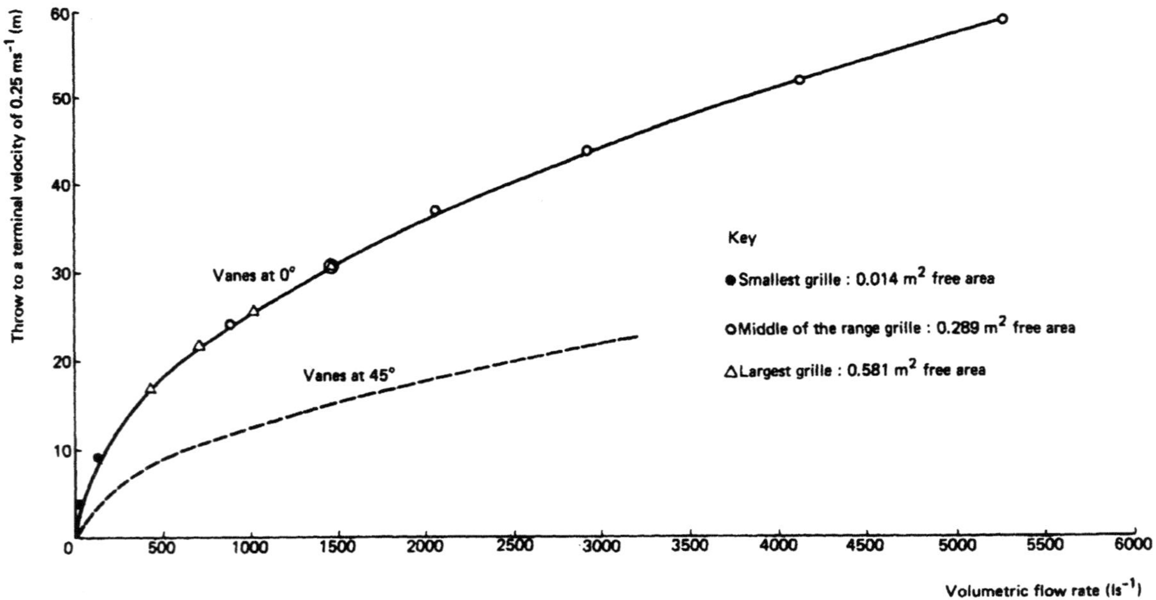

Figure 5.3 Relationship between flow rate and throw for a typical range of commercial side-wall grilles with vertical vanes set at 0°. The throw for a setting of 45° (90° included angle) is about half these values. Appropriate for room-to-supply air temperature differences from 8–13 K.

Data are sparse on the best separation, z, between grilles but one manufacturer suggests:

$$z = 0.20\,x, \text{ if the vanes are set at } 0° \tag{5.9}$$

$$z = 0.25\,x, \text{ if the vanes are set at } 22.5° \tag{5.10}$$

$$z = 0.30\,x, \text{ if the vanes are set at } 45° \tag{5.11}$$

When the grille is adjacent to a wall, half the above values of z should be taken as the minimum allowable distance from the wall.

Some general principles of grille selection are:

(1) A side-wall grille near a ceiling gives a longer throw than a square or circular ceiling diffuser.
(2) The effective use of grilles is limited to air change rates of less than about 12 per hour, with conventional ceiling heights.
(3) Stagnant air pockets may form if the air change rate is less than approximately four per hour, corresponding to specific supply rates of 3 l s^{-1} m^{-2}.
(4) The drop may be excessive, in the non-isothermal case, if the discharge velocity through the free area is below 0.2 m s^{-1}.
(5) The upper edge of the grille should be within 300 mm of the ceiling.
(6) Miller *et al.* (1971) consider that, for vertical vanes to be effective, the ratio of the depth of the vanes (in the direction of airflow) to their spacing, should lie between one and two.

EXAMPLE 5.4

Select suitable side-wall grilles for the supply of 0.4 m^3 s^{-1} to a room of size 7.2 m wide × 7.5 m long × 2.6 m high. The grilles are to be mounted in the wall that is 7.2 m wide and the height of the occupied zone is 1.8 m.

Answer

The air change rate is (0.4 × 3600)/(7.2 × 7.5 × 2.6) = 10.2 per hour, so side-wall grilles can be used. Required throw = 7.5 + (2.6 – 1.8) = 8.3 m. Figure 5.3 shows that a single grille would give a throw of about 17 m with its vanes at 0°, or 8.5 m with them at 45°, suggesting that two or even three grilles may be suitable. Table 5.2 shows that two grilles of size C can be used to give the desired throw with their vanes set at between 22.5–45°, which gives about NR 32, plus from 1 to 7 dB, depending on the vane setting. Supposing the vanes must be opened to 45° to get the throw with draughtless distribution then the desirable minimum separation between the grilles is z = 0.3 × 8.3 = 2.49 m, according to Equation (5.11). If the widest grille of size C is chosen, i.e. 508 × 102 mm, the minimum total width of wall required will be 2 × (2.49 + 0.508) = 5.996 m, which is less than the 7.2 m available, and is satisfactory.

If three grilles are chosen, each handling 133 l s^{-1}, a possibility is size B. Then, from Table 5.2 the required throw could be obtained with the vanes set at between 0° and 22.5°, with a probable noise level of about NR 26 + 1 dB for the diverging vanes. From Equation (5.10) the minimum permissible separation is z = 0.25 × 8.3 = 2.075 m. Selecting the widest grille, 457 × 102 mm, means that a distance of 3 × (2.075 + 0.457) = 7.596 m, which exceeds the available wall width of 7.2 m, and so three grilles is not possible, unless a smaller width is chosen, say 305 × 152 mm.

Manufacturers' tests show that the drop of an airstream depends on the air quantity as well as the temperature difference, a conclusion not readily inferred from Equation (5.3).

It follows that the risks of drop may be minimised by using many small outlets rather than few large ones. The relationship varies considerably with different types and reference should always be made to manufacturers' data for specific cases. For side-wall grilles, the risk of drop can be reduced by adjusting the horizontal vanes to curve the airstream towards the ceiling, if the top edge of the grille is more than 300 mm below it.

5.4 Circular ceiling diffusers

Circular ceiling diffusers distribute a radially expanding airstream over the ceiling that rapidly entrains air and gives a very good pattern of air movement in the room. Temperature differences of up to 14 K for cooling can usually be safely adopted with conventional ceiling heights, provided the diffusers are properly selected. Table 5.3 gives a selection of performance details for a commercial, circular ceiling diffuser with adjustable cones, and Figure 5.4 shows the overall performance for the complete range of sizes available for the type.

If the diffuser is mounted in an exposed duct the quoted radius of diffusion, to a terminal velocity of 0.25 m s^{-1}, must be multiplied by 0.7. A room-to-supply air temperature difference of 11 K for cooling is assumed and the NR values in dB are based on a room effect of zero. It should be noted that many makers quote NR values on the assumption of an 8 dB room effect.

It is claimed by some manufacturers that any pattern of air discharge from horizontally across the ceiling to vertically downwards is possible by adjusting the relative cone positions in the diffuser. Another view is that any distribution pattern between these two extremes is unstable and, at best, only transient. A small difference of pressure, inside to outside, across the cone of air will cause it to collapse to one or other of the two stable positions, on the ceiling or blowing vertically downwards.

As for all other air distribution terminals the makers' published data is for ideal conditions of installation and assumes smooth, uniform airflow normal to the plane of the cones with no upstream volume control damper. To achieve the tabulated performance these ideal conditions must prevail but they are only obtained with great difficulty in many actual installations and the presence of any turbulence in the neck of the diffuser will cause a departure from the expected behaviour, particularly from the NR value. Matters can be helped by having the longest possible straight duct feeding into the diffuser neck. Short lengths, which are the result of insufficient space above the suspended ceiling, are notorious for giving turbulent airflow into the diffuser cones and so causing noise. Projecting the diffuser neck into the duct itself makes things worse. The presence of proprietary volume control dampers in the diffuser neck often upsets the airflow enough to generate objectionable noise. It is best to use aerofoil section turning vanes, of short chord width, to assist smooth entry from the duct main to the branch feeding the diffuser and to size the main by static regain if at all possible. The neck velocity should be as low as possible, consistent with the desired throw. Where dampers must be used to balance airflow they should be located as far away from the diffuser cones as is practicable.

EXAMPLE 5.5

Select circular ceiling diffusers from Table 5.3 for the distribution of 0.4 m^3 s^{-1} of air in the room considered in Example 5.4.

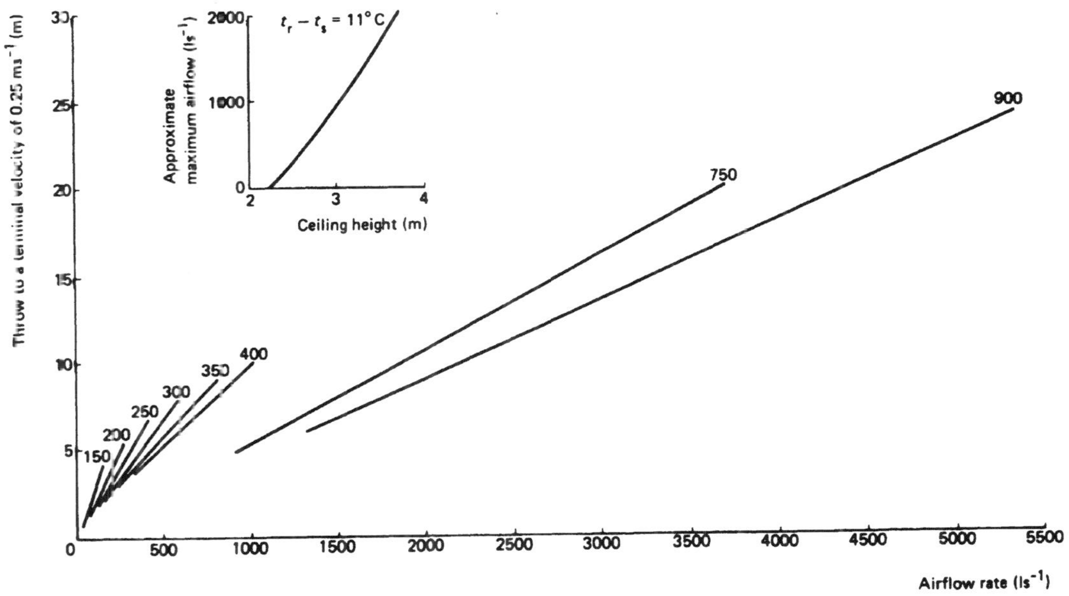

Figure 5.4 Overall performance of a range of sizes of a typical commercial ceiling diffuser. Sizes are nominal neck diameters in mm. Manufacturers' catalogues should always be referred to for specific cases.

Table 5.3 Performance of a commercial circular ceiling diffuser with adjustable cones

		Nominal size of ceiling diffuser (mm) 150			200			300			400		
Mean neck velocity ($m\ s^{-1}$)	Total pressure drop (Pa)	Flow rate ($l\ s^{-1}$)	Radius of diffusion (m)	NR (dB)	Flow rate ($l\ s^{-1}$)	Radius of diffusion (m)	NR (dB)	Flow rate ($l\ s^{-1}$)	Radius of diffusion (m)	NR (dB)	Flow rate ($l\ s^{-1}$)	Radius of diffusion (m)	NR (dB)
2.0	5	38	0.9		66	1.2		149	2.1		264	2.7	
2.5	8	47	1.2		83	1.5		184	2.4		330	3.4	22
3.0	12	57	1.5	23	99	2.1	24	222	3.0	26	396	4.0	27
3.5	16	66	1.8	28	116	2.4	29	260	3.4	31	462	4.6	33
4.0	21	76	2.1	32	132	2.7	34	297	4.0	35	529	5.2	37
4.5	27	85	2.1	35	149	3.0	37	333	4.6	39	595	6.1	40
5.0	33	94	2.4	39	165	3.6	40	370	4.9	42	661	6.7	44
6.0	47	111	3.0	44	198	4.0	46	444	5.8	48	793	7.9	49

Answer
Using a single, centrally-placed diffuser means that the throw should not exceed 3.6 + (2.6 – 1.8) = 4.4 m. Reference to Table 5.3 shows that the choice is a size of 400 mm, 4.0 m radius of diffusion and NR 27.

The practical difficulties of accommodating a large number of closely-located diffusers in a ceiling and the problems of ducting air to them in the void above, limit the air change rates that can be handled to a maximum of about 25 per hour, although more than this can be delivered in some instances. Not all manufacturers offer diffusers with neck velocities as low as 2.0 m s^{-1} (Table 5.3), a common minimum is often 3.0 m s^{-1}, and this provides a further limitation on the maximum practical air change rate.

5.5 Square ceiling diffusers

Square ceiling diffusers have performances that are similar to circular ones but the plan size of the square cones influences the throw for a given neck velocity. Thus, a diffuser of 200 mm square neck and a plan size of 300 mm square will deliver 99 l s^{-1} with a radius of diffusion of 3.4 m but the throw will be only 2.1 m when fed into a 600 mm square.

5.6 Linear slot diffusers

Linear slot diffusers are often used because they can be unobtrusively integrated with a suspended ceiling. They are commonly available in lengths up to about 2 m, greater distances needing several sections and shorter ones being cut to length on site. From one to 10 parallel slots may be combined in a single diffuser and arranged to blow in the same direction or outwards in opposing ways. The blow can also be directed vertically downwards, for spot cooling, although this is unlikely to give comfort under more normal conditions. It is essential that the duct connexions recommended by the diffuser manufacturer be followed and if good results are to be obtained, i.e. quiet, draughtless air distribution, the air velocity through the section of the plenum chamber or duct feeding the slots should be less than that through the slots themselves. The plenum chamber is usually called a diffuser boot.

The performance of a selection of commercial linear diffusers is given in Table 5.4 and the approximate performance over the entire range is shown in Figure 5.5. Generally speaking linear diffusers have a shorter throw than side-wall grilles but a longer one than circular or square diffusers. Outlets, like circular diffusers, that entrain air rapidly have short throws and rapid temperature equalisation, room-to-supply, and can, therefore, handle

Table 5.4 Performance of commercial linear ceiling diffusers

	Number of slots in diffuser 1			3			5		
Total pressure loss (Pa)	Flow rate (l s^{-1} m^{-1})	Throw (m)	NR (dB)	Flow rate (l s^{-1} m^{-1})	Throw (m)	NR (dB)	Flow rate (l s^{-1} m^{-1})	Throw (m)	NR (dB)
	12	1.2		37	3.4		62	4.8	
4	25	2.6		74	5.8		124	7.3	19
9	37	3.4	21	112	7.0	29	186	9.1	32
16	50	4.6	30	149	8.2	38	248	10.4	41

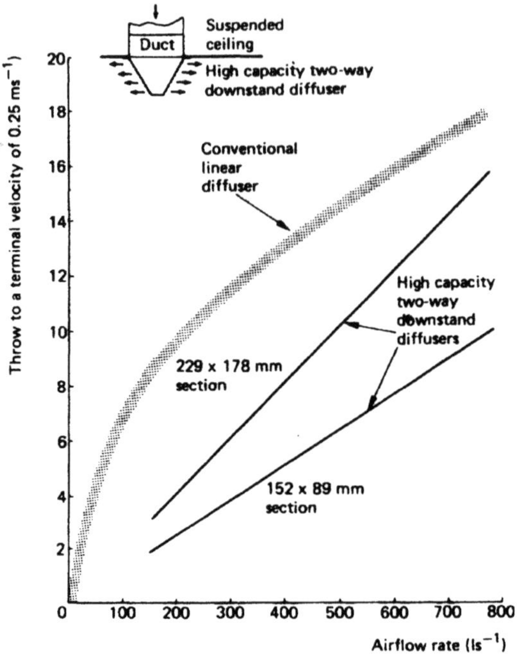

Figure 5.5 Typical performances of ranges of conventional and high capacity linear diffusers. The number of slots in a conventional linear diffuser does not affect the throw.

rather more air changes per hour than outlets like side-wall grilles which have longer throws. Linear diffusers lie in between these.

EXAMPLE 5.6

Select linear diffusers from Table 5.4 for the distribution of 400 l s^{-1} in the room considered in Example 5.1. Locate the diffuser in the centre of the ceiling parallel to the 7.2 m room dimension.

Answer

The required throw is 7.5/2 + (2.6 – 1.8) = 4.55 m, to a terminal velocity of 0.25 m s^{-1}. The specific airflow rate is 400/7.2 = 56 l s^{-1} m $^{-1}$. Figure 5.3 shows that a throw of about 4.6 m will not be exceeded with this rate. Table 5.4 shows that a three-slot diffuser could be used, interpolating between throws of 3.4 and 5.8 m.

A high capacity version of the linear diffuser is available for applications demanding a higher air change rate than can be conveniently handled by conventional air distribution terminals, and it is an alternative to a ventilated ceiling for duties up to about 40 air changes per hour. Figure 5.5 shows its triangular, downstand section and that the air is delivered parallel to the ceiling in opposing directions from a pair of perforated faces. Airflow must be smooth and uniform into the diffuser for good results. The performance lines for the two available sizes (152 × 89 mm and 229 × 178 mm, base and height, respectively) do not coincide, unlike the performance curve for conventional linear diffusers, where the sizes within the range overlap.

EXAMPLE 5.7

Determine the air change rate that can be supplied through high capacity, downstand, two-way diffusers, using the performance lines in Figure 5.5, to the room used in Example 5.1.

Answer

With the diffusers in the centre of the ceiling, parallel to the 7.2 m dimension, the tolerable throw is 4.55 m, as calculated earlier. Figure 5.5 shows that the larger diffuser will permit a delivery of about 225 $l\ s^{-1}\ m^{-1}$. The air change rate is, therefore, $(0.225 \times 7.2 \times 3600)/(7.5 \times 7.2 \times 2.6) = 42$ air changes per hour.

5.7 Swirl diffusers

Air leaves the diffuser through angled slots or over angled vanes that impart a twist to the air jet, increasing its momentum flow. As a result, the entrainment of air from the room is very rapid, the mean air velocity of the jet quickly decays and the jet temperature rapidly approaches room air temperature. Air change rates can be as high as 30 per hour under suitable circumstances. Adjustable and fixed vanes are possible and this offer scope for modifying the air distribution to suit changes in room partitions. When properly selected, such diffusers can be used in both ceiling and floor distribution systems. When mounted in ceilings the manufacturers claim that room-to-supply air temperature differences can lie between +10 and −10 K. If mounted in the floor, smaller temperature differences must be used for cooling applications, consistent with the perception of comfort at ankle level, according to the recommendations of the manufacturers. When floor-mounted, a dirt collection box in the diffuser assembly is needed, the diffuser must be flush with the floor surface and be sufficiently strong and durable to withstand the loads and wear likely in use. Total pressure drops appear to be about 2 or 3 Pa more than for conventional diffusers.

5.8 Permeable, textile, air distribution ducting

This comprises cylindrical, or semi-cylindrical, textile ducting of about 700 mm diameter in lengths up to 50 m. Conditioned air is supplied into one end and flows to the other end of the duct, diffusing uniformly through the fabric walls into the conditioned room. Individual ducts are attached to tracking fixed to the underside of the ceiling. The fabric materials available are polypropylene, polyester, nylon and Nomex. Other, non-inflammable materials can be provided. Cotton is not suitable because it is hygroscopic and because micro organisms can thrive in cotton. Coarse, medium and fine textile weaves are available and, although the usual colour is white, a range of other colours is available.

It is claimed that the materials used provide air filtration up to European standards EU6 or EU7, giving 99.2% removal of particles exceeding 5 μm and 75% removal of particles exceeding 0.5 μm in size. Inevitably the ducting material gets dirty and this is dealt with by periodic laundering or dry cleaning, without shrinkage. Zip fasteners facilitate installation and removal.

Air temperatures in the range from –30 to +50°C can be handled. Operation with static pressures between 60 and 200 Pa is possible but typical working pressures are from 120 Pa (clean) to 160 Pa (dirty). There is a risk of flapping if the static pressure is too low and the design must be arranged with this in mind. Typical mean air velocities within the textile duct are from 6 to 10 m s^{-1} with a face velocity through the circumferential fabric of about 0.1 m s^{-1}. Air change rates as high as 40 h^{-1} can be handled without discomfort.

Although applications are usually industrial, particularly in the food industry, textile ducting has also been used in commercial premises.

5.9 Smudging on walls and ceilings

Air distributed from a diffuser slot or a sidewall grille entrains air from the room and this air is dirty, the dirt having been brought in largely by people entering the room. Streaks of dirt are consequently deposited on ceilings and walls with the passage of time. If there is a gap between the flange of the grille or diffuser and the wall or ceiling in which it is fitted, dirt may occasionally be induced from the ceiling void and deposited on the surface in the room but this is unusual. Dirt streaks are invariably caused by people bringing in dirt on their footwear and clothing from outside. The only practical way of dealing with this is by routine cleaning and it is desirable that wall and ceiling surfaces in the vicinity of supply air outlets have a readily cleanable surface finish. Textured surface finishes are undesirable because of the difficulty of cleaning and this emphasises that smooth, cleanable finishes should be adopted for ceilings in which diffusers are fitted, particularly in places like airports, where passenger foot traffic is very dense. Dirt may also be deposited on the bars of grilles or the rings of diffusers and these must be cleaned from time to time. Some manufacturers offer anti-smudge rings around ceiling diffusers in order to keep the diffuser rings a small distance away from the ceiling itself, the aim being to prevent the dirty induced airstream from flowing over the ceiling surface in the immediate vicinity of the diffuser.

5.10 Ventilated ceilings

Although ventilated ceilings can be used for air change rates from about 7 h^{-1} upwards, they are unsuitable for small duties and only come into practical application for rates exceeding about 25 h^{-1}, where conventional air distribution terminals become difficult to select for draughtless conditions. Proprietary ceilings are available that will satisfactorily distribute from 30 to 520 l s^{-1} m^{-2} of live tile area with static pressures in the ceiling from 3 to 37 Pa, depending on the type of the tile selected. When unavoidable, it is sometimes possible to locate extract grilles in the ceiling itself but this is the least favoured position and it is much preferred to position them at low level in the walls, or in the floor. Rydberg (1962) has shown that if the whole of the ceiling is made live air distribution can be poor, particularly with a low air change rate, and it is even possible to get the downward flow of air moving obliquely across the room to a low-level extract point, giving draughts in part of the room and stagnant conditions elsewhere. The secret of good distribution is to achieve turbulent mixing, above the occupied zone, between the down-moving cold air and the rising, warm, convection currents. This is often done by making only part of the ceiling live, the live tiles being positioned over the major sources of sensible heat gain, e.g. computer cabinets. There is then ample opportunity for entrainment and good mixing.

In computer room applications many of the cabinets have in-built fans that forcibly eject air upwards, ensuring turbulent mixing with the down-coming air and it is then possible, and often essential, for a very much larger proportion of the ceiling to be live. It is generally not feasible to use 100% of the gross ceiling area because of the presence of light fittings and perhaps downstand beams above the ceiling and these dead areas must be subtracted from the gross area to establish the maximum usable net ceiling area. Air change rates as high as 250 h^{-1} have been successfully used with extract grilles at low level in the walls, but with rates up to 720 h^{-1}, as in clean rooms (Section 3.14), it may be necessary to use the whole of the floor area as an extract grille when laminar downflow is wanted.

To get good distribution beneath the ceiling it is first essential to arrange for good air distribution above it by introducing the air to the ceiling plenum chamber through a rudimentary duct system with a number of dampered outlet spigots blowing air horizontally across the top of the suspended ceiling (see Figure 5.6). Large ceiling plenum chambers must be divided into zones by downstand barriers from the soffit of the slab. Such zones should not exceed 500 m^3 and should be of fire-resistant material. The exit velocities from the supply spigots should not be greater than 5 m s^{-1} and the spigots ought to be as far above the suspended ceiling as possible, otherwise there is the risk that air may be entrained upward through the live tiles in the vicinity of the spigot. Access tiles are required near the spigots for damper adjustment during commissioning.

The exact design of the ceiling must be left to the manufacturer but a guide to safe blowing distances, based on manufacturers' literature, is given in Table 5.5.

There is a risk that small holes in the thicker (10–15 mm) non-metallic ceiling tiles may block with dirt, after a period of use. To minimise this risk it is recommended that the air handling plant includes a bag filter with a minimum atmospheric dust spot efficiency of 60% and a pre-filter having an arrestance of 80% (Holmes and Sachariewicz, 1973; ASHRAE, 1977).

It cannot be too strongly emphasised that the chamber above the ventilated ceiling must be properly sealed, insulated and vapour-sealed. If there are cracks or openings in the structure of the plenum chamber the system will be an abject failure because the conditioned air will leak away. If the soffit of the slab and the four side-walls, together with any downstand beams, are not properly insulated, and vapour-sealed, heat gains, or losses, will nullify the effectiveness of the installation. It should be remembered that the temperature difference

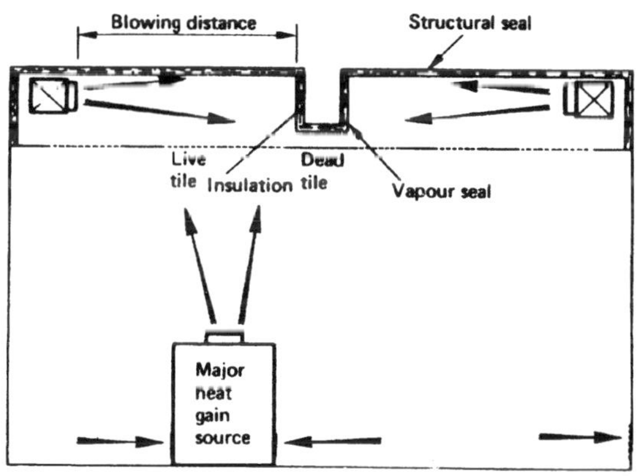

Figure 5.6 Air distribution for a ventilated ceiling.

Table 5.5 Safe blowing distances for a ventilated ceiling

	Safe blowing distance from a supply spigot (m)	
Clear depth of the ceiling plenum chamber (mm)	Parallel to the joists or downstand beams	At right angles to open-web trusses and joists
120–150	6.0–7.5	4.0–5.5
150–210	7.5–9.0	5.5–6.0
210–250	9.0–12	6.0–8.5
250–300	12–15	8.5–10.5
300–450	15–17	10.5–11.5
450–750	17–18	11.5–13.0
750–1000	18–20	13.0–14.0
> 1000	20–23	14.0–16.0

across the structure of the chamber is greater than usual, when calculating heat gains, because the air in the plenum is at about 14°C, instead of the more usual value of 22°C for conventional heat gains. A further cogent reason for insulating the slab, side-walls and beams is to isolate them thermally from the air temperature in the chamber. Otherwise the thermal inertia of the building structure will upset the response of the control system. For example, if a computer is switched off at weekends but the room is kept at 20°C, the supply air temperature to the plenum chamber above the ceiling will also be at about 20°C and the whole of the slab, side-walls and beams will have attained this temperature by start-up time on Monday morning. The outer 75 mm, or so, of the slab, etc. must then be cooled down to the required supply air temperature when the system then begins to try and meet the computer heat gains. It may be some time before the system is achieving good control again.

EXAMPLE 5.8

If the lower 75 mm of the concrete slab over a ventilated ceiling is at a temperature of 21°C determine the time taken for it to reach a value of 15°C if the air is supplied to the plenum chamber at a constant value of 14°C. Assume Newtonian cooling occurs and the slab has a specific heat capacity of 0.85 kJ kg^{-1} K^{-1} and a density of 2150 kg m^{-3}. Take the heat transfer coefficient at the surface of the slab was 9.5 W m^{-2} K^{-1}.

Answer

Newtonian cooling is defined by

$$\theta = \theta_o e^{-K/t} \tag{5.12}$$

where θ is the temperature of the material above a datum at time t, θ_o its initial temperature above the datum and $1/K$ is the time constant of the material, defined by

$$1/K = \text{Heat stored in the material (J kg}^{-1}\text{)}/ \text{Steady-state heat flow through the material (J kg}^{-1}\text{ s}^{-1}\text{)} \tag{5.13}$$

The value of the heat transfer coefficient is virtually independent of the air velocity across it for the velocities likely to be encountered. Considering 1 m^2 of the slab Equation (5.13) can be used to establish the structural time constant:

$$1/K = (0.075 \times 2150 \times 850)/(9.5 \times 3600) = 4 \text{ hours}$$

$$\theta_o = 21 - 14 = 7 \text{ K and } \theta = 15 - 14 = 1 \text{ K}$$

and so, from Equation (5.12)

$$1 = 7\,e^{-t/4} \text{ and } t = 7.8 \text{ hours}$$

5.11 Ventilated floors

Ventilated floors were dealt with in some detail towards the end of Section 3.18, which should be referred to.

5.12 Displacement ventilation

Conventional methods of supplying air for the removal of sensible heat gains deliberately arrange for turbulent mixing of the cold air supplied with the warmer room air. For comfort conditioning, this results in about five to 20 air changes per hour of air movement in the room, depending upon the heat gains. It is argued by the World Health Organization (1983) that this distributes polluting particles and gases generated by the structural materials and furnishings, and by the occupants themselves, producing a room environment that is less than fresh and may contribute to some of the symptoms of discomfort related to the sick building syndrome. An alternative to this method of mixing ventilation is to adopt what is termed displacement ventilation by ASHRAE (1993). Schultz (1993) points out that the sensible heat emitted by natural convection from a person in the room ascends the surface of the body and attains a mean velocity of about 0.25 m s^{-1} at head height, increasing in volume by thermal expansion and by entraining a small amount of the surrounding room air. This establishes a plume of warm air that rises out of the occupied zone and forms an upper layer of polluted air which is eventually extracted from the room. Displacement ventilation is designed to supply air at low level to form a shallow pool of cool, clean air that feeds the plume at ankle height (Figure 5.7). Air movement and temperature are critical factors at this position if comfort is to prevail: the air velocity approaching peoples' ankles should not exceed 0.15 m s^{-1} and the temperature should not be cooler than about one or two degrees less than the dry-bulb in the rest of the occupied zone. An advantage is that the plume of air around the individual is clean and has not been contaminated by mixing with any of the pollutants in the room. Other sources of convective heat gain in the room also generate plumes.

Schultz (1993) has suggested that displacement air distribution involves about 2.5–3.0 air changes per hour but the quantity supplied must be enough to feed the plumes in each case and this depends on the perimeter of the plume (because the air can only be supplied to a plume at its perimeter) and on the height of the convective heat source above the floor: sources near the top of the occupied zone require less air than those nearer the floor. The rate of air supply must also be enough to deal with any latent heat gains. Cool air may be supplied from wall-mounted diffusers located on peripheral walls at their junction with the floor. Even though the air velocity through the diffuser outlet area into the room is low, the air must descend onto the floor, under the influence of gravity, in order to form the pool

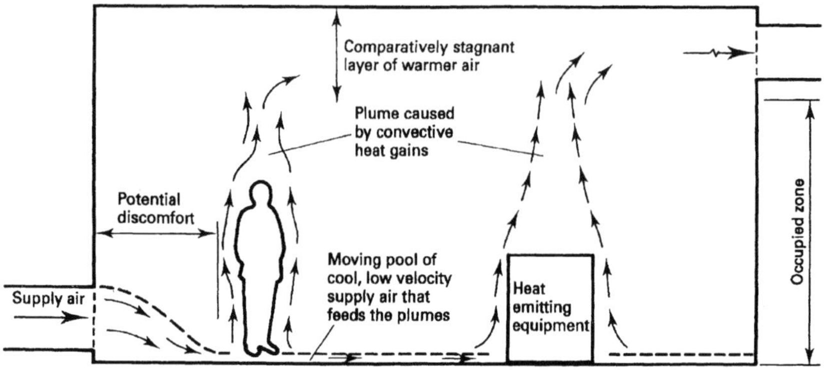

Figure 5.7 Displacement ventilation by supplying cool air from a wall-mounted diffuser at low level.

that spreads across the floor and feeds the plumes. It follows that there is likely to be a zone of discomfort that cannot be occupied, in the vicinity of a wall-mounted diffuser (Figure 5.7). If this form of air distribution is adopted it is essential that the extent of such an area of discomfort is established but not all manufacturers' catalogue data are clear and explicit on this matter. An alternative form of air distribution is from swirl diffusers mounted directly in the floor. Here also there is a local area around each diffuser where comfort may be lost but careful attention to the choice of supply air velocity and temperature, according to manufacturers' recommendations, is likely to solve the problem in many cases. Perforated floor tiles are used but the temperature differences adopted are very much smaller for comfort conditioning applications than those adopted with computer air conditioning (see Section 3.18). It is generally considered that the temperature difference from foot-to-head should not normally exceed about 1.5–2 degrees, with 3 degrees as an absolute limit. It follows that the supply air temperature used with displacement air distribution to deal with sensible heat gains in comfort air conditioning can only be about 2 degrees less than the room air dry-bulb in the occupied zone. This imposes a limit on the sensible cooling capacity provided by the air displacement method and supplementary radiant cooling by chilled ceilings may be needed. (It would not be a practical alternative to use fan coil units in the occupied zone to make up the necessary sensible cooling capacity because doing so would introduce local turbulent air mixing and nullify the displacement air distribution arrangement. Similarly, chilled beams might be unsuitable because of many of these give only a small proportion of their sensible cooling by radiation.) On the other hand, the comparatively stagnant, upper layer of air can be at a temperature that is much higher than that in the occupied zone and this allows the supply air to provide additional sensible cooling. Furthermore, the convective heat emissions from lighting at high level increase the air temperature in the stagnant layer, rather than in the occupied zone, particularly if the layer is 1.5 m or more deep. Radiant emission from the lights is a sensible heat gain in the occupied zone. Since the supply air temperature is usually only about 2 K less than the room dry-bulb, the sensible heat gain that can be dealt with by a system of displacement air distribution is about 25–40 W m^{-2}, referred to the treated floor area, according to ASHRAE (1993). This then gives a vertical temperature gradient of about 3 K in the occupied zone. (Engineering prudence suggests potential difficulties in attaining such sensible cooling capacities with air alone: for a similar arrangement but a higher air change rate with conventional, turbulent, mixing air distribution, room-to-supply air temperature differences are about 9 K and sensible cooling capacity is consequently about 90–100 W m^{-2}.) The use of air distribution terminals that induce some air from the room permit a lower supply air temperature from the diffuser and Jackman (1991) claims that as much as 50 W m^{-2} of sensible cooling capacity is then possible. The eddies and air patterns within the occupied zone when displacement air distribution is adopted are much gentler than with conventional, mixing distribution and Twinn (1994) points out that an opened door can upset a plume. Twinn (1994) also claims that displacement distribution is not suitable for winter heating – a perimeter heating system is needed.

5.13 The influence of obstructions on airflow

Accurate information for predicting the behaviour of an airstream upon encountering an obstruction is not available but Holmes and Sachariewicz (1973) have proposed a simple method for an approximate assessment. It is based on experimental results for airflow from a slot across a ceiling that meets a nearby, parallel, downstand beam. Both the beam and the slot extend over the full ceiling width and the influence of temperature is said to be negligible.

When an airstream moves over a surface and meets a barrier, such as a downstand beam or a light fitting (see Figure 5.8) it may behave in one of three different ways:

(1) The airstream may closely follow the contours of the barrier. This happens if the distance from the slot to the barrier, x_d, exceeds $8x_c$, where x_c is a critical distance that can be obtained approximately from Figure 5.9. To use the figure the nominal height of the slot, h, is determined by assuming that the total pressure drop across the slot equals to the velocity pressure of the airstream emitted from it. The downstand dimension of the barrier is d.

(2) The airstream may permanently separate from the ceiling, at the beam, and flow downwards in to the occupied zone. This happens when $x_d < x_c$.

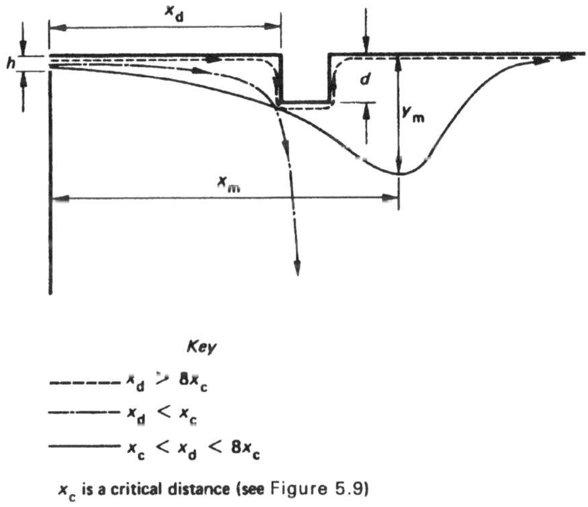

Figure 5.8 Behaviour of an airstream encountering an obstruction.

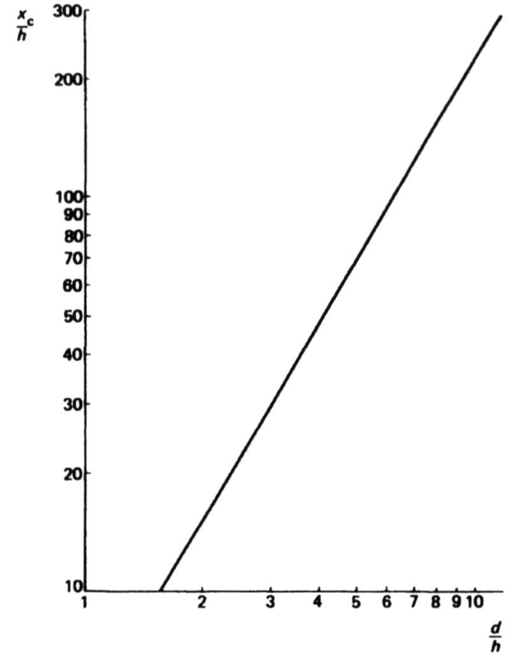

Figure 5.9 Relationship between critical distance and beam depth when an airstream follows the contours of the beam (after Holmes and Sachariewicz, 1973).

(3) The airstream may leave the ceiling at the beam but return to it on its downstream side. This happens if $x_c < x_d < 8x_c$. The maximum, vertical separation of the centre-line of the airstream from the ceiling, y_m, can be determined approximately from Figure 5.10.

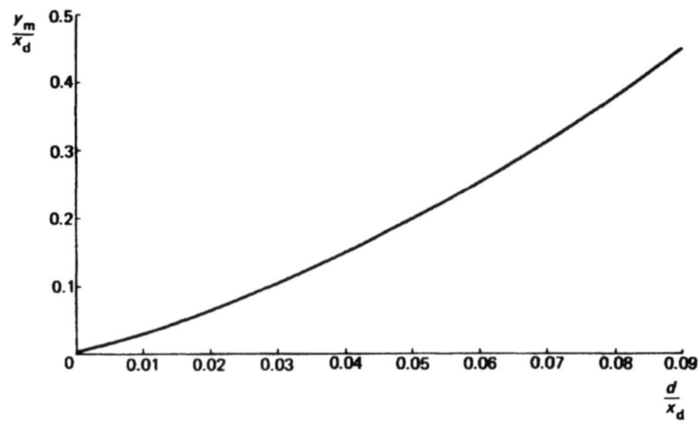

Figure 5.10 Maximum separation of an airstream from the ceiling on its downstream side (after Holmes and Sachariewicz, 1973).

EXAMPLE 5.9

A continuous slot along the full distance of the cornice at one end of a room is 7.2 m wide and delivers 400 l s^{-1} across the ceiling towards a continuous, downstand beam of 80 mm square section, parallel to the slot and 1.5 m distant from it. The total pressure drop through the slot is 10 Pa and the floor-to-ceiling height is 2.6 m. Assess the probable behaviour of the airstream.

Answer
Total pressure drop = velocity pressure, by assumption. Therefore,

$$10 = 0.6\, v_c^2$$
$$v_c = 4.082 \text{ m s}^{-1}$$

Nominal slot area = 0.4/4.082 = 0.098 m^2, and nominal slot height, h, = (0.098 × 1000)/7.2 = 13.6 mm.

$$d/h = 80/13.6 = 5.88$$

From Figure 5.9, $x_c/h = 90$, $x_c = 1.22$ m. $x_d < 8 \times 1.22$ m, therefore the airstream will not follow the contours of the beam. $x_d > 1.22$ m, therefore the airstream will return to the ceiling after a temporary excursion from its beyond the beam.

$$d/x_d = 0.08/1.5 = 0.0533$$

Therefore, from Figure 5.10

$$y_m/x_d = 0.22 \text{ and so } y_m = 0.33 \text{ m}$$

Remembering that the airstream expands and that it is the centre-line that is 330 mm below the ceiling, it is prudent to add 300 or 400 mm to the generally accepted value of 1.8 m for the height of the occupied zone. In this example, the centre-line of the airstream is 2.27 m from the floor and it is therefore probable that no discomfort will be felt in the occupied zone, near the beam.

The velocity of the airflow on the downstream side of the beam is difficult to predict with certainty. Holmes and Sachariewicz (1973) stated that "very large, low-frequency fluctu ations in velocity were experienced in the flow downstream of the barrier which made mean readings difficult to obtain". When the air supply opening does not extend over the full width of the ceiling but the downstand obstruction does, the critical distance, x_c, is increased and the method is not strictly applicable except with caution. It can only be regarded as giving a rough indication of behaviour. When the obstruction is less wide than the slot, the critical distance is decreased and the method may be applied to give approximate results. It is claimed by Holmes and Sachariewicz (1973) that when the span of the obstruction is less than half the span of the slot, the effect of barrier can be ignored, provided that $x_4 > x_c$.

5.14 Extract air distribution

Extract air distribution has been considered in Section 1.6 when extract light fittings were dealt with and needs little further comment, except to reiterate that the location of the extract opening has virtually no effect on air distribution in a room.

5.15 Air distribution performance index

Miller and Nash (1971), Nevins and Miller (1974) and ASHRAE (1993) have established experimentally that if the effective draught temperature, t_{ed}, lies between −1.5 and +1.0°C and if the local air velocity, v_x, is below 0.35 m s^{-1} then a large majority of the occupants in a room are likely to be comfortable. Effective draught temperature is defined by

$$t_{ed} = (t_x - \bar{t}_r) - 8(v_x - 0.15) \tag{5.14}$$

where t_x is the local air dry-bulb temperature in degrees centigrade and $\bar{t}_r$ is the mean room dry-bulb temperature in degrees centigrade. The air distribution performance index (ADPI) is the percentage of positions in a room where measurements of the effective draught temperature lie between −1.5 and +1.0°C and v_x is less than 0.35 m s^{-1}. The ideal would be an ADPI of 100% but, generally speaking, a value of 80% meets the most critical standards of comfort.

The ASHRAE handbook (1993) quotes characteristic room lengths, related to a terminal velocity of 0.25 m s^{-1} for all terminals except linear diffusers where 0.5 m s^{-1} pertains. These lengths are the horizontal distances from the air outlet to the nearest wall or obstruction, or half-way to the nearest outlet blowing in opposition plus the vertical distance downwards to the occupied zone, and if they are taken to be values of throw, the ADPI will be at a maximum. Measurements leading to the determination of an ADPI would only be relevant in the occupied zone.

5.16 Variable volume air distribution

The supply of air from grilles or diffusers in constant volume systems is designed to provide a throw to a terminal velocity that is consistent with human comfort in the occupied zone (see Section 5.3). This is achieved by the Coanda effect (Figure 5.1) which presses the jet of supply air against the ceiling. In the case of variable air volume (VAV) systems the momentum flow of the air jet supplied to the room is continually changing as the airflow rate supplied is intentionally varied, in response to thermostatic control as the sensible heat gains to the room alter. If the air distribution of a VAV system is to be successful proper attention must be paid to the characteristic behaviour of the supply air jet under all expected conditions of operation.

When the supply airflow rate is throttled the velocity reduces, the jet entrains less air from the room, the friction at the jet–ceiling interface (approximately proportional to the square of the air velocity) also reduces and the pressure difference between the air in the room and the ceiling tends to diminish rapidly. In due course, if the reduction in the sensible heat gains and the supply airflow rate continues, the jet of cold air leaves the ceiling and falls into the occupied part of the room, causing discomfort. This is called "dumping" and the airflow rate just before dumping occurs defines the minimum volumetric flow rate that a particular type of variable supply air terminal can provide.

There are many different types of supply air terminal used in VAV systems but one may generalise and refer to them as being of fixed or variable geometry form. See Figure 5.11.

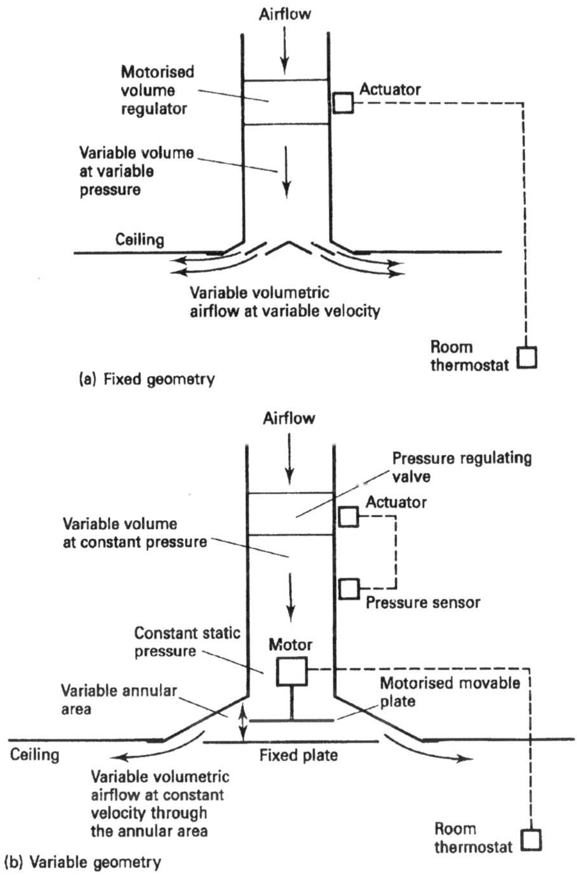

Figure 5.11 Fixed and variable geometry supply airflow terminals.

In Figure 11(a) room temperature is controlled using a fixed geometry ceiling diffuser. A motorised damper is regulated directly from a temperature sensor in the room. Hence the static pressure varies in the space above the diffuser rings and consequently the velocity through them varies as the volume is changed.

In Figure 11(b) a pressure regulator stabilises the pressure above the diffuser and the room temperature is separately controlled by a motorised plate that moves in the neck of the diffuser, providing an annulus of varying area in response to thermostatic demands. Since the pressure on the upstream side of the movable plate is constant the velocity of airflow through the variable annular area is also constant and hence airflow out of the diffuser is variable but at constant velocity.

Numerous different types of variable geometry supply terminals are available, some of which are motorised and some of which are self-acting, using the static pressure in the upstream duct as the power source for operation. All achieve much the same result: fixed geometry devices can turn down to about 40% of design airflow rate but the variable geometry versions can throttle to about 20–30% of design airflow.

Matters can be improved by the provision of a constant, minimum airflow from the central part of the diffuser, underneath the variable airflow. Figure 5.12 illustrates the principle of such diffusers. The constant airflow, representing the minimum percentage that the diffuser can handle, lies beneath the variable part as it spreads across the ceiling and helps to prevent the variable airflow from leaving the ceiling and falling into the room.

When a VAV terminal throttles to its minimum value and the jet leaves the ceiling, a natural rotary air circulation is set up in the far part of the room, beyond the dumped jet. Such air circulation represents kinetic energy and when more airflow is required from the VAV terminal it must have enough energy to be able to break up the pattern of rotary movement, before the jet can re-attach to the ceiling and normal air distribution can return. For example, dumping might occur when the airflow rate diminished to 30% of the design value but normal air distribution might not resume until the airflow had increased to 40%. This hysteresis effect is illustrated in Figure 5.13.

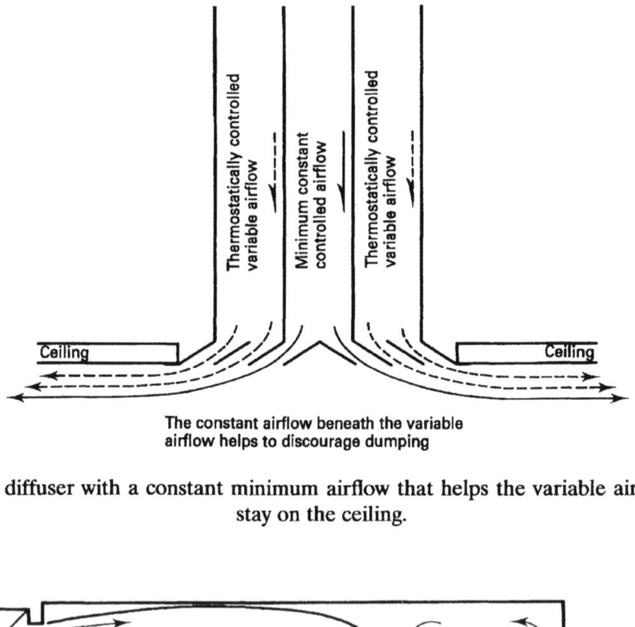

Figure 5.12 A diffuser with a constant minimum airflow that helps the variable airflow above it to stay on the ceiling.

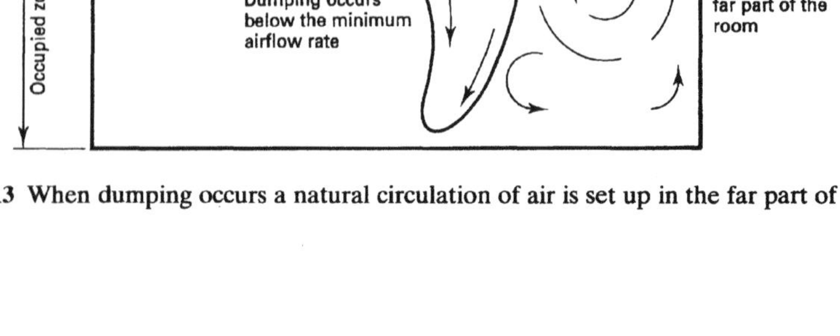

Figure 5.13 When dumping occurs a natural circulation of air is set up in the far part of the room.

Although the volumetric capacity of the extract fan that removes air from a building is varied in unison with the supply fan this only ensures that the total amount of air extracted matches the variable amount supplied, for the whole building. A possible consequence is that there may be variations in the static pressure in the various rooms themselves, perhaps with some difficulty in opening and closing doors. A case would then seem to emerge for the provision of extract VAV terminals on a modular basis, thermostatically controlled in unison with the supply terminals. There are obvious commercial objections to this. Fortunately, it appears that in the case of VAV systems which are competently designed, installed and commissioned, objectionable pressure variations in rooms are rare.

Notation

Symbols	Description	Unit
A_d	Cross-sectional area of a duct	m^2
A_g	Gross area of an outlet	m^2
A_s	Area of a slot in a linear diffuser	m^2
(Ar)	Archimedean number	–
B	Breadth of a room	m
D_h	Mean hydraulic diameter of a room	m
D_o	Diameter of an opening	m or mm
H	Floor-to-ceiling height in a room	m
K	Coefficient of proportionality	–
K'	Constant of proportionality	–
L	Length of a room or a horizontal distance from an outlet to the opposite wall	m
Q	Volumetric airflow rate	$m^3\ s^{-1}$ or $l\ s^{-1}$
R	Aspect ratio of an opening (b/h)	–
R_{fa}	Ratio of free to gross area of an outlet	–
T_r	Absolute room temperature	K
b	Breadth of an opening	mm or m
c_d	Coefficient of discharge of an outlet	–
d	Vertical dimension of a downstand obstruction in a ceiling	mm or m
g	Acceleration arising from the force of gravity	$m\ s^{-2}$
h	Height of an outlet	mm or m
p_s	Static pressure	Pa
p_v	Velocity pressure	Pa
t_{ed}	Effective drought temperature	°C
t	Time	s or h
t_o	Initial temperature of a non-isothermal jet	°C
t_r	Room temperature	°C
$\bar{t}_r$	Average room temperature	°C
t_w	Temperature of the heated surfaces in a room	°C
t_x	Mean temperature of an air jet at a distance x from an outlet or local air temperature	°C
u_r	Fictitious mean air velocity through the cross-section of a room	$m\ s^{-1}$
v_c	Nominal mean velocity through the area of an orifice or an open-ended duct	$m\ s^{-1}$

v_o	Effective velocity of an air jet over the section of the vena-contracta or the average velocity through an open-ended duct	m s^{-1}
v_x	Centre-line velocity in an air jet at a horizontal distance *x* from an outlet or local air velocity	m s^{-1}
x	Distance horizontally from an outlet	m
x_c	Critical distance of an obstruction from an air supply slot	m
x_d	Distance of an obstruction from an air supply slot	m
x_{max}	Maximum downward throw of a warm air jet	m
y	Vertical drop of the centre-line of a non-isothermal jet below the horizontal centre-line of an outlet	mm or m
y_m	Maximum vertical separation of the centre-line of an airstream from a ceiling	mm or m
z	Separation between adjoining side-wall grilles	mm or m
θ	Angle between a linear diffuser and the airstream flowing from it or temperature above a datum	angular degrees or °C
θ_o	Initial temperature above a datum	°C

References

ASHRAE Handbook, *Fundamentals*, SI Edition, Chapter 31, Space air diffusion, 1993.

Farquharson, I. M. C., The ventilating air jet: Part I. *JIHVE*, **19,** 449, 1952.

Frean, D. H. and Billington, N S., The ventilating air jet, Part II. *JIHVE*, **23,** 313, 1955.

Holmes, M. K. and Sachariewicz, E., The effect of ceiling beams and light fittings on ventilating jets. HVRA Laboratory Report No. 79, 1973.

Jackman, P. J., Air movement in rooms with side-wall mounted grilles – a design procedure. HVRA Laboratory Report No. 65, 1970.

Jackman, P. J., Displacement ventilation. CIBSE National Conference, University of Kent, Canterbury, 1991.

Koestel, A. and Young. C. Y., The control of airstreams from a long slot. ASHVE Research Report No. 1429, *ASHVE Trans.*, **57,** 407, 1951.

Koestel, A., Computing temperatures and velocities in vertical jets of hot or cold air. *ASHVE Transactions*, **60,** 385–410, 1954.

Koestel, A., Herman, P. and Tuve, G. L., Comparative studies of ventilating jets from various types of outlets. ASHVE Research Report No 1404, *ASHVE Trans.*, **56,** 459, 1952.

Koestel, A., Paths of horizontally projected heated or chilled air jets. *ASHVE Trans.*, **61,** 213–32, 1955.

Miller, P. L. and Nash, R. T., A further analysis of room air distribution performance. *ASHRAE Trans.*, 77, Part II, 205, 1971.

Müllejans, H., Uber die Ähnlichkeit der nichisotherm Strömung und den Wärmeübergang in Raümen mit Strahllüftung, Forschungsberichte des Lauds Nordrhein–Westfalen, Nr 1656, Westdeutscher Verlag, Koln und Opladen, 1966.

Nevins, R. G. and Miller, P. L., ADPI – an index for design and evaluation. *Australian Refrigeration, Air Conditioning and Heating*, **28,** 7, 26, 1974.

Parkinson, J T L and Billington, N S., The ventilating air jet, Part III. *JIHVE*, **24,** 415, 1957.

Rydberg, J., Introduction of air to perforated ceilings. HVRA Translation No. 45, June 1962.

Schultz, U., A new eye for indoor climate. *Heating and Air Conditioning*, 1 July, 24–26, 1993.

Tuve, G. L., Air velocities in ventilating jets. ASHVE Research Report No 1476, *ASHVE Trans.*, **59,** 261–82, 1953.

Twinn, C., Displacement ventilation – fact or fiction? *Building Services*, June, 39–41, 1994.

World Health Organization, Indoor air pollutants, exposure and health effects. *European Reports and Studies*, **78,** 23–26, 1983.

6

Plant Location and Space Requirements

6.1 Plant location

Jones (1970) suggests that the art of plant location is to minimise the lengths of piping and ductwork that link the plant with ancillary components or terminal units, while ensuring proper performance. The implications of this are seen by considering some common examples.

COOLING TOWERS

Because copious quantities of outside air are essential and its free flow to and from the tower must not be compromised by the proximity of neighbouring structures, the best location for a cooling tower is often on a roof. Towers should be sited on the prevailing upwind side of any flues, to minimise the risk of cooling water contamination and the pond water level must always be sufficiently above the suction branch of the cooling water pump to keep it primed at all times (see Section 4.8). If need be, the tower should be erected on stilts.

AIR-COOLED CONDENSERS

Similar considerations apply about airflow through air-cooled condensers, with the additional constraint that the practical maximum length of the refrigerant lines leading to the compressor and evaporator is often about 15 m. Furthermore, when evaporators are above condensers there is the risk that the loss of static head in the rising liquid line may cause flashing to vapour before the expansion valve is reached.

REFRIGERATION PLANT

Where the roof can bear the dynamic loads imposed during operation and there is no risk of noise and vibration transmission to the building, the proper precautions being taken (see Sections 7.12, 7.20 and 7.26), refrigeration plant can be successfully erected on roofs. There is then the advantage that the lengths of cooling and chilled water pipelines to cooling towers and air handling units are minimised. Nevertheless, the problems of roof installation are sometimes misunderstood or underestimated by engineers and architects and it is frequently better to put water-cooled water-chillers in the basement on a solid foundation, where noise and vibration will pose fewer difficulties. In the UK it is not permissible to have refrigeration plant in the same room as the boilers because of the inflammable or toxic nature of the refrigerants generally used.

AIR-HANDLING PLANT

With smaller plant using direct-expansion air-cooler coils the restrictions on the lengths of refrigerant lines, as with air-cooled condensers usually, apply but with larger plant having chilled water cooler coils there is no such restraint on location. Air-handling plants can be put at basement level but this is often a poor choice because space must then be found for the vertical duct mains needed for supply air up the building, extract air downwards, discharge air upwards if discharge to waste locally is impossible, and fresh air down the building if a local outside air inlet is out of the question. On the other hand, when plant is on the roof only two vertical ducts are required: for supply and extract air. With high-rise buildings the cost of duct-work and the space it occupies can be minimised by putting air-handling units in a plant room on an intermediate floor, specially intended and constructed for the purpose. Between 10 and 16 storeys above and below can be conveniently dealt with.

6.2 Cooling tower space

Broadly speaking, towers of the same type occupy the same volume, per kW of refrigeration, when working under the same conditions, although induced draught towers tend to need more space than forced draught towers, particularly at the lower end of the range of duties. Data published in catalogues suggest a range of specific volumes required for commercially available induced and forced draught towers: 3–4 m^3 per 100 kW for the former and 2–3 m^3 per 100 kW for the latter.

Shapes, heights and types of cooling towers differ and there is consequently less correlation between specific plan area and refrigeration load (Figure 6.1). Small towers are less efficient than larger ones and induced draught versions require more space than forced draught. Generalisation is not as easy as with specific volumes but it can be inferred that forced draught towers require from 0.6 to 0.8 m^2 per 100 kW, whereas induced draught need between 1.0–1.5 m^2 per 100 kW, for loads over about 1000 kW of refrigeration. Increasing the ambient wet-bulb from 20°C, assumed for Figure 6.1, to 25.6°C means that the specific plan area rises by about 15% for the induced draught type and by about 20% for the forced draught tower.

As well as the plan area of the tower itself, a walkway space of at least 1.5 m must be provided all round, for maintenance, and, in addition, sufficient extra plan area must be allowed for free airflow into the tower. The total plan area is unlikely to be less than twice the plan area of the tower alone; for a provisional estimate of the plan area needed it is suggested that 5 m^2 plus 2.5 m^2 per 100 kW of refrigeration be allowed.

6.3 Air-cooled condensers

Horizontally arranged condenser coils are always to be preferred to vertical coils because the vagaries of wind effect can then be ignored. Since air-cooled condensers are multi-row, finned-tube, heat exchangers the volume occupied is much less significant than the plan area, which depends on the refrigerant used, e.g. R12 needs slightly more than R22, where the duty lies in the commercial range of sizes available and a typical difference between the condensing refrigerant and the ambient air. However, Figure 6.2 shows that the

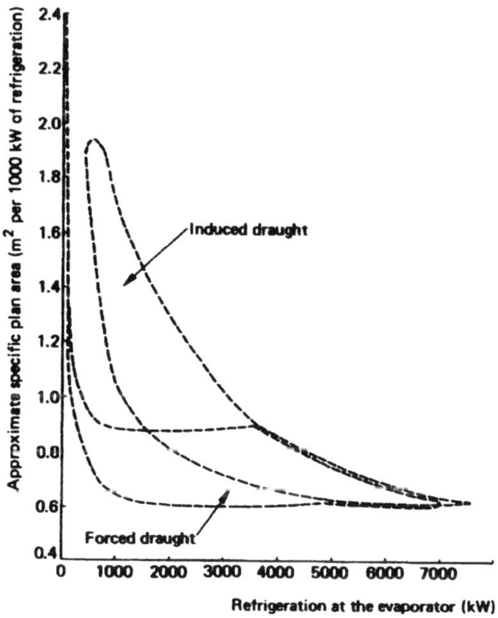

Figure 6.1 Comparative approximate plan areas occupied by typical commercial cooling towers. Based on 20°C wet-bulb (screen) and water cooled from 32.2 to 26.7°C.

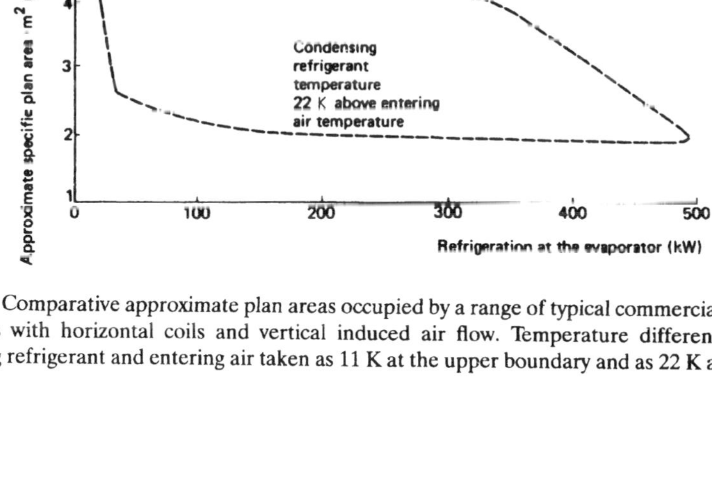

Figure 6.2 Comparative approximate plan areas occupied by a range of typical commercial air-cooled condensers with horizontal coils and vertical induced air flow. Temperature difference between condensing refrigerant and entering air taken as 11 K at the upper boundary and as 22 K at the lower.

approximate specific plan area required is from 2.5 to 5.0 m^2 per 100 kW for duties exceeding about 100 kW of refrigeration, but below this the requirement increases to 8 m^2 per 100 kW, or more, of refrigeration. At least twice these plan areas should be allowed for preliminary planning purposes.

6.4 Water chillers

Figure 6.3 shows the results of a survey of manufacturers' literature for typical, commercial, centrifugal and screw chillers. The specific space allowance includes 1 m all round for access, plus tube withdrawal space at one end. Curve A is for a commonly used hermetic, centrifugal machine with a single shell to house condenser and evaporator tubes and the motor-compressor assembly mounted on top. Curve C is for machines having separate condenser and evaporator shells and their hermetic compressors positioned alongside. The end of the curve, for the larger duties, includes open centrifugal machines intended to be driven by steam turbines. Curve B is for a middle range, single shell, hermetic, centrifugal machine that is shorter than the others but somewhat wider. The dashed line is for screw compressors, which tend to be rather longer but narrower, than centrifugals. Reciprocating chillers are used in the lower span of duties, i.e. less than about 550 kW of refrigeration, and in general, use more space. They are not shown in Figure 6.3, but can need as much as 10 m^2 per 100 kW of refrigeration, for small duties.

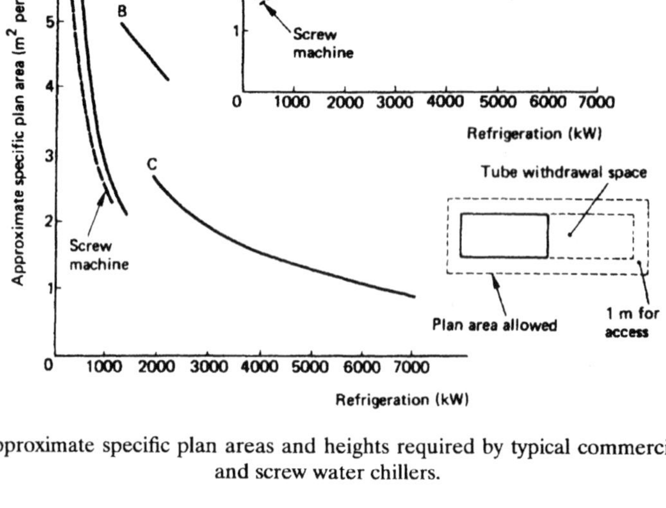

Figure 6.3 Approximate specific plan areas and heights required by typical commercial centrifugal and screw water chillers.

The specific space requirements given in Figure 6.3 are the bare minimum for the machines. Additions must be made to cover the height of concrete bases and antivibration mountings, and overhead clearance is essential to accommodate large diameter, lagged chilled water and unlagged cooling water pipes. Piping connections also have to be made at the ends and pumps are usually located nearby, often with stand-by. Switchgear and controls occupy space as well. Furthermore, the cooling duty is frequently shared among several machines to give a measure of stand-by capacity. To cover all these items it is recommended that, at the preliminary formative design stage, i.e. before any layout drawings are prepared, an allowance of 8 m^2 plus 8 m^2 per 100 kW of refrigeration be made, for centrifugal and screw machines. Thus, for a cooling load of 1000 kW, the provisional estimate of space would be $8 + 8 \times 1000/10 = 88$ m^2 of plan area. For reciprocating chillers it is recommended that manufacturers' dimensions be obtained before any estimate of the space needed is done.

Finally, it is fatal to underestimate the clear height required in plantrooms, of any sort. Figure 6.3 shows that up to 4 m can be needed. Less than 3.5 m often results in difficulties.

6.5 Air-handling plant

Commercial, packaged, air-handling units occupy between 0.4 and 0.1 m^2 per m^3 s^{-1} of air delivered and their heights are from 0.5 to 2.0 m over the range of duties from 0.5 to 20 m^3 s^{-1}, the better quality, quieter units requiring more space and being higher than those that are cheaper and noisier. Individual, forward-curved, centrifugal fans need from 0.06 to 0.03 m^2 per m^3 s^{-1} of air, over a similar span of airflow rates.

Air-handling plantrooms may accommodate a packaged air-handling unit for supply air; an extract fan and motor set; ducts for supply, recirculated, discharge and fresh air; silencers; duct-mounted heated batteries; controls; switch-gear; pumps and lagged pipes. The ducts are large and of a complicated shape, usually having to cross over one another, and plant is best erected on concrete plinths. Taking account of these factors and of the need for enough space to give access for maintenance suggests that, for preliminary planning purposes, it is wise to allow 12 m^2 plan area plus 13 m^2 per m^3 s^{-1} of air handled, with a clear internal height of 2.7 m plus 0.1 m per m^3 s^{-1} of airflow, up to a maximum of 4.5 m.

6.6 Systems

With the exception of Swain *et al.* (1964) little information giving details of the space occupied by different systems has been published. Table 6.1 gives typical space requirements for the mechanical plant of air-conditioning systems as a percentage of the treated area. Some latitude is necessary when interpreting the data, bearing in mind the figures relating space occupied with plant duty. The tabulated information is perhaps best used for comparative purposes, the actual plan area needed depending greatly on the system design and the building shape and size. The clear ceiling space required by different systems is a function of the disposition of the terminal units, grilles or diffusers, their selection and the ductwork arrangement (see Figure 6.4). Suggestions are given in Table 6.2. Putting induction units above suspended ceilings is really a misapplication (see Section 3.2) and although successful results have been achieved, air distribution patterns and unit capacities cannot be determined from catalogue data but must always be established by tests in a full-scale mock-up of the building module.

The plan area occupied by terminal units should be determined by actually using dimensional details of the units, properly laid out on architect's drawings. However, Table 6.3 gives

Table 6.1 Typical space requirements for the mechanical plant of air-conditioning systems

System	Total area of plant space as a percentage of treated floor area: From Swain *et al.* (1964)	Suggested allowance
Terminal heat-recovery units, plus ducted fresh air	–	5
Perimeter-induction	4.5	5
Fan coil units, plus ducted fresh air	4.5	5
Chilled ceiling, plus ducted fresh air	5.1	5
Single duct, all-air, low or high velocity	5.9	7
Dual duct, high velocity	6.5	8

Table 6.2 Clear ceiling space requirements of different air-conditioning systems

System	Clear ceiling space needed (mm)
Chilled ceiling. plus ducted fresh air fed from side-wall grilles in a bulkhead	120*
Variable air volume (see Figure 6.4)	370–600
Ceiling-mounted induction units	450
Ceiling-mounted fan coil units	500
Chilled ceiling, plus ducted fresh air fed from ceiling diffusers	500†‡
Water loop air conditioning/heat pump units	550
Dual duct feeding ceiling diffusers	600
Minimum return air space (Section 1.6)	150

*More clear space must be allowed if pipes have to cross over one another

†Smooth air flow into the necks of the diffusers is essential if quiet operating conditions are to be achieved. More than 500 mm may, therefore, be needed to secure this.

‡Reducing the ceiling space to a minimum for this type of diffuser, or any system using ceiling diffusers, is inviting noise problems

Table 6.3 Floor area needed by terminal units of air-conditioning systems

Terminal unit	Floor area occupied as a percentage of the total treated floor area
Perimeter-induction, conventional	1.3–2.8
Perimeter-induction, low silhouette	1.6–2.7
Fan coil	1.3–2.0
Water loop air conditioning/heat pump units	1.5–2.1

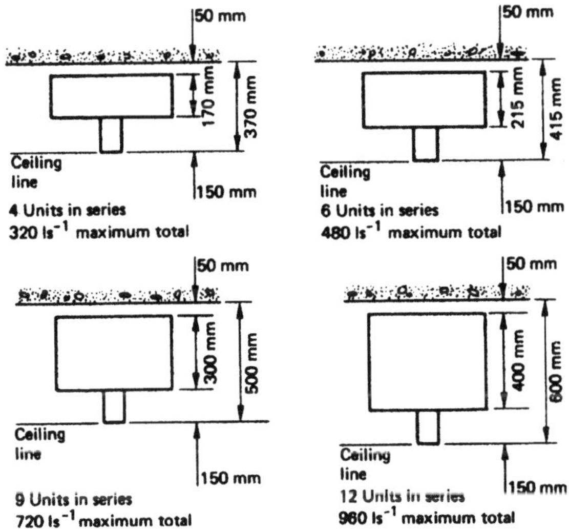

Figure 6.4 Ceiling space requirements for a variable air volume system with variable area linear slots and the possibility of partitions running along the slots, independent temperature control being retained in the offices on each side (see Section 2.8 and Figure 2.10).

an indication of the floor area used up. The actual floor area required by the units depends on the building design, the sensible heat gains and the unit selection in the commercial range available. Four-pipe induction units are at the upper end of the range quoted in Table 6.3 but four-pipe fan coil units do not seem to need significantly more space than two-pipe units, a split four-row coil with three rows for cooling and one for heating, is no bigger than an ordinary four-row coil which is used entirely for cooling.

6.7 Duct space

At a very early stage in the design process an estimate must often be made of probable space to be occupied by the ductwork. It is never wise to underestimate space requirements at the conceptual phase and Table 6.4, therefore, lists conservative suggestions for the supply air quantities needed for office blocks. Generalisation for other applications is best made by

Table 6.4 Supply air quantities required for office blocks with different air-conditioning systems

System	Specific supply air quantity ($l\ s^{-1}\ m^{-2}$)
Perimeter induction	3.25
Variable air volume	8.5
Double duct	12.5
Ducted, uncooled, fresh air for a fan coil system	1.8–2.7*
Ducted, cooled and dehumidified air for a fan coil system	3.0–5.0†

*Depends on the amount of smoking
†Depends on the moisture contents of the supply air and room air.

means of a rough estimate of the sensible heat gain and the use of Equation (2.3), bearing in mind that the choice of conventional air distribution terminals (Chapter 5) is limited to about 25 air changes per hour or 12.5 $l\ s^{-1}\ m^{-2}$. The figures in Table 6.4 refer to the total treated floor area, i.e. the modular area plus the area of the central corridor.

6.8 Miscellaneous items

Although air conditioning is often regarded as merely comprising refrigeration plant and cooling towers, etc. with systems of air and water distribution to deploy cooling capacity, there are other mechanical services that must be provided for an air-conditioned building if it is to be habitable. Table 6.5 gives the approximate plantroom space required for most of these other services, specifically applied to office blocks. The designer must exercise judgement if he or she wishes to refer the data to different applications. The figures in Table 6.5 include an allowance for maintenance and, where relevant, tube withdrawal or burner removal. Plantrooms should be of rectangular shape with a plan aspect ratio of 2:1. An excessive departure from this, an unusual shape, or the presence of columns, will increase the area needed. An air conditioned building seems to require between 5–15% of the treated floor area as space for the plant of the mechanical services and vertical shafts to house pipes and ducts. Small buildings need more plant space than large ones, proportionally.

EXAMPLE 6.1

Make a preliminary estimate of the plantroom floor area that should be allocated for air conditioning the hypothetical office block considered in Chapter 1, assuming the following:

- System. Four pipe, fan coil with a low velocity ducted supply and extract system to deal with the latent heat gains.
- Refrigeration load. 1616 kW (Example 1.11).
- Boiler power. 1500 kW.
- HWS – 7000 l.
- CWS – 84 000 l, to cover domestic needs and the storage required by most water companies in the UK for cooling towers, i.e. half a day's usage or four hours consumption at design duty, whichever is the greater.

Answer

(1) Air handling plant (see Table 6.4 and Section 6.5):
$(3.5 \times 13\,997)/1000 = 49.0\ m^3\ s^{-1}$. A duty of $49.0\ m^3\ s^{-1}$ would be too much for a single plant and well beyond the range of packaged plants. Four packages would probably be chosen, each to handle $12.25\ m^3\ s^{-1}$.
Plant area needed (see end of Section 6.5) $4[12 + 13 \times 12.25] = 685\ m^2$
Clear height required: $(2.7 + 0.1 \times 12.25) = 3.925$ m

(2) Refrigeration plant (see Section 6.4): two centrifugal water chillers, each of 808 kW capacity.
Plant area needed for each machine: $2(8 + 8 \times 808/100) = 146\ m^2$
Clear height required (Figure 6.3): 3.5 m

(3) Cooling towers (see Section 6.2): Assume two induced draught towers, each for a refrigeration duty of 808 kW.
Volume required: $(4 \times 808)/100 = 32.32\ m^3$

Plan area needed for two towers: $2(5 + 2.5 \times 808/100) = 50\ m^2$
Overall height. required: difficult to say with certainty but probably wise to allow 5 m.

(4) CWS (see Table 6.5): Assume the lower tank heights are desirable.
Plant area needed: $9 + 1.5 \times 84 = 135\ m^2$
Clear height required: 2.2 m

(5) Boilers (see Table 6.5): Assume three boilers are to be provided.
Plant area needed: $3(9 + 3 \times 5.0) = 72\ m^2$
Clear height required: 3.9 m

(6) HWS (see Table 6.5): Assume two vertical storage vessels.
Plant area needed: $(6 + 1.65 \times 7) = 18\ m^2$
Clear height required: $(3.5 + 0.25 \times 3.5\ m) = 4.5\ m$

(7) Oil tanks (see Table 6.5): Assume two equal tanks.
Plant area needed: $(15 + 2.9 \times 1500/100) = 59\ m^2$
Clear height required: $(3 + 0.1 \times 1500/100) = 4.5\ m$

(8) Space for vertical supply and extract low velocity mains: Assume that two air handling plants are located on the roof of the services block at the north end of the building (Figure 1.1) and two similar plants on the roof of the services block at the south end Four equally sized ducts run vertically downwards in each services block. Each duct handles $12.25\ m^2\ s^{-1}$ of supply or extract air The ducts would be spirally-wound of circular section and sized by the equal pressure drop method at a rate of $0.8\ Pa\ m^{-1}$, but subject to a maximum velocity of $8.5\ m\ s^{-1}$ (see Jones (1994)) Reference to the CIBSE duct-sizing chart (1986) shows that a main duct handling $12.25\ m^3\ s^{-1}$ must be sized for the limiting velocity Then, since

Table 6.5 Approximate plantroom space required for the services provided for office blocks

Item	Approximate space needed	Suggested minimum clear height
Boiler	$9\ m^2$ per boiler + $3\ m^2$ per 100 kW of boiler power	3.9 m
Oil tanks:		
1 tank	$10\ m^2 + 2.9\ m^2$ per 100 kW of boiler power	3 m + 0.1 m per 100 kW, up to 4 m maximum
2 tanks	$15\ m^2 + 2.9\ m^2$ per 100 kW of boiler power	3 m + 0.1 m per 100 kW, up to 4 m maximum
HWS (hot water service) storage:		
Vertical vessels:		
1 vessel	$3.5\ m^2 + 1.65\ m^2$ per 1000 l	3.5 m + 0.25 m per 1000 l
2 vessels	$6\ m^2 + 1.65\ m^2$ per 1000 l	3.5 m + 0.25 m per 1000 l
Horizontal vessels:		
1 vessel	$7\ m^2 + 3.5\ m^2$ per 1000 l	2.5 m + 0.1 m per 1000 l
2 vessels	$10\ m^2 + 3.5\ m^2$ per 1000 l	2.5 m + 0.1 m per 1000 l
CWS (cold water service) storage:	$9\ m^2 + 1.5\ m^2$ per 1000 l	2.2 m
	or	
	$9\ m^2 + 0.75\ m^2$ per 1000 l	3.5 m

(Reproduced by kind permission of Haden Young Ltd.)

$$d = 1.128\sqrt{A} \qquad (6.1)$$

where d is the inside duct diameter in m and A is the cross-sectional area in m^2, each of the main ducts has a cross-sectional area of 12.25/8.5 = 1.441 m^2 and a diameter of 1.354 m. The supply ducts must be lagged with 50 mm of insulation and hence their overall diameter is 1.454 m The extract ducts are unlagged. Fifty mm of clearance must be allowed between ducts, and between ducts and the neighbouring walls of the vertical builder's shaft. Adding the clearance to the outside diameter of each supply duct gives a gross diameter of 1.554 m and the horizontal space occupied by the ducts at the top of each main builder's shaft is a rectangle, of size 1.554 × 5.866 m, having a horizontal area of about 9 m^2. This space reduces as the ducts proceed down the shaft to deal with the lower floors but little advantage can be taken of the extra area available On each floor, ducts branch horizontally to feed half the building from the north services block and half from the south services block. It is probable that the supply and extract ducts from each pair of air-handling plants would be cross-connected so that, in the event of the failure of one air-handling unit, the other could provide a reduced supply to both sides of the building. At each floor, branches from the supply and extract ducts would travel horizontally, above the suspended ceiling in the central corridor, to provide supply and extract air for each module. The horizontal supply ducts, at. their largest where they leave the builder's shaft, would each be handling 1.021 $m^3 s^{-1}$ to feed their half of the relevant face of the building. The CIBSE duct sizing chart (1986) shows that their size would be about 470 mm internal diameter (for a pressure drop rate of 0.8 Pa m^{-1}), or 570 mm externally, including lagging. The extract duct would be the same size but unlagged. To accommodate these ducts the suspended ceiling in the corridor would have to be lower than in the adjoining modules, giving a little less than the 2.6 m headroom provided therein.

To summarise:

	m^2	%
Air handling plant:	685	4.9
Refrigeration plant:	146	1.0
Cooling towers:	50	0.4
CWS tanks;	135	1.0
Boilers	72	0.5
HWS vessels:	18	0.1
Oil tanks	59	0.4
Vertical supply and extract ducts (2 × 9)	18	0.2
Totals:	1174	8.5

One might have multiplied the duct area by 12, since the two builder's shafts pass through each floor and this would make the total space used about 11%. The percentages quoted refer to the total treated floor area of 13,997 m^2. The estimate is liberal, as it should be in the conceptual stage of a design.

References

CIBSE guide,C4, Flow of fluids in pipes and ducts, Fig C4.2, Flow of air in round ducts, CIBSE, 1986.

Jones, W. P., The why and when of air conditioning-planning and building design. *RIBAJ*, **77,** 36–5, August 1970.

Jones, W. P., *Air Conditioning Engineering*, 4th Edition, p. 398, Edward Arnold, London, 1994.

Swain, C. P., Thornley, D. L. and Wensley, R., The choice of air conditioning systems. *JIHVE*, **32,** 1–41; 307–20, 1964.

7

Applied Acoustics

7.1 Simple sound waves

Sound consists of progressive wave trains, each comprising a series of alternate compressions and rarefactions travelling past a point in an elastic medium such as air, water, or a building material. In a simple imaginary case, a point source of sound, e.g. a very small pulsating balloon, imparts energy to the ambient air at a rate defining its sound power and produces an expanding, spherical, travelling wave system that is non-directional. An alternative simple illustration is a reciprocating piston at the end of a tube, producing a directional plane wave train, moving along the axis of the tube (Figure 7.1). The distance between the crests is the wavelength, λ the speed of propagation is c (344 m s^{-1} in air at 22°C) and f is the frequency, defined by

$$f = c/\lambda \tag{7.1}$$

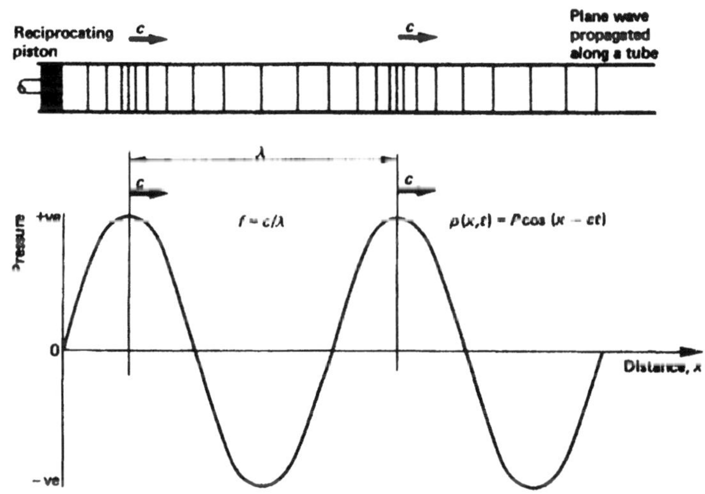

Figure 7.1 A directional plane wave train.

Real wave trains are more complicated and contain a large number of tones, not necessarily related harmonically. When a sound consists of many frequencies that are harmonically-related, the prime frequency is called the first or fundamental harmonic and the others the second, third, etc. Simple, rotating objects such as the impellers of centrifugal compressors, tend to produce harmonically-related wave trains, but random sources like air turbulence do not. Most plant items generate a complicated wave train with perhaps a few, discernible harmonics imposed on an aperiodic background.

7.2 Simple wave equations

In a wave train, the variation of sound pressure about the mean barometric pressure, at any time t, and distance x from the sound source, can be expressed by

$$p(x, t) = P \cos k(x - ct) \tag{7.2}$$

where k is a constant termed the wave number and equals $2\pi/\lambda$ and P is the maximum amplitude of the pressure. When a forward-travelling wave (see Figure 7.1) is reflected from a non-absorptive surface, a backward-travelling wave is produced, which is expressed by

$$p(x, t) = P \cos k(x + ct) \tag{7.3}$$

The combined acoustic field in the tube is then expressed by

$$\begin{aligned} p(x, t) &= P \cos k(x - ct) + P \cos k(x + ct) \\ &= 2P \cos kx \cos 2\pi ft \end{aligned} \tag{7 4}$$

This is the equation for a standing wave (Figure 7.2). At a distance from the source, $x = 0$, $p(0, t) = 2P \cos 2\pi ft$ and the sound pressure varies from $+2P$ to $-2P$ with time. This is also true for distances $x = \lambda/2$, λ, $3\lambda/2$, etc., defining areas of permanent sound in the tube. On the other hand, at distance $x = \lambda/4$ from the source, $p(\lambda/4, t) = 2P\cos k\lambda/4\cos 2\pi ft = 2P \cos \pi/2\cos 2\pi ft$, which always equals zero, regardless of the value of t. The same is true for distances $x = 3\lambda/4$, $5\lambda/4$, and so on, defining places of permanent silence along the axis, alternating with the areas of permanent sound.

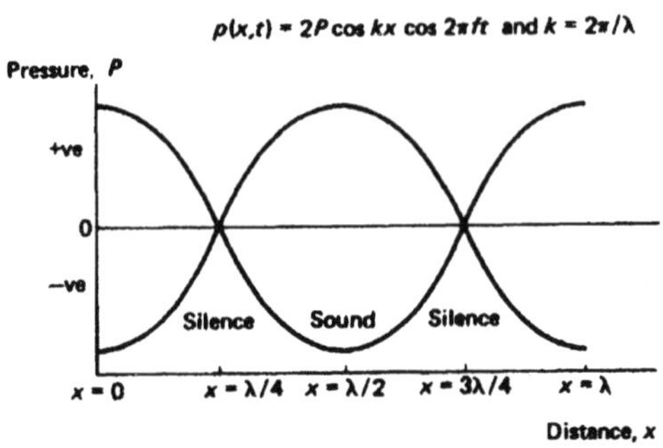

Figure 7.2 A standing wave.

Standing waves are characteristic of sound fields in any enclosure with reflecting or partially-reflecting surfaces when the waves from the acoustic source contain harmonics.

7.3 Root mean square pressure

Since both positive and negative pressures are the same to a listener and both contribute in equal measure to the sound power, measurement must be done in a way that avoids cancellation. This is achieved by measuring the square of the pressures and taking the square root of the average. Thus the measured indication of the strength of a sound is the root mean square pressure (P_{rms}). Since, at any time, the average of a cosine squared is half the amplitude of the cosine (Figure 7.3)

$$P_{rms} = P/\sqrt{2} = 0.707\,P \tag{7.5}$$

7.4 Intensity, power and pressure

If the molecules in a progressive wave train vibrate with a velocity u, it can be shown that $u = p/\rho c$ for a plane wave train, where ρ is the density of the air and p is defined by Equation (7.2). By substitution, therefore

$$u = (P/\rho\,c)\cos k(x - ct) \tag{7.6}$$

The intensity (I) of a sound is defined as the sound power (W) per unit area of wavefront and, for a spherical wave train, this is

$$I = W/4\pi r^2 \tag{7.7}$$

It can also be shown that the intensity is the product, pu, and hence, at any instant

$$\begin{aligned} I &= (P^2/\rho\,c)\cos^2 k(x - ct) \\ &= (P^2/\rho\,c)[1/2 + 1/2\cos^2 2k(x - ct)] \end{aligned} \tag{7.8}$$

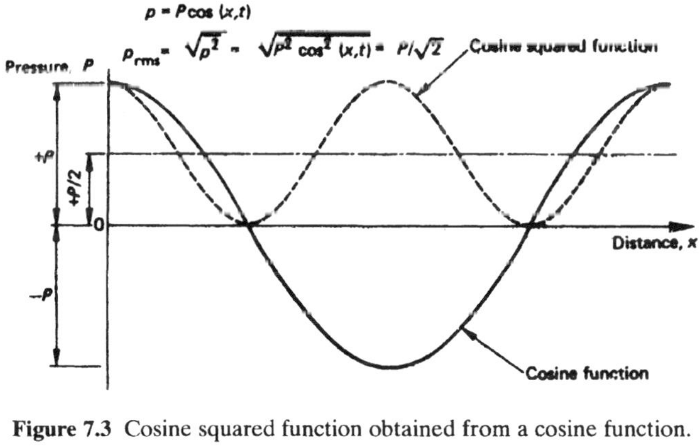

Figure 7.3 Cosine squared function obtained from a cosine function.

There is an analogy with fan power for $I = pu$. The volumetric flow rate per unit area corresponds to u, and the fan total pressure corresponds to p, implying the product corresponds to fan power per unit area.

Since the average of a cosine function with respect to time, at any place, is zero we can see that the average intensity is

$$\begin{aligned} I_{av} &= (P^2/2\rho c) \\ &= (P/\sqrt{2}c)^2(1/\rho c) \\ &= p_{rms}^2/\rho c) \end{aligned} \tag{7.9}$$

from Equation (7.5).

For a spherical wave train an analysis of the propagation yields

$$p(r, t) = (P'/r) \cos k(r - ct) \tag{7.10}$$

Therefore, p is inversely proportional to r, the radial distance from the source, by analogy with the circumferential stress in a thin-shelled boiler, and P' is a pressure amplitude factor. Equations (7.7) and (7.9) show that the sound power is proportional to the square of the sound pressure, and so

$$W = (4\pi r^2 p_{rms}^2)/\rho c \tag{7.11 }$$

which is the expression for the sound power in a spherical wave train. The expression ρc is termed the characteristic resistance of the medium and for air at 22°C and 101.325 kPa barometric pressure it has a value of 406 N s m^{-2}.

EXAMPLE 7.1

The peak sound pressure in a spherical wave train, measured 2 m from the source, is 2 Pa. Determine the root mean square pressure, the average intensity and the sound power.

Answer

$P = 2$ Pa, $\rho c = 406$ N s m^{-2} and $r = 2$ m, therefore

$$P_{rms} = P/\sqrt{2} = 2/\sqrt{2} = 1.4142 \text{ Pa}$$

$$I_{av} = p^2{}_{rms}/\rho c = 2/406 = 0.004926 \text{ W m}^{-2}$$

$$\begin{aligned} W &= 4\pi r^2 p_{rms}/\rho c \\ &= 4\pi \times 2^2 \times 1.4142^2/406 = 0.2476 \text{ W} \end{aligned}$$

Alternatively, by Equation (7.7)

$$W = 4\pi r^2 I = 4\pi \times 2^2 \times 0.004926 = 0.2476 \text{ W}$$

7.5 Decibels

Because of the vast range of values considered, linear measurement scales are not used to express acoustic pressure, intensity and power, logarithmic scales to base 10 being used instead. In each case the logarithm used is the ratio of the measured quantity to a reference quantity. Although, strictly speaking, such ratios are dimensionless, their logarithms are termed bels or, when one-tenth of the size for convenience, decibels.

Sound pressure level (L_p) is defined in terms of the ratio of the squares of the root mean square pressures:

$$\begin{aligned} L_p &= \log (p^2/p_{ref}^2) \text{ bels} \\ &= 20 \log (p^2/p_{ref}^2) \text{ dB} \end{aligned} \qquad (7.12)$$

In this equation the subscript 'rms' is dropped, for convenience, as it will be for the majority of the following text. The reference pressure adopted by international agreement is 2×10^{-5} Pa, regarded as the threshold of hearing for a healthy young adult. The use of the word level implies, in an acoustic context, the use of a reference value and for this reason the reference value should always be stated, although this principle is not always followed in practice or where usage makes the matter clear. For example, Equation (7.12) should have the expression 're 2×10^{-5} Pa' added to it.

EXAMPLE 7.2

Determine the sound pressure level at a distance of (a) 2 m and (b) 4 m from the source mentioned in Example 7.1.

Answer

(a) At $r = 2$ m, $P = 2$ Pa and $p_{rms} = 1.4142$ Pa; $p_{ref} = 2 \times 10^{-5}$ Pa
According to Equation (7.10) pressure is inversely proportional to radius in a spherical wave train, hence

$$L_p = 20 \log 1.4142/(2 \times 10^{-5}) = 96.99 \text{ dB: re } 2 \times 10^{-5} \text{ Pa}$$

(b) At $r = 4$ m, $p_{rms} = 1.4142 \times 2/4 = 0.7071$ Pa

$$\begin{aligned} L_p &= 20 \log(0.7071/(2 \times 10^{-5})) \\ &= 20[5 + \log (0.7071/2)] = 90.97 \text{ dB: re } 2 \times 10^{-5} \text{ Pa} \end{aligned}$$

This is an important result, showing that doubling the distance from a source of sound in a direct, free field reduces the sound pressure level by about 6 dB. This is not true for the reverberant field which tends to develop in a room some distance from the acoustic source.

Sound power level (L_w) and sound intensity level (L_I) are defined by international agreement in relation to 10^{-12} W and 10^{-12} W m^{-12}, respectively

$$L_W = 10 \log(W/W_{ref}) \text{ dB: re } 10^{-12} \text{ W} \qquad (7.13)$$

$$L_I = 10 \log (I/I_{ref}) \text{ dB: re } 10^{-12} \text{ W m}^{-2} \qquad (7.14)$$

EXAMPLE 7.3

Determine the sound power level and the sound intensity level at a distance of (a) 2 m and (b) 4 m from the source mentioned in Example 7.1.

Answer

(a) At r = 2 m, P = 2 Pa, p_{rms} = 1.4142 Pa and W = 0.2476 W

$$\begin{aligned} L_W &= 10 \log (W/W_{ref}) \\ &= 10 \log(0.2476/10^{-12}) \\ &= 10\,(12 + \log 0.2476) \\ &= 113.94 \text{ dB: re } 10^{-12} \text{ W} \end{aligned}$$

and, since I = 0.004926 W m^{-2} (from Example 7.1)

$$\begin{aligned} L_I &= 10 \log (I/I_{ref}) \\ &= 10 \log(0.004926/10^{-12}) \\ &= 10(12 + \log 0.004926) \\ &= 96.92 \text{ dB: re } 10^{-12} \text{ W m}^{-2} \end{aligned}$$

(b) At r = 4 m

$$L_W = 113.94 \text{ dB: re } 10^{-12} \text{ W}$$

as before, because sound power level is an absolute quantity, independent of distance from the source

$$I = (2/4)^2 \times 0.004926 = 0.0012315 \text{ W m}^{-2}$$

$$\begin{aligned} L_I &= 10 \log(0.0012315/10^{-12}) \\ &= 10(12 + \log 0.0012315) \\ &= 90.90 \text{ dB: re } 10^{-12} \text{ W m}^{-2} \end{aligned}$$

It can be shown that $L_I \simeq L_p$ by considering Equations (7.7) and (7.9): $I = W/4\pi r^2 = p^2/\rho c$, where the rms subscript for pressure has been dropped. Then, noting that ρc = 406 N s m^{-2}

$$\begin{aligned} L_I &= 10 \log (p^2/\rho c I_{ref}) \\ &= 10 \log (p^2/p^2_{ref})(p^2_{ref}/\rho c I_{ref}) \\ &= L_p + 10 \log(p^2_{ref}/\rho c I_{ref}) \text{ dB} \\ &= L_p + 10 \log [(2 \times 10^{-5})^2/(406 \times 10^{-12})] \\ &= L_p + 10 \log [(4 \times 10^{-10})/(406 \times 10^{-12})] \\ &= L_p + 10 \log [(4 \times 10^{2})/406] \\ &= L_p - 0.06 \text{ dB} \end{aligned}$$

The second term on the right is trivial and hence $L_p = L_I$, for all practical purposes.

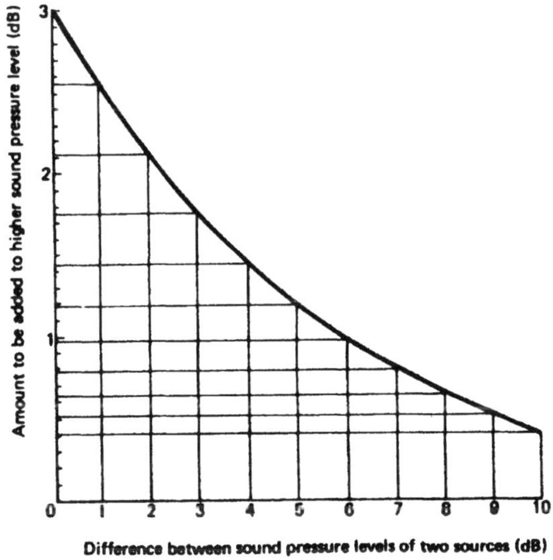

Figure 7.4 Addition of sound pressure levels from two sources.

When two levels in dB are added a straightforward arithmetical addition cannot be followed. Instead the levels must be converted to bels, the antilogarithms taken, the numbers produced added, the logarithm of the sum taken and the result converted to decibels.

EXAMPLE 7.4

Determine the sound pressure level in dB resulting from the addition of two sources, each of 20 dB: re 2×10^{-5} Pa.

Answer

20 dB = 2 B: antilog 2 = 100. Hence 2 B+2 B becomes 200. Therefore, log 200 = 2.301 B = 23.01 dB: re 2×10^{-5} Pa.

In the same way, 2 B + 2 B + 2 B becomes 300 with a logarithm of 2.477 B (≃ 25 dB). Thus, the addition of two equal sources gives a combined level about 3 dB higher than either and the addition of three, a level 5 dB higher, and so on. The same approach can be adopted for the addition of unequal sources and the results of doing this are shown in Figure 7.4.

A change of 3 dB is just noticeable, a change of 5 dB is distinctly noticeable and one of 10 dB appears to correspond to a doubling or halving of the noise.

7.6 Sound fields and absorption coefficients

Large rooms are those of most concern in air-conditioning applications because they are defined acoustically as having dimensions many times larger than the wavelength of concern. Using Equation (7.1) gives the figures shown in Table 7.1.

Unfortunately, simple calculations cannot be used for large rooms and statistical methods must be adopted instead. There are two sound fields to consider: near a source the field has

Table 7.1 Values of frequency and wavelength

f (Hz)	63	125	250	500	1000	2000	4000	8000
λ (m)	5.33	2.67	1.33	0.67	0.33	0.17	0.08	0.04

a directional quality but further away this direct field decays (see Example 7.2) and multiple reflections from the room surfaces (see Equations (7.2), (7.3) and (7.4)) produce a background reverberant field which swamps the direct field. The reverberant field strength may be described by the mean of the square of the sound pressure, averaged with respect to distance, or by the mean intensity, in the vicinity of the point of measurement.

Sound waves meeting a surface will not be reflected with the same amplitude if the surface has a sound-absorbing quality. Such a quality is expressed by the sound absorption coefficient (α) which has at least three forms:

(1) Coefficient for a given angle of incidence (θ)

$$\alpha_\theta = \frac{\text{Sound energy absorbed by the surface}}{\text{Sound energy incident on the surface at an angle } \theta}$$

(2) Statistical coefficient

$$\alpha = \frac{\text{Sound energy absorbed by the surface}}{\text{Sound energy incident on the surface when the field is perfectly diffuse}}$$

(3) Sabine coefficient (α_{sab}): this is measured in a laboratory and requires a nearly perfect, diffuse field with a large specimen of the material being tested laid on the floor (10–12 m^2 in Europe, 6.7 m^2 in USA). It is to be noted that a diffuse field is one in which, at any point of measurement, the net flow of acoustic energy is independent of direction.

A random incidence coefficient, similar to (2) and (3) may be obtained theoretically from the normally incident coefficient, α_n, which is a special case of α_θ. Values of α_n can be obtained more readily than values of α_{sab}, since a much smaller specimen is used and located at the end of a tube in which an exploration of the standing waves set up yields the desired information. Values of α and α_{sab} are often very similar, except for occasional, inexplicable anomalies. Values of α_{sab} are often merely written as α. Figure 7.5 shows how α varies with frequency for some common materials, although it is misleading since it conceals a certain amount of information because the method of mounting the material greatly influences its characteristics. For example, the presence of a 50 mm air gap behind the polystyrene panels improves their acoustic absorption properties, and the peak at 500 Hz is not necessarily a feature of the material-in all probability it is the result of mounting the panels on battens spaced at 600 mm intervals (see Table 7.1). Tuning effects of this kind are not uncommon. There is a further example in the case of the perforated aluminium ceiling tiles which may owe part of the peak at 500 Hz to the dimensions of the panels (600 mm square).

The term noise reduction coefficient (NRC) is sometimes used, being the average of α_{sab} over 250, 500, 1000 and 2000 Hz. More often, the average absorption coefficient, $\bar{\alpha}_{sab}$, must

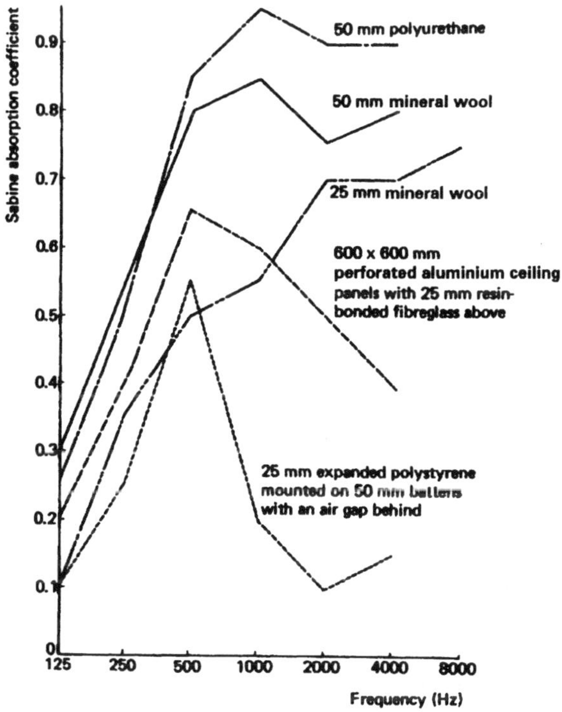

Figure 7.5 Variation of the Sabine absorption coefficient with frequency for some common materials.

be calculated as follows, to establish the acoustic quality of the room

$$\bar{\alpha}_{sab} = (S_1\alpha_{sab1} + S_1\alpha_{sab2} + S_n\alpha_{sabn})/S \tag{7.15}$$

where S is the total room surface area and equals $S_1 + S_2 + ..., S_n$, the sum of the individual surface elements, having absorption coefficients α_{sab1}, α_{sab2}, ..., α_{sabn}. Sound-absorbing objects, such as people and furniture, must be included in the calculation but because such objects are often of ill-defined area they are generally given a value of $S\alpha_{sab}$, in units termed sabines, for inclusion in the numerator of Equation (7.15). Some typical values are given in Table 7.2. No modification is made to the value of S when using the values given in Table 7.2 for objects, and the room constant, R, is then defined by

$$R = S\bar{\alpha}_{sab}/(1 - \bar{\alpha}_{sab}) \tag{7.16}$$

7.7 Octave bands

The human ear can, at best, detect changes of frequency over a spectrum between 20 and 20 000 Hz. Since the energy of most sounds is distributed over part or all of this range, a spectral analysis is necessary to describe the nature of a noise. Because the response of the human ear is more logarithmic, to base two, than linear, an octave scale is used to divide

Table 7.2 The sound-absorption of objects in a room

	Sound absorption (Sasab) (sabines)					
	Frequency (Hz)					
Object	125	250	500	1000	2000	4000
Desk or table	0.01	0.02	0.02	0.04	0.04	0.03
Upholstered chair	0.11	0.18	0.28	0.35	0.45	0.42
Clothed person	0.17	0.36	0.47	0.50	0.50	0.46
Glass-fronted cabinet	0	0.01	0.01	0.02	0.02	0.02
Wooden door	0.15	0.20	0.10	0.10	0.10	0.10

the spectrum so that successive frequencies are doubled, the span between them denoting octave bands: 62.5–125 Hz, 125–250 Hz, and so on, usually up to 8000 Hz, for most practical purposes. The mid-frequency, f_c, in each band equals $\sqrt{(f_1 \times f_2)}$, where f_1 and f_2 are the extreme frequencies of the band. The sound pressure level in a band is termed the octave band level.

An octave band analyser does not always yield sufficient information for engineering needs and so one-third octave bands are also used. Whereas with an octave band the frequency bands could be represented by 2^n and 2^{n+1}, with a one-third octave band they would be given by 2^n and $2^{n+1/3}$, the mid-frequency still being the root of their product. The narrower the frequency band adopted, the less the sound pressure level in it: for instance, three equal sound pressure levels of 40 dB in contiguous one-third octave bands are added to give 45 dB in a whole octave band (see Section 7.5 and Figure 7.4).

For the extension of acoustics into infra-sound (less than 20 Hz) and ultra-sound (greater than 20 kHz) it has become convenient to use powers of ten for the separation of frequency bands, rather than octaves. The consequent incompatibility has been resolved by noting that $2^{1/3}$ (= 1.2599) is almost the same as $10^{1/10}$ (= 1.2589) British Standards (1963) describe a range of preferred frequencies that has been established over the entire span of the spectrum used in acoustics, but with 1000 Hz retained unchanged. The values of preferred frequencies have been rounded off to whole numbers, as listed in Table 7.3

The centre frequency of an octave band still equals the square root of the product of its limits, but this is only approximate for one-third octave bands: one-third octave band frequencies cannot be obtained exactly by applying a factor $2^{n+1/3}$. For example, $250 \times 2^{1/3} = 250 \times 1.2599 = 314.98$, but $250 \times 10^{1/10} = 250 \times 1.2589 = 314.72$. The preferred frequency is 315 Hz.

Table 7.3 Preferred frequencies (Hz) for measurement

Octaves:	16	31.5	63
1/3 Octaves:	16 20 25	31.5 40 50	63 80 100
Octaves:	125	250	500
1/3 Octaves:	125 160 200	250 315 400	500 630 800
Octaves:	1000	2000	4000
1/3 Octaves:	1000 1250 1600	2000 2500 3150	4000 5000 6300
Octaves:	8000	16000	
1/3 Octaves:	8000 10000 12500	16000	

When a source of sound radiates acoustic energy into a room it does so over the audio range of frequency and. since the absorption coefficients of the surfaces are frequency-dependent, it is necessary to analyse the noise of the source in octave, or one-third octave, bands. The resultant sound pressure level in the room can then be established with respect to the octave bands and some idea formed of the subjective quality of the sound field in the room.

7.8 Room effect

If a steady source of sound exists in a room the sound waves will suffer multiple reflections from the surfaces and, provided there are no highly absorptive areas, a complicated pattern of standing waves will be established, constituting the reverberant field. The energy lost by absorption at the surfaces equals the input by the source and the waves travel with equal probability in all directions.

In such a field, according to Beranek (1957), the concept of energy density, per unit volume, D_r, is useful. The average intensity over a small area in the field is the rate of energy flow through it at speed c, expressed in W m^{-2} or J s^{-1} m^{-2}. Dividing the intensity by the speed of sound then gives the average energy density in J m^{-3}

$$D_r = I_{av}/c = p_{av}^2/\rho c^2 \qquad (7.17)$$

If a small area δS, within the field, is considered, half the sound waves will, on average, enter one side of it and half the other, at various angles of incidence, θ. Any particular wave, with intensity I, may be resolved in a direction normal to the surface to yield an expression $I \cos \theta \delta S$ for the energy flow rate into one side of the surface. Over the solid angle subtended by this side the average value for such an expression is $\frac{1}{2}I\delta S$. Since, statistically, the average energy flow through one side of a surface in the field is also half that through both sides, the average energy flow through the surface in one direction only is $\frac{1}{4}I\delta S$. From Equation (7.17) the flow of power through one side of a surface can be expressed as $\frac{1}{4}D_r c\delta S = \frac{1}{4}I_{av}\delta S$. If the surface has an absorption coefficient α, the power loss from the reverberant field is $\frac{1}{4}D_r c\alpha\delta S$. If the power output from the source is W, then after each complete reflection of sound waves has occurred on the room surfaces, the power absorbed by them, on average, is $W\bar{\alpha}$, where $\bar{\alpha}$ is the average absorption coefficient for all the room surfaces. The remaining power, $W(1 - \bar{\alpha})$, is what the source contributes to the reverberant field and is balanced by the power flow expression referred to all the surfaces in the room, of area S, having an average absorption coefficient $\bar{\alpha}$. Thus $\frac{1}{4}D_r c\bar{\alpha}S = W(1 - \bar{\alpha})$ and the energy density in the reverberant field can be expressed as

$$D_r = 4(W/c\bar{\alpha}S)\left(\frac{1-\bar{\alpha}}{\bar{\alpha}}\right) \qquad (7.18)$$

Equations (7.7) and (7.9) give an expression for the sound pressure in the direct field at distance r from the source:

$$p^2 = (\rho cW)/(4\pi r^2)$$

If the source is in the centre of a large open space it is radiating all its energy into a sphere and it is said to have a directivity factor, Q, of unity. If it is in the middle of a large flat ceiling

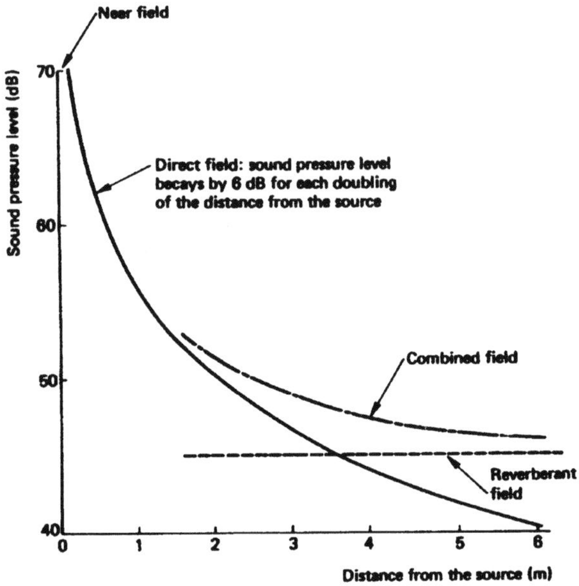

Figure 7.6 Graphical representation of the sound field in a room with a source of noise radiating acoustic power from a point at one side.

its energy is radiated into a hemisphere and Q has a value of two. Values of four and eight correspond to sources of sound located in the junction of a ceiling and wall, and in a corner, respectively. Strongly directive sources, which beam the sound, can have factors exceeding eight.

At a point in the direct field, distant r from the source, $I = QW/4\pi r^2$ (Equation (7.7)) and from Equation (7.9)

$$p^2 = \rho cQW/4\pi r^2 \tag{7.19}$$

At any point, sound pressures in the direct and reverberant fields can be added and so, from Equations (7.16)–(7.19) an expression for the combined mean square sound pressure is obtained

$$p^2 = \left[\frac{\rho cQW}{4\pi r^2}\right] + \left[\frac{\rho c^2 4W}{c\overline{\alpha}S}\frac{(1-\overline{\alpha})}{\overline{\alpha}}\right]$$

$$= \rho cW\left[\frac{Q}{4\pi r^2} + \left(\frac{4}{\overline{\alpha}S}\right)\left(\frac{1-\overline{\alpha}}{\alpha}\right)\right]$$

$$= \rho cW\left[\frac{Q}{W\pi r^2} + \frac{4}{R}\right]$$

where $\bar{\alpha}$ is used instead of $\bar{\alpha}_{sab}$

$$L_p = 10 \log \frac{\rho c W}{(2 \times 10^{-5})^2} \left(\frac{Q}{4\pi r^2} + \frac{4}{R} \right) \text{dB: re } 2 \times 10^{-5} \text{ Pa}$$

since $(2 \times 10^{-5})^2/\rho c \simeq 10^{-12}$, numerically

$$L_p = 10 \log (W/10^{-12}) + 10 \log [(Q/4\pi r^2) + 4/R] \text{ dB: re } 2 \times 10^{-5} \text{ Pa}$$
$$= L_W + 10 \log [(Q/4\pi r^2) + 4/R] \text{ dB: re } 2 \times 10^{-5} \text{ Pa}$$

whence

$$L_W - L_p = -10 \log [Q/4\pi r^2 + 4/R]$$
$$= -10 \log [Q/4\pi r^2 + 4(1 - \bar{\alpha}_{sab})/S\,\bar{\alpha}_{sab}] \quad (7.20)$$

The difference between the sound power level, L_w, and the sound pressure level of the sound field in the room, $L_{p,}$ is termed the room effect.

Figure 7.6 shows a graphical representation of the sound field in a room with a source of noise radiating acoustic power from a point at one side. Immediately next to the source anomalous and inconsistent readings are often obtained but, after the ill-definable area of this near field, the far field is established. This comprises the direct field and the reverberant field, as shown in Figure 7.6.

EXAMPLE 7.5

A double module office has dimensions 4.8 m wide × 2.6 m high × 6.0 m deep. One wall (2.6 × 4.8 m) contains two windows of total area 7.92 m^2 and the opposite wall contains a door of 1.61 m^2 area. A single air conditioning terminal unit is located in the centre of the ceiling. Making use of the information given, establish the room effect in each octave band from 125–4000 Hz. Take *r* as 1.769 m, the distance between the measuring microphone and the terminal unit.

The furnishings and statistical absorption coefficients are as follows:

Item	Area, S_n (m^2)	Absorption coefficients at the mid-octave bands					
		125 Hz	250 Hz	500 Hz	1000 Hz	2000 Hz	4000 Hz
Carpet and underlay	28.8	0.05	0.25	0.50	0.50	0.60	0.65
Plastered walls	46.63	0.03	0.03	0.02	0.03	0.04	0.05
Curtains	0.99	0.05	0.15	0.35	0.55	0.65	0.65
Acoustic ceiling	28.8	0.25	0.70	0.85	0.85	0.85	0.40
Door	1.61	0.19	0.25	0.12	0.12	0.12	0.12
Window	7.92	0.10	0.07	0.04	0.03	0.02	0.02
Total area, S	114.75						
$\Sigma(S_n\alpha_n)$		11.19	29.87	40.67	41.25	44.62	33.56

Two people are present and the total absorptions of the occupants and furniture are as follows:

Absorption units at the mid-octave bands

Item	125 Hz	250 Hz	500 Hz	1000 Hz	2000 Hz	4000 Hz
Hard-topped desk	0.01	0.02	0.02	0.04	0.04	0.03
Two soft chairs	0.22	0.36	0.56	0.70	0.90	0.84
Two hard chairs			0.03	0.04	0.08	
Two people	0.34	0.72	0.94	1.04	1.00	0.92
Glass-fronted bookcase	0.01	0.02	0.02	0.04	0.04	0.03
Totals:	0.58	1.12	1.57	1.86	2.06	1.82

Answer

It should be noted that because the terminal unit is in the centre of the ceiling the directivity factor, Q, is 2.0, and the component of the direct field can be evaluated as follows:

$$Q/4\pi r^2 = 2/(4 \times \pi \times 1.769^2) = 0.051$$

From the tabulated information above it was determined that S is 114.75 (including curtains) and it can be calculated that

Mid-octave bands	125 Hz	250 Hz	500 Hz	1000 Hz	2000 Hz	4000 Hz
$\Sigma(S_n\alpha_n)$	11.19	29.87	40.67	41.25	44.62	33.56
$\Sigma(S_n\alpha_n)$ + absorption units = total absorption	11.77	30.99	42.24	43.11	46.68	35.38
$\bar{\alpha}$ = total absorption/S	0.103	0.270	0.368	0.376	0.407	0.308
$(1 - \bar{\alpha})$	0.897	0.730	0.632	0.624	0.593	0.692
$R = S\bar{\alpha}/(1 - \bar{\alpha})$	13.18	42.44	66.82	69.14	78.76	51.07
$4/R$	0.303	0.094	0.060	0.058	0.051	0.078
$Q/4\pi r^2$	0.051	0.051	0.051	0.051	0.051	0.051
$4/R+Q/4\pi r^2$	0.354	0.145	0.111	0.109	0.102	0.129
$L_w - L_p$ (dB) (Equation (4.20))	−4.5	−8.4	−9.5	−9.6	−9.9	−8.9

Therefore, it can be seen that even a very soft room only gives a room effect of between 4 and 10 dB. It is unrealistic to quote sound pressures levels to fractions of a decibel since they cannot be measured that accurately. A hard room, however, sometimes can give positive values for the room effect.

If the room effect is known a given sound power level for a terminal unit can be added to it to yield the sound pressure level likely in the room at the position considered. Alternatively, a reverse procedure may be followed. If the desired sound pressure level in each octave band is known, the room effect can be subtracted from it to give the maximum acceptable sound power level in each octave band for the terminal unit.

7.9 Noise criteria, noise ratings and room criteria

The response of the human ear to sound is frequency-dependent and various indices have been proposed to express this subjective quality. One of the earliest was the concept of loudness level, expressed in phons and defined as the sound pressure level of a pure tone at 1000 Hz, presented centrally in front of the listener, that seems equal in magnitude to the tone at the frequency considered. A slightly different index, the perceived noise level (expressed in PNdB), and also the effective perceived noise level (expressed in EPNdB), adopts a band of noise between 910 and 1090 Hz as a reference and is commonly used to express the annoyance value of steady noises, particularly aircraft.

An attempt to define noises of subjectively equal loudness level indoors, has led to noise criterion (NC) curves, originally proposed by Beranek (1957). These refer to steady noises and specify acceptable band levels, usually between 63 and 8000 Hz. Since then, the improved standards of preferred noise criteria (PNC) curves have been put forward by Beranek (1971). Developments in Europe induced Kosten and Van Os (1962) to suggest noise rating (NR) curves and these are gaining some acceptance internationally. Although based on the idea of the acceptability of a noise to listeners, equations have been derived for the curves and they can therefore be used beyond the range of human hearing if need be. Interpolation can also be done. Figure 7.7 shows NC and NR curves with a single RC curve superimposed. There does not seem to be a great deal of difference between NC and NR values and, given the subjective nature of the curves and the practical difficulties of exact measurement, they can probably be regarded as the same (see Figures A.1 and A.2 in the Appendix).

If the spectrum of the sound field in a room is plotted on the same coordinate system as NR curves, the NR value for the room is determined from the highest value of the NR curve cut by the spectrum of the sound field for the room. The shape of the spectrum influences the subjective appreciation of the noise in the room: if the spectrum is like that of curve A in Figure 7.8 the subjective feeling will be quite different from that of curve B, even though both sound fields give NR 40 for the room. Curve A contains a lot of energy in the low frequencies and this would be apparent to the listener whereas, on the other hand, spectrum B has much of its energy in the high frequencies and the noise heard in the room would suggest such a quality. This is a criticism of the use of NC and NR curves for describing the

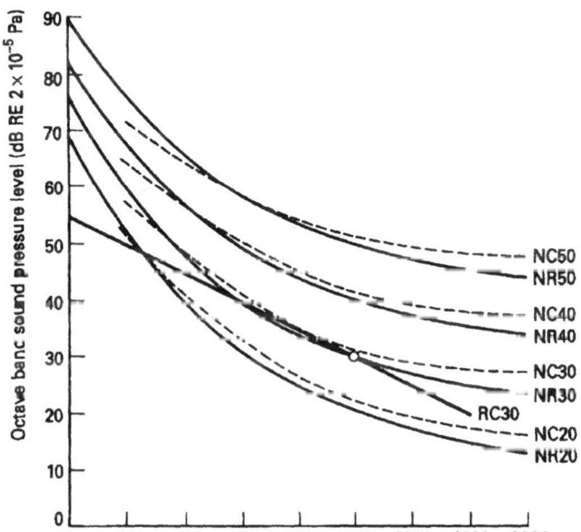

Figure 7.7 NC and NR curves with a single PNC curve superimposed. RC curves are straight and parallel with a slope of –5 dB per octave. The numerical value of an NR curve or an RC curve is the same as the sound pressure level at 1000 Hz.

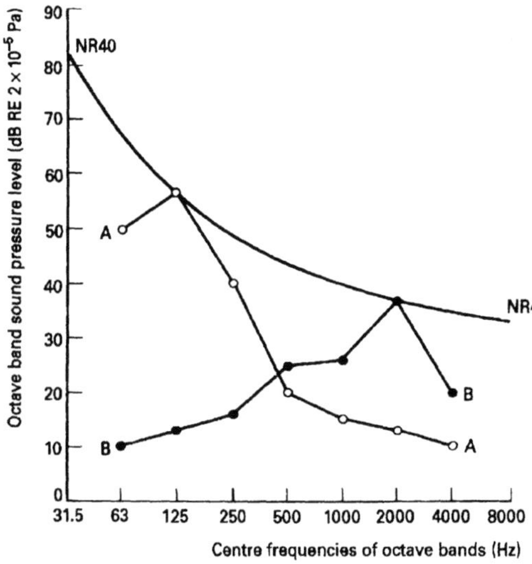

Figure 7.8 Examples of sound field spectra having the same NR value but different subjective qualities: curve A has a low frequency quality but curve B has a high frequency quality.

acoustic quality of an air-conditioned room. To be acceptable, the sound in an air-conditioned room should be bland and unobtrusive, without discernible low or high frequency characteristics. The sound pressure level should not be so high that it interferes with communication between the occupants, nor should it be so low that it fails to provide privacy by masking (see Section 7.11).

A room criterion (RC), meeting some of the criticism, was introduced by ASHRAE (1980) in the United States and subsequently developed (1991) as an alternative to the use of NC and NR values. The RC curves are based on a study of office environments and offer a more balanced, and hence more acceptable, indication of a sound field than do the NC or NR curves. Referring to a particular spectrum in a room, the numerical value of the room criterion is the preferred speech interference level (PSIL) in dB (see Section 7.11). The RC curve is then drawn as a straight line through the point defined by the PSIL value in dB and 1000 Hz, with a slope of –5 dB per octave, from 31.5 to 4000 Hz. This is shown in Figure 7.9. There is an alternative form that includes 16 Hz.

To express a low frequency character in the spectrum, a straight line is drawn from 31.5 to 500 Hz, parallel to the RC line but at 5 dB above it. Similarly, to show a high frequency character, a straight line is drawn from 1000 to 4000 Hz at 3 dB above and parallel to the RC line (Figure 7.9). To denote a spectrum having a neutral character the sound pressures levels of all its components between 31.5 and 4000 Hz must not exceed the two limiting lines for low and high frequencies. mentioned above. The spectrum of the noise in the room would then be described by an appropriate RC number and the letter N, for example, RC 30(N), 30 dB being the measured speech interference level. A spectrum with a low frequency rumble is defined as having one or more sound pressure levels exceeding the relevant low-frequency tolerance line and would be described by adding the letter R to the RC number, say RC 30(R). Similarly, a spectrum characterised by a high-frequency quality has one or more sound pressure levels greater than the relevant high-frequency tolerance line and is

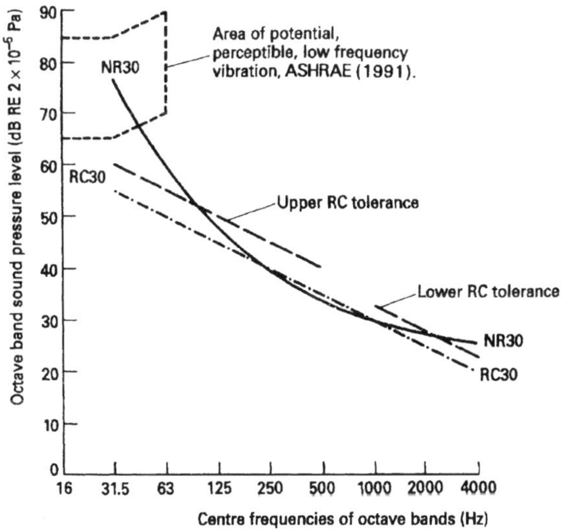

Figure 7.9 An RC curve is established by measuring the PSIL of a spectrum and locating this as a value in dB on the 1000 Hz octave line. The RC curve is then a straight line with a slope of –5 dB per octave. An NR30 curve is shown here for comparison with RC30.

described by adding the letter H, for example, RC 30(H). There are also prescriptions for spectra having discernible pure tones (indicated by adding the letter T) and for spectra at high sound pressure levels and low frequencies, that can induce perceptible vibration in lightweight building components (described by adding the letters RV to the RC number).

Another method of describing the subjective quality of a sound field is to quote a dB(A) value (see Section 7.23). Although this gives a broad idea of the loudness of a sound field, being roughly 7 dB more than an NR value, it is less satisfactory than the use of NR or RC values because it does not have a unique spectrum.

7.10 Traffic noise and windows

Traffic noise varies considerably with its density, the gear ratios used to climb the inclines and the distance from the listener. In an attempt to define traffic noise a quantity L_{10}, has been proposed. This is the arithmetic mean hourly value of the sound pressure level

Table 7.4 The attenuation given by different windows

Window type	Mean attenuation over 160–3150 Hz (dB)
Ordinary single, openable, without weatherstrip	20
Fixed or openable, single, with weatherstrip	25
Sealed, double-glazed unit	30
Weatherstripped, openable, double, with 200 mm air gap	40
Weatherstripped, openable, double, with 400 mm air gap and sound absorbant reveals	45

in dB (A) (see Section 7.23), at a distance of one metre from the building facade, that is exceeded for 10% of the time between the hours of 6 a.m. and midnight. It is termed the 18-hour value of L_{10}. Values of 55 dB (A) for general offices and 45 dB (A) for private offices, inside the building, are suggested as reasonable. More than 60 dB (A) leads to considerable complaint. Clearly, the attenuation given by the windows is highly significant (Table 7.4).

7.11 Privacy of speech

The PSIL has been developed as an indication of the intelligibility of speech in a noisy environment. It is defined as the arithmetic mean of the sound pressure levels in the three octave bands having mid-frequencies of 500, 1000 and 2000 Hz. Table 7.5 gives the level corresponding to a communication reliability of 60% for words and numbers, out of doors. For example, if two people are 1.8 m apart a PSIL of 70 dB will mask more than 40% of their conversation, even if they are shouting. However, if they are 0.60 m apart, it needs 80 dB to achieve the same effect.

Masking noise may be deliberately introduced in open-plan offices, at times, to achieve a level of privacy. Such noise should lie within an acceptable spectrum, although it may be provided by mechanical air-conditioning equipment and, at, the higher frequencies, by electronic devices. However, the designer does not have the freedom to offer a carelessly

Table 7.5 The preferred speech interference level (PSIL)

	PSIL (dB)			
Distance between two people (m)	Normal speech	Raised voice	Very loud speech	Shouting
0.3	68	74	80	86
0.6	62	68	74	80
1.2	56	62	68	74
1.8	52	58	64	70

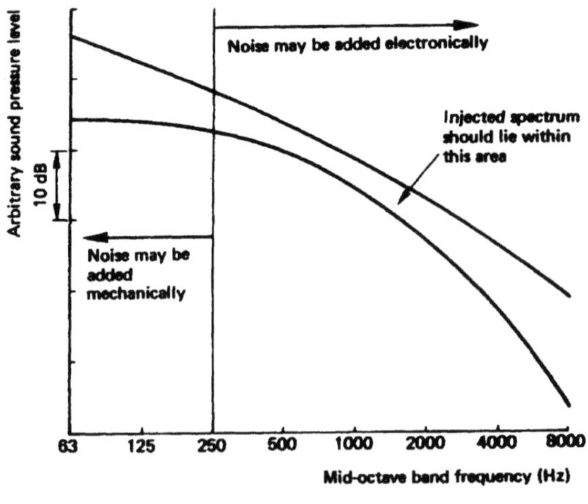

Figure 7.10 Recommended methods and spectrum for injecting noise (after Beranek, 1975).

noisy system. Beranek (1957) has suggested that the injected noise ought to lie within the pair of curved lines shown in Figure 7.10.

Typically acceptable levels for steady background noise, according to Beranek, are: concert halls NR10–20; private offices NR30–40; large offices NR35–45; general offices NR40–50; computer rooms NR45–55.

In quiet rooms, it is only possible to achieve privacy if intruding noises are excluded and this is best done by having massive walls, extending up above the suspended ceiling and sealed tightly to the soffit of the slab. Any break in the wall negates the isolating effect of its mass and it is therefore necessary to provide well-fitted packing around ducts and pipes where they pass through a wall or partition. For ducts, their walls may offer less acoustic impedance than does the partition penetrated and it then becomes necessary to reduce the flanking path so provided by covering the duct with a dense material and possibly also lining it with a sound-absorbing quilt.

7.12 Sound transmission through building structures

The equipment working inside a plant room sets up a reverberant field which loses energy continuously through the walls, roof and floor (Section 7.8). Further, the direct fields in the vicinity of the plant items also impart energy to the surfaces nearby and, also, some sound waves travel directly from the equipment through the building material. These effects excite the plant room structure, producing compression and shear waves in it that combine to make the surfaces bend, periodically, in various modes and harmonics. Such structural flexures produce sources of acoustic radiation on the outer surfaces of the plant room.

To reduce the transmission of sound through the plant room surfaces, one first step might be to lower the strength of the reverberant field within by covering the inner surface with sound-absorbing material. This is seldom the most effective approach, although it may sometimes help, in conjunction with other measures. The first and most important step is always to construct the plantroom walls, roof and floor, of massive materials. Theory suggests that the mass law transmission loss for homogeneous panels should be 6 dB per doubling of mass but experimental evidence indicates about 5 dB, which is the figure usually taken.

In practice, a sound reduction index (R) is measured for a building material and BS 2750 (1987) specifies how this should be done, according to

$$R = L_{p1} - L_{p2} + 10 \log (S/A) \text{ dB: re } 2 \times 10^{-5} \text{ Pa} \tag{7.21}$$

in which L_{p1} and L_{p2} are the sound pressure levels in the rooms on each side of the specimen, S is its area and A is the total number of absorption units in the receiving room. The sound transmission loss (R) may be used as an alternative term for the sound reduction index and defined in terms of mean squared sound pressures, $|p_1|^2$ and $|p_2|^2$, in the transmitting and receiving rooms, respectively

$$|p_2|^2 = (\tau s/S)\,|p_1|^2 \tag{7.22}$$

leading to a concept of τ, the sound transmittance coefficient. The sound power entering a room through a partition is then directly proportional to the product τS, in m^2. BS 661 (1969) defines the sound reduction index as

$$R = 10 \log 1/\tau \text{ dB: re } 2 \times 10^{-5} \text{ Pa} \tag{7.23}$$

Given an experimental value of R, τ and hence τS can be deduced by Equation (7.23). Thus the overall value of R for a parallel combination of building elements can be calculated.

EXAMPLE 7.6

A 6.0 × 2.6 m partition, constructed of Thermalite blocks 100 mm in thickness contains a door, 2.0 m × 0.9 m× 62 mm thick, fitted with a rubber gasket. Determine the overall average sound transmission loss.

Answer
From Table 7.6 the average values of R for the partition and the door are 41 and 30 dB, respectively. Using Equation (7.23), the following may be calculated:

Item	R (dB)	$1/\tau$	τ	S (m^2)	τS (m^2)
Partition	41	12 590	7.9428×10^{-5}	13.8	109.65×10^{-5}
Door	30	1000	100×10^{-5}	1.8	180.00×10^{-5}
Total				15.6	289.65×10^{-5}

Therefore, $\bar{\tau} = (289.65 \times 10^{-5})/15.6 = 18.567 \times 10^{-5}$ and $\bar{R} = 10 \log 1/\bar{\tau} = 37.3$ dB.

The presence of a hole or a crack in a wall significantly reduces its transmission loss. For the case of a hole or a slit, the diameter or width of which is large compared with the thickness of the wall, $\tau \simeq 1.0$. This is usually the case for open windows, provided that the opening is a large percentage of the total window-wall area. When the opened window is 10% or less of the total area, the loss is generally taken as 10 dB. If the width of the hole is not large in comparison with the wall thickness, τ has a value of less than 1.0. For a slit width of from 5 to 25 mm, in a 50 mm thick wall, or door frame, the value of τ lies between about 0.1 at 2000 Hz and about 1.0 at 125 Hz. A reasonable average is 0.3 for speech frequencies. τ can have a value of greater than 1.0, at low frequencies. An approximate equation, proposed by Cook and Chrzanowski (1957) gives a value for the mean transmission loss in accordance with the mass law mentioned earlier:

$$\bar{R} = 13 + 14.5 \log m \text{ (dB)} \tag{7.24}$$

where m is the mass of the barrier in kg m^{-2}, referred to the surface area.

When acoustic energy travels through a medium some of it is dispersed in damped vibrations of the material and so its stiffness is important. Preferably, most building materials, with natural frequencies of 5–20 Hz, should be limp, like lead. However, when a sound wave passes through different, successive media, especially an air gap in a solid structure, it encounters discontinuities and a reflection occurs at each interface, constituting an extra transmission loss. Table 7.6 shows that a pair of 3 mm glass sheets, separated by a 50 mm air space, has a mean loss of 33 dB whereas a single 6 mm sheet only gives 26 dB. The greater the width of the air gap the better; less than about 100 mm is seldom worthwhile.

To prevent noise travelling through a floor it is sometimes uneconomical to thicken the slab to increase its mass. Instead it is often better to use a more conventional structural slab and to float a floor on top of it. The floating floor should cover the entire floor area and frequently comprises 100 mm of concrete, reinforced as necessary, resting on pads of resilient material to give an air space of about 50 mm. The air space is often filled loosely

Table 7.6 Average values of R for various objects

		Approximate transmission loss (R) (dB)						
	Mass	Mid-octave frequency (Hz)						
Item	(kg m^{-2})	Av	125	250	500	1000	2000	4000
6 mm plywood glued to both sides of 25 mm × 75 mm studs at 400 mm centres; 75 mm thick	12	24	16	18	26	28	37	33
18 swg corrugated steel sheet	21	24	30	20	22	30	28	31
Glass fibre board; 50 mm thick	26	30	27	23	27	34	39	41
6 mm plywood glued to both sides of 25 mm × 75 mm studs at 400 mm centres plus 12 mm plasterboard nailed to each side; 99 mm thick	32	40	26	33	39	46	50	50
Heavy wooden door plus rubber gasket around edges; 62 mm thick	61	30	30	30	24	26	37	36
Common brick; 100 mm thick	187	44	30	36	40	50	54	60
Thermalite block; 110 mm thick	200	41	27	33	40	44	56	57
Breeze block, plastered both sides; 125 mm thick	205	42	27	33	40	50	57	56
Common brick, plastered both sides; 135 mm thick	220	45	31	36	41	51	55	61
Cavity wall, 2 × 110 mm brick plus 2 × 12 mm plaster plus 50 mm air gap plus butterfly ties	407	29	29	40	45	62	72	84
Reinforced concrete; 150 mm thick	366	46	37	42	47	51	56	60
Reinforced concrete; 200 mm thick	488	48	38	44	49	54	58	62
Reinforced concrete; 300 mm thick	732	50	39	46	51	56	60	63
3 mm glass in a light frame	7	21	15	15	20	23	29	27
3 mm glass in a heavy frame	7	23	14	20	22	28	30	30
2 × 3 mm glass sheets plus 50 mm air gap	14	33	19	24	29	39	48	53
2 × 3 mm glass sheets plus 100 mm air gap	14	37	19	25	37	45	51	53
6 mm plate glass in a heavy frame	17	26	16	21	27	29	29	37
10 mm plate glass in a heavy frame	25	29	21	26	31	32	32	39
16 mm plate glass in a heavy frame	42	32	23	28	33	31	38	45

with sound absorbing material. The conventional housekeeping concrete pads, anti-vibration mountings and inertia blocks are still very necessary and lie on top of the floating floor. Resilient material is placed around the edge of the floating floor to isolate it from the walls and it is most important that the floor be laid properly, under competent supervision, to make sure there are no solid bridges across it to the building structure.

7.13 Sources of noise in mechanical systems

No single element in a mechanical system fails to contribute to the production or transmission of noise. The plant itself generates noise, ducts and pipes help to distribute it about the building, the air and water flowing create further noise and, finally, the terminal units in the rooms provide their quota. Even silencers can produce more noise than they are supposed to remove, under some circumstances. It is clearly necessary to know something of the noise generating characteristics of the components in a system.

7.14 Fan noise

When air passes over a fan blade a pressure gradient is developed across it that is uniform if laminar flow occurs, generating little noise. This case seldom prevails in practice and the airstream is not laminar but separates in a random way from the curvature of the blade, forming eddies in a fluctuating pattern with vortices that are shed from the trailing edge. With centrifugal fans, vortices may also be formed at the leading edges of the blades. Noise over a wide spectrum is produced, its power depending on fan selection, efficiency, size and speed. Further noise is created by turbulence within the casing and matters may be aggravated by obstructions at the fan outlet or inlet and also by ill-conceived or badly installed duct connexions. Although it does not seem possible to relate fan efficiency directly with noise, some approximate generalisations can be made. A given fan will be noisier if it works at a higher fan total pressure or if it handles more air. Further, fans with higher discharge velocities tend to be noisier than those with lower. With axial flow fans, tip clearance should be minimal since much of the noise generated is at the blade tips. Reducing fan capacity by throttling at a damper produces noise. It is best to use an inverter to regulate the speed of the driving motor, or variable pitch blades in the case of axial flow fans. It is possible to overdo the attempt to specify a quiet fan. If the manufacturers are restricted in terms of impeller diameter, tip speed, outlet velocity and efficiency, they may find it impossible to offer a fan that will operate on the stable part of its characteristic where pressure increases with volume reduction.

The scroll of a centrifugal fan is designed like an involute, its radius of curvature increasing with the angle turned through from the cut-off. Consequently, the static pressure in the casing increases from the cut-off to the fan discharge but there is a sharp fall in static pressure as the cut-off is passed again. The result is that noise is produced here at the blade-passage frequency.

Non-aerodynamic noises are generated by out-of-balance impellers; bearings, particularly ball or roller; driving motors, giving noises largely of magnetic origin that are notably bad with single phase motors of more than the smallest size. Couplings and vee-belt drives seldom cause complaint, if properly selected.

Figure 7.11 compares the noise spectra of two fans, each selected to deliver 2830 $l\ s^{-1}$ at a fan total pressure of 0.623 kPa. It can be seen that the axial flow fan is the noisier and this emphasises that its selection and installation requires care. It is not merely a matter of fitting silencers but also of ensuring that noise does not break out through the flexible couplings and is not radiated objectionably from the casing. Silencers should be bolted directly onto the flanges of axial flow fans at both ends, the flexible couplings being fitted at the extremes of the silencer–fan–silencer combination. Radiation from the casing may be dealt with by installing the fan in a plant room with walls having adequate sound reduction indices. All holes in the walls, etc. must be made good with grout and the annular holes where ducts and pipes pass through the walls tightly plugged with fibreglass, or similar material. It is often acoustically fatal to locate a fan above a suspended ceiling, even if silencers are fitted. A partial remedy for such a difficult situation is to cover the fan casing completely with a layer of sound-absorbing material and, most important, to put a layer of heavy, barrier matting on top. The covering must be done very carefully, with 75 mm lap joints for the outer barrier mat. Leaving small gaps in the coverage will nullify the effort and expense (see Figure 7.12). A further, if expensive remedy

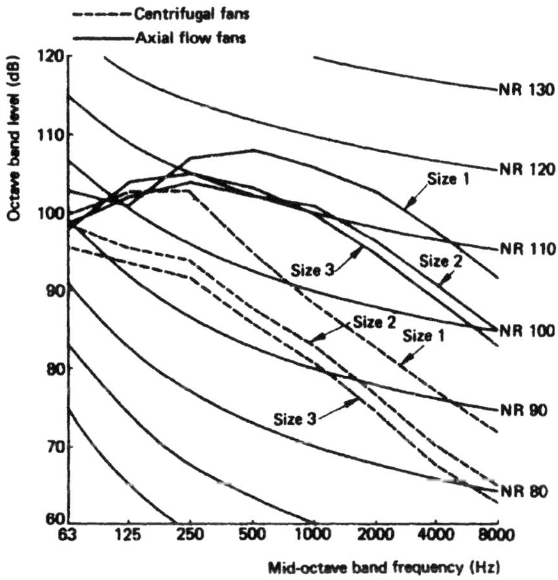

Figure 7.11 Comparison of the noise spectra of two fans.

is to install a massive subceiling, suspended from resilient hangers. Again it is important that there is no gap around the edges and that the ceiling is free from holes or cracks.

7.15 Noise in ducts

Ducts should be sized for the lowest practicable velocities and bends and fittings designed to give smooth airflow with the minimum of turbulence. The characteristic roar from turbulent airflow is very difficult to remove after a duct system is installed. It is not easy to lay down hard and fast rules but Tables 7.7 and 7.8 suggest maximum air velocities and quantities for high velocity systems in comfort applications. The acoustic power emitted by the fan is partly absorbed by the silencer, reduced in intensity by being spread over wavefronts of increasing area as the duct system branches out, augmented by locally generated noise at fittings, further augmented by turbulence at the air distribution terminal and slightly reduced in the lower frequencies by an end reflection when the air finally enters the room. Some attenuation is provided by the unlined duct system as the duct walls are excited to vibrate and bends also give a loss by reflection, but these effects, together with the end reflection at the terminal opening are somewhat unpredictable and comparatively small. It is usually best to ignore any benefit from them. The terminal unit itself may have some added sound-absorbing material to deal with residual noise.

The starting point in the assessment of the silencing to be provided is the fan, and an accepted method of measuring its noise is set out in BS 848 (1985) on fan noise testing. The test method yields sound power levels at the mid-octave frequencies from 125 to 4000 Hz. It is important to note that the test uses good inlet and outlet duct connexions and up to 5 dB should be added to the test results to cater for disturbed airflow in a real case. The sound power levels in the inlet and outlet duct connexions are of most interest and it is these in-duct levels that are most often quoted. Since sound pressure level is almost equal to sound

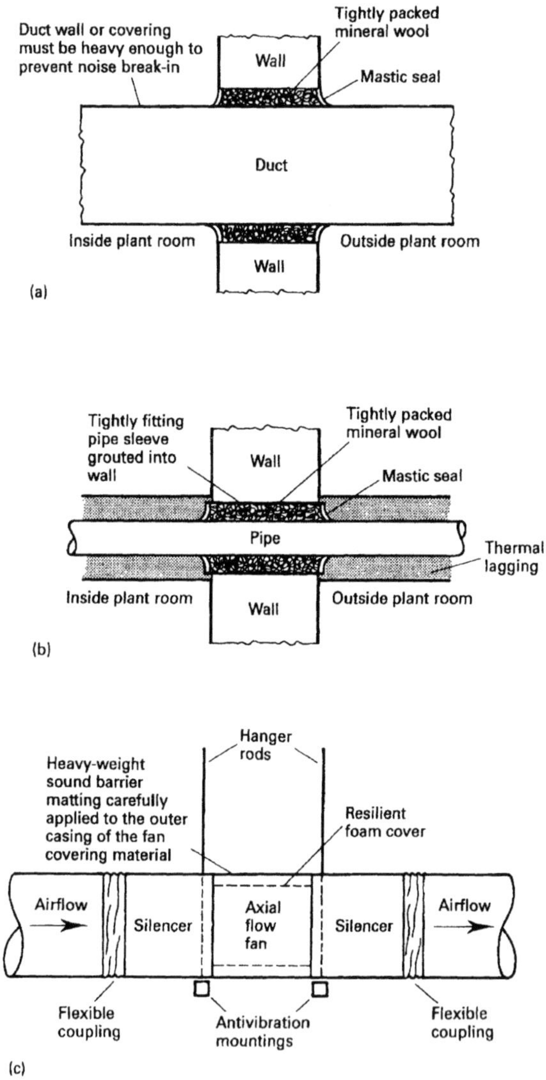

Figure 7.12 Minimising noise break-out from a plant room. (a) A duct passing through a wall. (b) A pipe passing through a wall. (c) Silencing installation for an axial flow fan.

intensity level (see Section 7.5), differing only by 10 log $[p_{\text{ref}}^2/\rho c I_{\text{ref}}]$, sound level intensity can be calculated. Multiplying by the duct cross-sectional area gives the in-duct sound power level:

$$L_w = L_p + 10 \log S \tag{7.25}$$

where L_p is the mean sound pressure level across the duct section and S is the area of the cross-section in m^2.

Table 7.7 Suggested maximum air velocities for duct. systems close to occupied areas.

High velocity, spirally-wound ducts	
Main risers:	18 m/s
Branch ducts (having conical tees or the equivalent):	12 m/s
Low velocity rectangular ducts	
Main risers:	10 m/s
Branch ducts:	8 m/s
Ducts passing through quiet areas:	5 m/s
Ducts supplying to or extracting from quiet areas:	3 m/s

Table 7.8 Maximum air quantities for a high velocity system

Duct diameter (mm)	Maximum volume ($l\ s^{-1}$)
Flexible ducting	
75	40
100	90
Rigid ducting	
75	45
100	100

An approximate formula for the overall sound power level generated by ducts and fittings according to the CIBSE (1986) is

$$I_w = C + 10 \log A + 60 \log v \text{ (dB)} \tag{7.26}$$

in which C is a constant related to the type of fitting, A is the minimum cross sectional area of the fitting and v is the maximum velocity. The formula is said to be roughly correct for a range of velocity from 10 to 30 m s^{-1}. Values for C are tabulated and corrections given to be applied to yield sound power levels at the mid-octave frequencies. Thus, the value of C is -10 for a straight duct, 0 for a conventional bend, i.e. aspect ratio $\leqslant 2$, throat radius $\geqslant$ half the width, and $+10$ for a square-mitred bend with turning vanes, i.e. closely spaced, short-radius, single skin. In each case the mid-octave band corrections are given as -2, -7, -8, -10, -12, -15 and -19 dB, respectively, over the range of 125–8000 Hz.

EXAMPLE 7.7

Using the above data and Equation (7.26) calculate the sound power level produced by a square-mitred bend fitted with short-radius, closely spaced, single skin turning vanes if the section is 300 mm square and the velocity is (a) 10 m s^{-1} and (b) 30 m s^{-1}.

Answer
From Equation (7.26)

(a) $L_w = 10 + 10 \log (0.3 \times 0.3) + 60 \log 10 = 60$ dB
(b) $L_w = 10 + 10 \log (0.3 \times 0.3) + 60 \log 30 = 88$ dB

Hence, applying the quoted corrections the octave band analysis is

Overall L_w	125 Hz	250 Hz	500 Hz	1000 Hz	2000 Hz	4000 Hz	8000 Hz
(a) 60 dB	58 dB	53 dB	52 dB	50 dB	48 dB	45 dB	41 dB
(b) 88 dB	86 dB	81 dB	80 dB	78 dB	76 dB	73 dB	69 dB

It is to be noted that large radius turning vanes in high velocity ducts can produce pure tones, of objectionable strength, generated by vortices formed when the airstream breaks away from the vanes.

Tie rods in ducts produce noise at a frequency given by

$$f = 0.2\, v/d \text{ (Hz)} \tag{7.27}$$

where d is the rod diameter and v the velocity of airflow. The noise produced is much amplified by tuning effects when f equals $nc/2D$, n being an integer, c the velocity of sound and D the transverse duct dimension.

7.16 Silencers

The attenuation of noise in ducts lined with sound-absorbing material has been the subject of a lot of experimental and theoretical research. No simple, accurate expression is possible but some acceptance has been gained for the equation due to Sabine (1940), despite the severe restrictions on its use:

$$R/L = 1.052(P/A)\alpha^{1.4} \text{ (dB m}^{-1}\text{)} \tag{7.28}$$

where R/L is the attenuation per unit length of duct, P is the duct perimeter, A its cross-sectional area and α the sound-absorption coefficient (statistical or Sabine) for its lining. Sabine originally referred it to frequencies between 128 and 2048 Hz, for ducts of dimensions from 225 to 450 mm with aspect ratios of 1:1 to 2:1. It is evident from the form of the equation that the duct shape should provide a large perimeter for a given cross-sectional area. It is at the perimeter of each wavefront, where it meets the sound-absorbing material, that acoustic energy can flow out of the wave train and attenuation occur. For this reason, proprietary silencers often take the form of an array of splitters, covered with sound-absorbing material and offering multiple, parallel paths to the wavefronts, with a large value of P/A. The thicker a sound-absorbing material the better its ability to absorb low-frequency components. Less than 25 mm is scarcely worth bothering with and splitters are usually from 50 to 200 mm in thickness. At the very least they should be covered with a scrim to minimise erosion but the most desirable cover is perforated metal, not least because of the extra structural strength it gives when being man-handled on site.

Low-frequency noise is not very directional, but high-frequency components do tend to travel in straight lines and follow the laws of reflection from surfaces. For this reason it is possible for high-frequency noise to "beam" straight through a silencer, between splitters. To deal with this, one proprietary silencer has curved pathways between the splitters, ensuring high-frequency reflections from the sound-absorbing surfaces. Typical silencers achieve some selectivity over the spectrum by varying the thickness of the splitters and the spacing between them. Air turbulence occurs at entry and exit and to minimise this the ends of the splitters are often streamlined. Figure 7.13 shows typical silencer performances; all

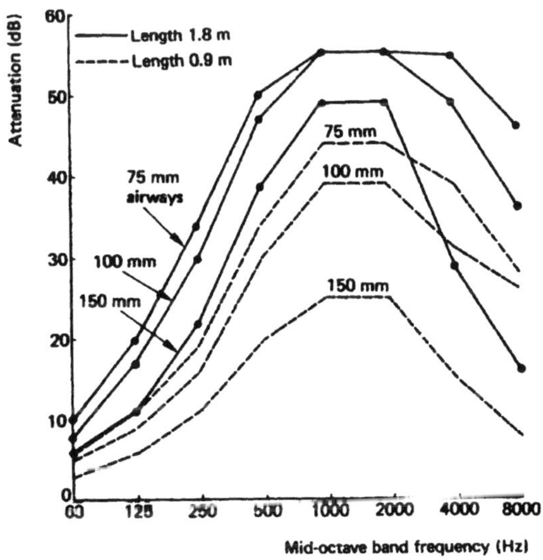

Figure 7.13 Comparison of the performance of two typical silencers with different airway widths.

of them have splitters 100 mm thick but each has a different airway width: 75, 100 and 150 mm. As expected, a width of 75 mm gives the best performance and a silencer 1.8 m long is nearly twice as effective as one of 0.9 m. The cross-sectional area and the air pressure drop are influenced by the splitter configuration. Excessively high velocities and pressure drops should not be adopted in an effort to economise in the space occupied. All silencers generate some noise, half of which travels downstream and half upstream. For the silencer to be useful it must absorb significantly more noise than it produces. Note that according to Equation (7.26), each doubling of the airflow velocity adds 18 dB to the self-generated noise. Not all silencers use splitters; for axial flow fans, lined circular ducts are used, with or without coaxial pods of sound-absorbing material. Other silencers can be purpose-made to provide a bend for the airflow and give extra silencing by reflection. Yet another possibility is to construct silencers with resonant chambers, tuned by design to absorb sound selectively at particular frequencies.

As a general guide it is suggested that self-generated noise is not likely to be troublesome if the face velocity is less than 7 m s^{-1}, the velocity through the airways is less than 15 m s^{-1} and the air pressure drop does not exceed 75 Pa. Special consideration, beyond these suggestions, must be given to low-noise areas such as television studios, etc.

EXAMPLE 7.8

By means of Equation (7.28) estimate the attenuation likely in the range of 125–4000 Hz for a silencer 1200 mm wide × 600 mm high × 1800 mm long, fitted with five splitters, parallel to the 600 mm dimension, each 100 mm thick to form six, equally-spaced airways 100 mm apart. Assume the lining is mineral wool with absorption coefficients as shown in Figure 7.5.

Answer
If each splitter is 100 mm wide it may be assumed that this is equivalent to a 50 mm thick lining for each airway of cross-sectional dimensions 100 × 600 mm.

$$P/A = [(0.6 + 0.6 + 0.1 + 0.1)(0.6 \times 0.1)] = 23.33$$

for one airway or for several airways in parallel L is 1.8 m and by Equation (7.28) $R = 1.052 \times 23.33 \times 1.8\ \alpha^{1.4} = 44.18\ \alpha^{1.4}$. The results can, therefore, be tabulated as follows:

f (Hz)	125	250	500	1000	2000	4000
α	0.30	0.53	0.80	0.84	0.75	0.80
R (dB)	8	18	32	35	30	32

Figure 7.13 shows that at the corresponding frequencies the attenuations for a proprietary silencer of a similar type are 16, 29, 47, 55, 55 and 48 dB. This seems to confirm that the Sabine equation underestimates attenuation. ASHRAE (1991) quote other equations for the attenuation of noise in lined ducts that yield more accurate answers.

It is no good installing a silencer if the noise from the plant breaks out of the duct before the silencer and re-enters after it. Silencers should therefore be bolted directly onto the plant item, the flexible coupling being after the combination (see Figure 7.12a), or be fitted in the duct where it leaves the plant room, butting tightly against the wall. Running ducts through places where the noise level is high should be avoided, otherwise break-in may nullify the effect of the silencer Covering the duct is effective in minimising break-in and break-out. The rules for the cover, i.e. mass, discontinuity, homogeneity, still apply but it is worth noting that an 18 gauge steel duct of 1.2 mm nominal thickness, lagged with 25 mm of glass-fibre and covered with a hard-setting cement plaster, trowelled smooth, may give a fairly good sound reduction index (R) (Table 7.9), provided that at least 20 mm of hard-setting cover is used. Lack of uniformity in the thickness of the cement finish nullifies the effect.

It is possible to generate electronically a wavetrain that is 180 degrees out of phase with another wavetrain. If the two wavetrains are merged, silence should ensue. This well-established principle has been developed with the aid of small computers to deal with both pure tones and broad band noise. Silencers of this sort are termed active silencers, in distinction to the passive silencers, involving the use of absorptive material, described earlier. As applied to fans, an active silencer comprises a microphone, located in the duct, near to the fan, followed by a loudspeaker and another microphone, further downstream. The first microphone measures the noise from the fan in the ducted airflow and passes this information to a processor which analyses it, modifies its phases and feeds this as an output to the loudspeaker. A selectively amplified wavetrain, 180 degrees out of phase with the fan noise,

Table 7.9 Sound reduction index (R)

	Sound reduction index (R) (dB)							
	Frequency (Hz)							
	63	125	250	500	1000	2000	4000	8000
Bare duct	3	10	17	22	28	33	38	40
Duct plus cover	18	24	30	34	38	42	45	45

is then delivered by the loudspeaker into the duct. The second, downstream microphone measures the results of the interference of the original and the injected wavetrains and provides any negative feedback that may be necessary to achieve an improvement in the attenuation. Active silencers are most effective for frequencies less than 400 Hz and ASHRAE (1991) report that attenuations of from 20 to 38 dB have been achieved over a spectrum from 40 to 400 Hz. The performance of an active silencer can be automatically modified if the fan duty changes, for example, as in the case of variable air volume systems.

7.17 End reflection

When a wave train encounters a change in the impedance of its acoustic path, a reflection occurs (see Section 7.2 and Figure 7.2). A change in the section of a duct, as at the entry to a plenum chamber, or at the terminal diffuser delivering air to a room, are instances of impedance variations that produce significant end reflections, mostly in the lower frequencies. The square root of the area of the opening or of the duct cross-sectional area is an approximate index of the attenuation occurring in this way, although this depends also on the shape of the opening or duct, its position at entry to the chamber or room, and the sound-absorption properties of the surfaces of the latter. Figure 7.14 illustrates typical end reflections and shows that these are most effective at the lower frequencies. Even so, it is the low frequency components of system noise that are the most difficult to deal with. For example, the broad spectrum noise produced by air turbulence often has objectionable peaks at 125, 250 and 500 Hz.

7.18 Duct branches

When a duct divides into branches so does the wavefront and it is usually assumed that the acoustic energy is distributed accordingly. If the main duct after the branch has a cross-

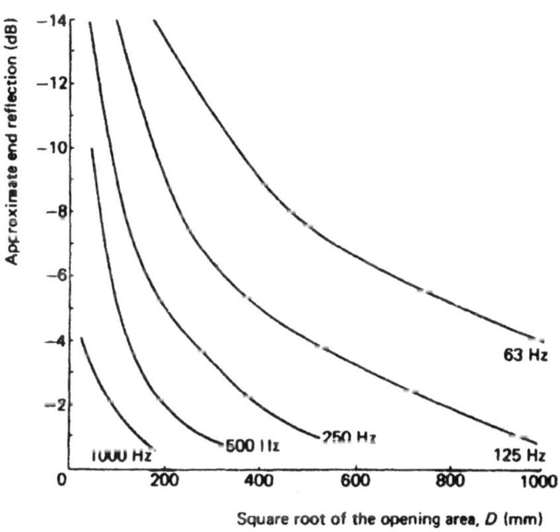

Figure 7.14 Typical end reflections.

sectional area a_1 and the area of the branch is a_2, then the attenuation in the main, because of the loss of acoustic energy down the branch, is given by

$$R = 10 \log [a_1/(a_1 + a_2)] \text{ (dB)} \tag{7.29}$$

EXAMPLE 7.9

If 2830 l s^{-1} flows down a duct and is distributed equally between a pair of ducts each conveying the air at 7.5 m s^{-1}, determine the branch attentuations.

Answer

$$a_1 = a_2 = 2.830/7.5 = 0.377 \text{ m}^2$$

and from Equation (7.29)

$$R = 10 \log [0.377/(0.377 + 0.377)] = -3 \text{ dB}$$

Although it is common to ignore the natural attenuation of unlined ductwork, which is really because of break-out, the effect of branches is considerable and should not be ignored.

EXAMPLE 7.10

A centrifugal fan has an in-duct sound power level as given by the curve for size 2 in Figure 7.11 and is directly connected to a silencer 1.8 m long with properties of attenuation as in Figure 7.13 for 75 mm airways. If the air quantity is ultimately distributed by successive branchings amongst eight subducts of equal area before being delivered through a side-wall grille having an area of 0.16 m^2, determine the sound power level emitted at the grille. If the grille supplies air to a room similar to that forming the basis of Example 7.5, determine the noise level therein. Use Figure 7.14 to establish the end reflection.

Answer
The square root of the area of the supply grille is 400 mm and this is used in Figure 7.14 to find the end reflection. The first branching produces two subducts and gives 3 dB attenuation for each, since they are of equal area. The second branching gives four subducts and a further 3 dB. The third provides the eight duct branches and another 3 dB. The following can now be tabulated.

Frequency (Hz)	63	125	250	500	1000	2000	4000	8000
Sound power level at fan outlet	99	96	94	88	83	77	70	65
Attenuation from silencer	–10	–20	–34	–50	–55	–55	–54	–46
Subtotal	89	76	60	38	28	22	16	19
Branches attenuation	–9	–9	–9	–9	–9	–9	–9	–9
Subtotal	80	67	51	29	19	13	7	10
End reflection	–9	–5	–2	0	0	0	0	0
Sound power level entering room	71	62	49	29	19	13	7	10
Room effect		–4	–8	–10	–10	–10	–9	–
Sound pressure level in room:	71	58	41	19	9	3	–	10
L_p values for NR 50	75	66	58	54	50	47	45	44
L_p values for NR 40	67	57	49	44	40	37	35	33

If the 63 Hz component is ignored, the level in the room is just above NR 40.

7.19 Noise from pumps and pipes

The major component of pump noise is the blade passage frequency but the noise output is related in practice to the pump speed. Pumps that operate against a high head must run at a high speed and those running at 47.5 rev s^{-1} (2850 rpm) are undoubtedly much noisier than at 23.8 rev s^{-1} (1425 rpm). The former often constitute a major noise source in plant rooms and are to be avoided.

Water flow in pipes is generally turbulent and, therefore, generates some noise, according to Rogers (1954). Fortunately, tube walls are massive and rigid, so noise radiation from them is seldom a problem, provided that the water is free of air and velocities are not too high. Pipes with rough and corroded internal walls tend to create more noise than do smooth bores but noise-production in air-free water is not the limiting criterion of velocity. The limit is generally set at 2.4 m s^{-1} for constant flow to minimise erosion in the tubes, especially at the heels of elbows. The commonest source of noise in pipes arises from the air present in improperly vented systems. Bubbles are conveyed by the water and cause a characteristic tinkling at modest velocities, or a broader spectrum, whiter noise at high velocities. Cavitation can also occur in systems when the static head is insufficient for the prevailing temperature, and is another origin of noise, particularly at the pump, which may give a pronounced roar when cavitation is well established. Quieter running pumps can be obtained by selecting an impeller diameter that is only about 85% of the maximum impeller diameter, in the process using a larger pump to achieve the duty, and paying the small extra cost for a dynamically balanced electric driving motor.

7.20 Refrigeration plant

Sound power levels for refrigeration compressors are difficult to obtain because of the intrusion of noise from auxiliary plant, notably pumps. However, Blazier (1972) has established that compressor noise is dominant within 1 m of the machine, allowing representative sound pressure levels to be measured. When assessing the transmission of noise through floors and walls that are near to the compressor, such values should be used without correction, but at distances greater than 3 m from the machine the reverberant field is fairly constant at about 5 dB less than the sound pressure levels measured within 1 m.

Hermetic and open-drive, centrifugal and reciprocating chillers, over the range from 135 to 14 000 kW (40 to 4000 tons of refrigeration), show no significant correlation between sound pressure level and machine size, but centrifugal compressors running at partial load produce some 3 dB more noise than at full duty. Open-drive machines are approximately 5 dB louder than hermetics. Figure 7.15 shows the results of a survey of 34 centrifugal and 14 reciprocating chillers, available in the North American market. Reciprocating compressors are clearly quieter than centrifugals and reach a peak at 500 Hz, probably arising from piston stroke frequency or its harmonics. The centrifugal spectrum is more uniform and flatter, although noisier. The broadening between 500 and 4000 Hz is caused by the larger, open-drive compressors.

Some sound power levels have been measured and Figure 7.16 compares a single-stage, semi-hermetic chiller with speed-increasing gear and a semi hermitic, screw machine. Each plant has a nominal refrigeration capacity of 1370 kW. The screw machine is up to 20 dB quieter than the centrifugal at full load and we also see that the centrifugal generates more

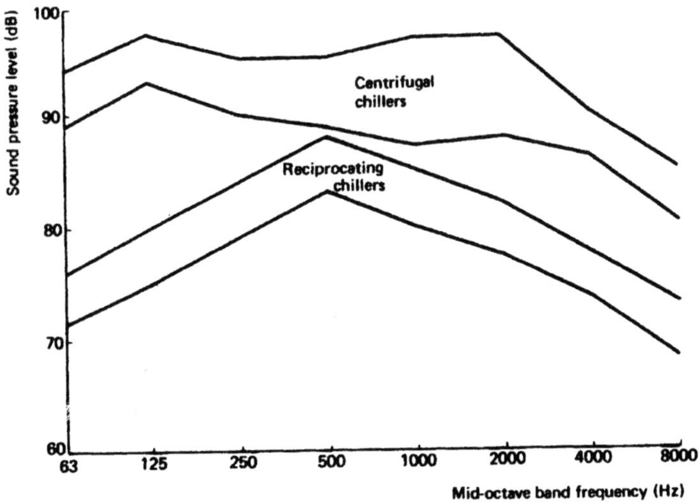

Figure 7.15 Comparison of the sound pressure levels of centrifugal and reciprocating chillers.

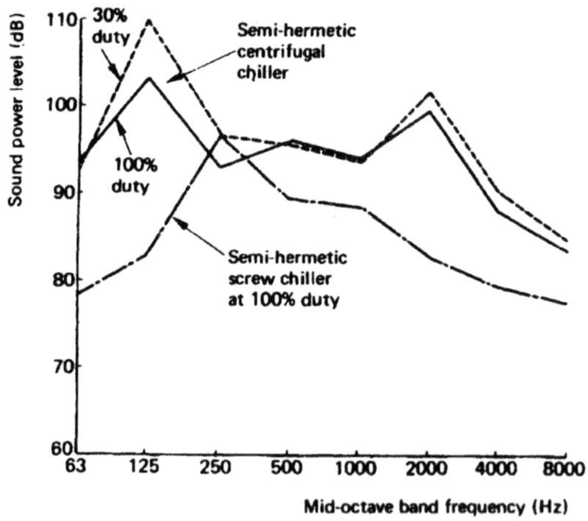

Figure 7.16 Comparison of the sound power levels of a single-stage, semi-hermetic centrifugal chiller with speed-increasing gear and a semi-hermetic, screw machine.

noise, mostly at 125 Hz, when its duty is turned down to 30%. This is a characteristic of centrifugals but not of screws.

7.21 Cooling towers and air-cooled condensers

A straightforward comparison of in-duct sound power levels for two cooling towers, one with an axial flow fan and the other with a centrifugal chosen for the same duty from the same

Table 7.10 Comparison of in-duct sound power levels for two cooling towers

	Sound power level (dB)					
	Frequency (Hz)					
Cooling tower	125	250	500	1000	2000	4000
With axial flow fan	100	99	95	93	89	83
With centrifugal fan	88	85	81	75	69	63

manufacturer (Table 7.10), verifies that the forced draught centrifugal arrangement is much the quieter.

Whether the fan is inside or outside the tower casing and the type of tower itself, also have some bearing on the noise created, as the comparative sound pressure levels, measured on the same basis, show (Figure 7.17). There is little to choose among them, except that the ejector tower, which has no fans, relying on the water jets to induce the airflow, is significantly quieter at the low frequencies. Air-cooled condensers have the reputation of being noisy but this is because they generally use axial flow or propeller fans. The greatest nuisance to the surroundings is sometimes caused by the intermittent operation of condenser or cooling tower fans under automatic control, particularly at night-time. Solid-state speed controllers on air-cooled condensers make significantly more noise when they reduce the speed of the fans. One other cause of noise from cooling towers is water dripping from the fill into the pond. This can be partly overcome by fitting screens beneath the fill to break up the pattern of drips.

Silencing towers after the event is a difficult and expensive business. It is better to select a quiet tower initially. Screening can be effective, 20 dB(A) being achieved with barriers of surface density 10–20 kg m^{-2}, but as much as 10 dB(A) is possible with as little as 5 kg m^{-2}, according to the BRE (1973a). The screen obtains its results by casting an acoustic shadow and it is pointless to try for more than about 20 dB(A) because of diffraction effects. Barriers should be homogeneous and as close as possible to the source of noise, remembering the need for adequate airflow paths into the tower or condenser.

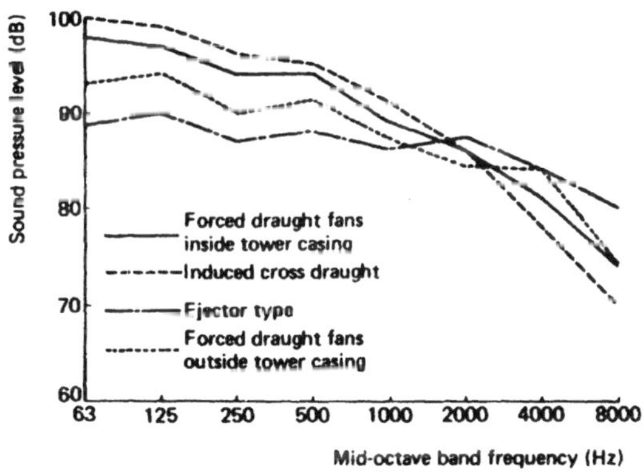

Figure 7.17 Comparative sound pressure levels for various cooling towers.

7.22 Noise radiated to areas outside a building

The noise from louvred air inlets and discharge air outlets in plant room walls, cooling towers, air-cooled condensers, roof-mounted air handling units, boiler houses, flues, etc. may cause a noise nuisance to neighbouring, occupied property. The extent of the nuisance depends on the existing ambient noise level, the time of the day or night that the noisy plant operates, the duration of the operation, the intensity of the noise emitted, its spectrum, and the distance from the source of the noise to the listener.

To deal with the problem the procedure is to start at the property receiving the noise. Table 7.11 gives some approximate details of the noise reductions provided by typical exterior window-wall constructions and these can be used with an appropriate inside sound pressure level to establish the maximum sound pressure level outside the premises, during the time that the plant will be running. Site measurements are taken at the appropriate time of the day or night, over 24 hours, and compared with the values determined above. The higher of the two sets of sound pressure levels should be regarded as the maximum sound pressure levels that are tolerable, immediately outside the property considered.

Alternatively, and less satisfactorily, reference may be made to techniques for predicting traffic noise in various localities, provided by the Building Research Establishment (1973a, 1976).

Reference to Equation (7.20) shows that, in the absence of any room effect (that is, in a free field), the acoustic decay from the sound power level at the noise source, L_w, to the sound pressure level at the receiver, L_p, is given by the directional effect only:

Table 7.11 Typical approximate noise reductions for various external window–wall constructions.

Construction	Noise reductions in dB over the audio spectrum							
	63 Hz	125 Hz	250 Hz	500 Hz	1 kHz	2 kHz	4 kHz	8 kHz
Any typical wall with windows open	9	9	11	14	16	17	17	17
Any typical wall with windows shut but small air vents open	15	15	17	19	21	23	23	23
Any typical wall with no cracks or openings and windows shut	19	19	22	25	28	31	31	31
100 kg m^{-2} wall, with no windows, no cracks and no openings	25	26	30	34	38	42	44	46
250 kg m^{-2} wall, with no windows, no cracks and no openings	31	32	36	40	44	48	51	52

$$L_w - L_p = -10 \log (Q/4\pi\ r^2)$$

whence

$$L_w = L_p + 10 \log (4\pi\ r^2/Q) \tag{7.30}$$

in which r is the distance between the plant noise source and the receiving property, and Q is the directivity factor (see Section 7.8). When the noise source is a louvred opening in a plant room wall Q is 2 and Equation (7.30) becomes

$$\begin{aligned} L_w &= L_p + 10 \log (2\pi r^2) \\ &= L_p + 10 \log (2\pi) + 20 \log r \\ &= L_p + 8 + 20 \log r \end{aligned} \tag{7.31}$$

Since L_p is the tolerable sound pressure level immediately outside the property receiving the noise and is known, the sound power level, L_w, immediately outside the louvred opening, can be calculated. This is done for each octave over the audio spectrum and establishes the extent of the silencing treatment that must be provided for the plant room behind the louvred opening.

7.23 Terminal units

These can be separated into three classes: mechanical units, high-velocity air terminals and low-velocity air diffusers or grilles, roughly in order of noisiness. In the first class, air-cooled, self-contained, 'unit conditioners' or 'window units' are undoubtedly the noisiest but differences of price and quality are reflected in the sound created. Water-cooled units are invariably quieter than air-cooled. In the same class, fan coil units can be very much quieter, particularly if selected for operation at medium or low speed, when they can then fall into the lowest noise category.

High-velocity terminals selected to deliver large quantities of air at high pressure will be noisier than those chosen for reduced duties. This feature is to the advantage of variable air volume units (see Figure 7.18) which operate for much of the time at a partial load and consequently at much less than their design noise level. High-velocity terminals always radiate some noise from their casings and are often located above suspended ceilings that scarcely impede the flow of noise into the conditioned space beneath at all. The presence of a silencer in the low-velocity side of the terminal does nothing to prevent such radiation.

The sound power levels claimed for low-velocity air distribution devices always assume smooth, uniform airflow from a silent duct system into the cones of the diffuser or the bars of the grille. This is a most unlikely situation in reality. The presence of a damper disturbs the airflow over the cones or bars and more noise than anticipated is generated. Further, airflows are frequently distorted before entering the terminal by the duct system and it follows that the noise levels claimed should never be taken at their face value. Caution in selection is advisable. Regardless of the sound produced at the terminal, turbulent airflow noise in a poorly-designed duct system feeding it may give additional noise.

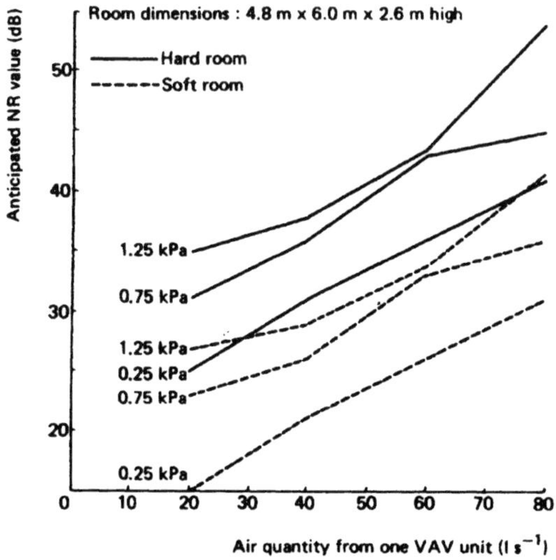

Figure 7.18 Relationship between NR value and duty for a VAV terminal unit.

7.24 Measurement of sound

Sound pressure level is measured by means of a sound level meter which comprises microphone, preamplifier, filters, amplifier, rms rectifier and indicating meter. An undamped measurement is often difficult because the needle on the meter oscillates rapidly as the sound pressure level fluctuates; it is usually easier to damp the response of the meter and so give averaged values over a short time period. This is most conveniently done as a digital display.

There are four types of microphone: crystal, condenser, electret and dynamic, of which the first two are the most commonly used. The cheaper and more rugged crystal microphone relies on the piezo-electric effect of a crystal placed against a diaphragm, where small changes of air pressure strain the atomic lattice of the crystal, generating a proportional voltage. The condenser microphone has greater acoustic sensitivity. Alterations in sound pressure deflect a diaphragm and produce variations in the capacitance with an adjacent plate. A constant charge is usually maintained on the capacitor assembly so that the voltage output varies with the change in capacity, that is, with change in sound pressure. All microphones, especially condenser instruments, need careful treatment.

Measurements should never be made close to a strongly absorbing or reflecting surface and microphones used in moving airstreams must be shielded from the airflow to avoid the irrelevant noise from the air turbulence.

The purpose of the filter in the meter is to simulate the subjective response of the human ear. An A-weighted network (see Figure 7.19) includes filters that modify the output signal to the meter, according to frequency, in such a way that there is a high degree of correlation with speech interference levels, loudness level (see Section 7.9) and NC or NR levels. In a rough and ready way, an NR value is about 7 dB less than the dB(A) value. It follows that sound pressure levels in dB(A) are most useful. B-weighted networks are seldom used and

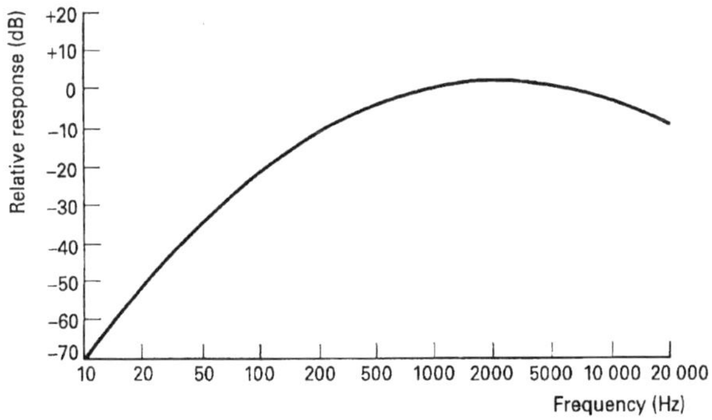

Figure 7.19 Modified output to a meter for an A-weighted network.

the C network gives a flat response. Other weightings are useful for special purposes. Where measurements in dB(A) are not sufficiently accurate or informative, a frequency analyser must be used, in conjunction with a sound level meter, to yield an octave band or even a one-third octave band analysis. Real time displays of sound pressure level against one-third octave bands are useful for the diagnosis of difficult problems.

7.25 Vibration transmission

Because of imperfect construction and balance, rotational and reciprocating machines impose oscillatory, out-of-balance forces on their mountings, producing linear, rocking, rolling and yawing movements. If simple harmonic motion, giving only linear, vertical displacements is assumed, elementary analysis leads to some convenient, if optimistic, answers. When a resilient mounting of stiffness k, is placed between the machine and the supporting building structure, the ratio of the force transmitted through the mounting to the force imposed on it by the machine is termed the transmissibility (τ) of the mounting and, ignoring damping effects, is defined by

$$\tau = 1/[1 - (f/f_o)^2] \qquad (7.32)$$

whence, as a useful alternative,

$$f_o = f\sqrt{[\tau(1 + \tau)]} \qquad (7.33)$$

in which f is the frequency of the imposed force and f_o is the natural frequency of the mounting, defined by

$$f_o = 15.8/\sqrt{d} \qquad (7.34)$$

where d is the static deflection in mm of the machine at rest on its mountings. Since the static deflection depends on the stiffness of the mounting, i.e. the force needed to give a unit static deflection, it follows that the choice of suitable mountings depends on the static deflection, this requiring to be large if low frequencies are involved.

EXAMPLE 7.11

Out-of-balance forces have frequencies of (a) 100 Hz and (b) 50 Hz. What static deflection must be specified for the resilient mountings in each case, to give a transmissibility of 0.005?

Answer

(a) Using Equation (7.33) $f_o = 100\sqrt{(0.005/1.005)} = 7$ Hz and from Equation (7.34) $d = 249.6/49 = 5.1$ mm.

(b) Similarly, $f_o = 50\sqrt{(0.005/1.005)} = 3.5$ Hz and $d = 249.6/12.25 = 20.4$ mm.

Resonance occurs when f/f_0 equals unity. If a machine runs continuously under this condition, successive amplitudes of vibration increase until there is a mechanical failure or damping forces restrict further amplitude increases. When a machine starts, its speed builds up and it passes through the resonant frequency of the mountings. If the run-up time is short, resonance is only a transient effect but with slow accelerations it may be necessary to introduce extra damping to minimise resonant amplitudes.

There is the risk of a beat frequency developing when two machines, running at slightly different speeds, are connected, as by a pipe between a chiller and its pump. The beat frequency is the difference of the component frequencies and may be near to the natural frequency of the link.

7.26 Damping

When a mass, m, freely supported on a resilient mounting, is displaced vertically by an amount, y, from its position of equilibrium, it will oscillate indefinitely with a frequency, f_o, and maximum amplitudes $\pm y$, in the absence of damping forces. In practice, air resistance and the internal friction of the material of the mounting will reduce the movement, which will eventually cease. The reduction may be helped by adding external damping, as by a dash pot in which the viscous resistance to the flow of oil through an orifice retards the movement, but it does not follow that the oscillations will always then occur. If c is the damping force per unit of velocity, oscillations will only take place when $c/m < 4\pi f_0$; above this the movement is dead beat and decays slowly to zero without oscillation. The value $c_0 = 4\pi f_0$ is the critical damping force per unit of velocity and the fraction c/c_0 is termed the damping ratio, D.

If forced vibrations of frequency f are imposed on a damped, resilient mounting, a new expression for the transmissibility can be derived from elementary theory

$$\tau = \sqrt{\frac{1 + 4D^2(f/f_o)^2}{[1-(f/f_o)^2]^2 + 4D^2(f/f_o)^2}} \tag{7.35}$$

EXAMPLE 7.12

A machine at rest on a damped, resilient mounting gives a static deflection of 25 mm. If the damping ratio is 0.25 and the machine runs up slowly from rest to a normal operating frequency of 25 Hz, determine (a) the transmissibility as the machine passes through resonance and (b) the transmissibility at normal running.

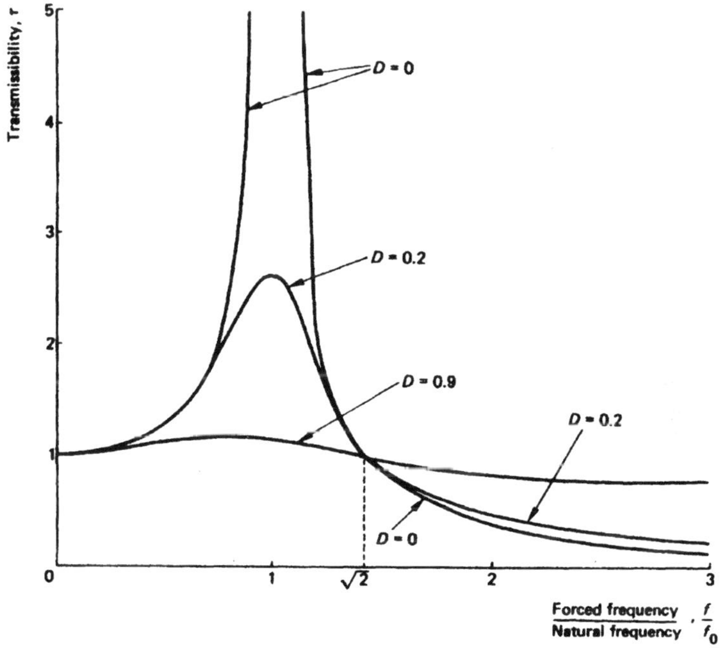

Figure 7.20 The effect of damping ratio upon transmissibility.

Answer

(a) $(f/f_o) = 1$ at resonance and so from Equation (7.35)

$$\tau = \sqrt{\frac{1 + 4 \times 0.25^2}{0 + 4 \times 0.25^2}} = 2.24$$

(b) From Equation (7.34) $f_o = 15.8/\sqrt{25} = 3.16$ Hz and from Equation (7.35)

$$\tau = \sqrt{\frac{1 + 4 \times 0.25^2 \times (25/3.16)^2}{[1 - (25/3.16)^2]^2 + 4 \times 0.25^2 \times (25/3.16)^2}} = 0.066$$

In general, a damping ratio greater than zero reduces the transmissibility for $f/f_o = \sqrt{2}$, but increases it for $f/f_o - \sqrt{2}$ (Figure 7.20).

7.27 Anti-vibration mountings

Pads or mats in compression, synthetic or natural, ribbed or studded rubber, glass-fibre, cork, felt, etc., are used for dealing with high frequencies where static deflections are less than about 5 mm. It is essential that they are uniformly and correctly loaded, as underloading gives insufficient static deflection and overloading causes distortion and spoils resilient properties. Rubber-in-shear fittings may be used for comparatively small loads with static deflections up to about 8 mm. They are affected by ageing, oil or other solvents, and temperature. Internal damping ratios are of the order of 0.03.

Helical steel springs must be used when static deflections greater than 8 mm are desired. Since they have relatively low stiffness and hence large static deflections they tend to be unstable if badly installed or subjected to lateral movement. They have very little internal damping and transmit high frequencies unattenuated. It is therefore customary for proprietary spring isolators to incorporate levelling screws, to be fitted with rubber snubbers to restrict excessive lateral and vertical movement, especially during run-up or run-down, and to be provided with synthetic rubber pads for attenuating the high frequencies.

Machines and their anti-vibration mountings should be located on level plinths, or house-keeping pads, 100 mm or more in thickness. The resilient mountings must be positioned to carry equal loads and the machine lowered carefully onto them, so that lateral displacements of the mountings or torsional forces on them, are avoided. The machine base-frame should then be levelled, using shims if necessary.

It is important that there are no rigid bridges, e.g. by electrical conduit, across the anti-vibration mountings. Flexible couplings with lower stiffnesses than the mountings should be fitted in duct and conduit. Flexible couplings in pipes on each side of a pump or a chiller, to limit the transmission of vibration, are largely a waste of time as they become rigid and ineffectual under hydraulic pressure. It is much better to bolt the pipe directly to the machine flanges and to use Victaulic couplings for the pipe joints to take up any vibrational movement; such joints can accommodate from 1°, for large pipes, to 5°, for small pipes, of angular displacement. The pipes themselves should be supported by spring hangers with a static deflection of at least 25 mm for a distance of about 30 m, horizontally, from the machine. In this way they will attenuate any transmitted noise and vibration. Thereafter, conventional pipe brackets may be used. It should be noted that unrestricted heavy valves or other fittings may impose a rotational force on pipework with Victaulic joints. The valves or fittings then tend to rotate into an undesirable stable position, hanging down beneath the pipe.

Pumps should be either bolted down rigidly to floors with solid ground beneath, with their pipe connexions as above, or, otherwise, fitted on a strong, rigid, steel base-frame, loaded with concrete to form a substantial inertia block and mounted on steel springs.

When two machines are coupled, e.g. a fan, belt-driven from a motor, it is essential that they are both mounted on a common steel base-frame and that this is supported as a whole by anti-vibration fittings. The lowest frequency component in the pair should be used for the selection of the resilient mountings. In this connexion it is bad to fit a stand-by motor, that is normally not running, on a common base frame with the running plant, as transmitted vibrations will cause bearing damage by brinelling. When stand-by motors are used they should be belted up to the fan, for example, and driven, along with the fan, by the duty motor.

Inertia blocks are very desirable. They are often provided by filling with concrete the space within the steel channel base-frame, to which the machine is bolted and to which the resilient mountings are fitted. The increased mass so provided lowers the centre of gravity of the assembly and so adds stability, increases the static deflection on the mountings, reduces the amplitude of movement, reduces the transient disturbances as the machine runs up or down, and minimises the bad effects of any unequal loading on the resilient mountings caused by imprecise knowledge of the weight disposition in the machine itself. Ideally, the centre of gravity should be below the upper surface of the anti-vibration mountings and for this reason they are often fitted on extension members from the upper part of the inertia block.

Table 7.12 Suitable static deflections for various locations and different dynamic loadings

Dynamic load	Static deflections (mm)				
	On ground	Span of floor (m)			
		6	9	12	15
Centrifugal chillers					
Open:	10	15	45	45	90
Hermetic or semi-hermetic	10	25	45	45	65
Absorption chillers	5	25	25	45	45
Reciprocating compressors					
8.5–12.5 Hz	25	45	65	65	90
> 12.5 Hz	25	25	45	65	65
Air compressors					
8.5–12.5 Hz	25	45	65	65	90
> 12.5 Hz	25	25	45	65	65
Cooling towers					
⩽ 8.5 Hz	10	10	45	65	90
> 8.5 Hz	10	10	25	45	65
Boilers and steam generators	0	0	25	45	65
Centrifugal pumps					
close-coupled ⩽ 4 kW	10	25	25	25	25
close-coupled 5–30 kW	25	25	45	45	45
close-coupled > 30 kW	25	25	45	65	65
Air-handling units					
suspended, ⩽ 4 kW	0	25	25	25	25
suspended, > 4 kW and ⩽ 8.5 Hz	0	45	45	45	45
suspended, > 8.5 Hz	0	25	25	45	45
floor-mounted, ⩽ 4 kW	10	25	25	25	25
floor-mounted, > 4 kW, ⩽ 8.5 Hz	10	45	45	45	45
floor-mounted, > 4 kW, > 8.5 Hz	10	25	25	45	45
Fans					
suspended, axial flow, ⩽ 40 kW, ⩽ 5Hz	0	65	65	65	90
suspended, axial flow, < 40 kW, 5–8.5 Hz	0	45	45	65	65
suspended, axial flow, < 40 kW, > 8.5 Hz	0	25	25	45	65
suspended, axial flow, > 40 kW, 5–8.5 Hz	0	45	50	65	90
suspended, axial flow, > 40 kW, > 8.5 Hz	0	45	45	45	65
floor-mounted fans, ⩽ 40 kW, ⩽ 3.5 Hz	10	65	65	90	*
floor-mounted fans, < 40 kW, 3.5–5 Hz	10	45	65	65	90
floor-mounted fans, < 40 kW, 5–8.5 Hz	10	45	45	65	90
floor-mounted fans, < 40 kW, > 8.5 Hz	10	25	25	45	90
floor-mounted fans, > 40 kW, ⩽ 5 Hz	45	65	90	90	*
floor-mounted fans, > 40 kW, 5–8.5 Hz	45	45	65	90	90
floor-mounted fans, > 40 kW, > 8.5 Hz	25	45	45	65	65
Internal combustion engines					
< 25 kW	10	10	15	65	65
25–75 kW	10	45	65	90	90
> 75 kW	10	65	90	110	110

*Avoid this situation.

With fans, in particular, it is sometimes possible for a reactive movement to occur which means the fan base tilts upon start-up. This may be minimised by the addition of a heavy inertia block and a lengthening of the base frame in the direction of reaction. Alternatively, the assembly can be designed to have a reverse tilt that is corrected when the fan runs.

Locating anti-vibration mountings on a suspended floor may yield unpredictable results. The natural frequency of the resilient mountings should be as far as possible from the natural frequency of the floor, which itself is difficult to predict. The approximate natural frequency of a concrete floor slab is the shorter span in metres divided by 600. The safest plan is to use large static deflections, never less than 30 mm, for large machines such as refrigeration compressors. Lightweight floors and cantilevered slabs are very likely to cause trouble and mounting dynamic loads on them is to be avoided. When machines are located on suspended floors it is wise to position them close to load-bearing walls and columns. Better still, they should be entirely off the floor, supported on resilient mountings fixed to steel channels or joists which are in turn supported by load-bearing walls and columns. Even so, there is the risk of noise penetrating the slab and passing to spaces beneath. Table 7.12 suggests suitable static deflections for various locations and different dynamic loadings. In spite of the suggestions made in Table 7.12, it is best to locate boilers and refrigeration plants in basements, on solid non-suspended floors with ground beneath. The same should apply, as far as possible, to pumps.

Exercises

1 Evaluate the room effect in dB in each octave band from 125 to 4000 Hz for an acoustically hard room of the same dimensions as in Example 7.5. Take the same absorption coefficients for the walls, window and door. Use the following coefficients for the hard floor and ceiling:

Frequency (Hz)	125	250	500	1000	2000	4000
Floor	0.02	0.04	0.05	0.05	0.10	0.05
Ceiling	0.03	0.03	0.02	0.03	0.04	0.05

Assume that there is no furniture, that two people are present measuring the sound pressure levels, that the terminal unit is in the middle of the ceiling and that $r = 1.769$ m.
(*Answer* –0.1, +0.3, –0.7, +0.1, +1.8, +1.3 dB)

2 If the room in Exercise 7.1 is fitted with a variable air volume terminal having a sound power level of

Frequency (Hz)	125	250	500	1000	2000	4000
dB: re 10^{-12} W	10.0	11.0	16.5	25.0	32.0	35.5

determine the NR level in the room, with the aid of Figure 7.6.
(*Answer* Approximately NR 38)

3 A wall 4.6 m long × 2.7 m high has a mean sound transmission loss of 45 dB. If a door of dimensions 0.9 × 2.1 m having a loss of 31 dB is fitted in the wall, what will be the overall loss?
(*Answer* 38 dB)

Notation

Symbols	Description	Unit
A	Cross-sectional area	m^2
	Total number of absorption units	m^2
C	Constant	dB
D	Damping ratio	–
	Transverse duct dimension	m
D_r	Energy density in a reverberant field	$J\ m^{-3}$
I	Intensity of a sound	$W\ m^{-2}$
I_{av}	Average intensity of a sound	$W\ m^{-2}$
L	Duct length	m
L_p	Sound pressure level	dB
L_I	Sound intensity level	dB
L_w	Sound power level	dB
P	Maximum sound pressure in a wave train	Pa
	Duct internal perimeter	m
P'	Pressure amplitude factor	m Pa
Q	Directivity factor	–
R	Sound reduction factor or index	dB
	Sound transmission loss	dB
	Room constant	dB
	Attenuation in a duct	dB
S	Surface of an element in a room	m^2
	Total surface area in a room	m^2
	Cross-sectional area of a duct	m^2
S_n	Surface area of an element in a room	m^2
W	Sound power	W
a	Cross-sectional area of a duct	m^2
c	Speed of sound	$m\ s^{-1}$
c	Damping force per unit velocity	$N\ s\ m^{-1}$
c_o	Critical damping force per unit velocity	$N\ s\ m^{-1}$
	Diameter of a rod	m
	Static deflection of a machine at rest	mm
f	Frequency of a sound	Hz
	Frequency of an imposed force on a mounting	Hz
f_c	Mid frequency of an octave band	Hz
f_o	Natural frequency	Hz
k	Wave number	m^{-1}
	Stiffness of a resilient mounting	$N\ m^{-1}$
m	Mass	kg
n	Integer	–
p_{rms}	Root mean square pressure	Pa
p	Root mean square pressure	Pa
$p(x, t)$	Sound pressure at distance x and time t	Pa
$p(r, t)$	Sound pressure at distance r and time t	Pa
r	Radial distance	m
t	Time	s

u	Velocity of vibrating molecules in a wave train	m s^{-1}
v	Air velocity	m s^{-1}
v	Maximum air velocity	m s^{-1}
x	Distance	m
y	Vertical displacement	m
α	Sound absorption coefficient	–
	Sabine sound absorption coefficient	–
	Statistical sound absorption coefficient	–
α_{sab}	Sabine sound absorption coefficient	–
$\bar{\alpha}$	Average sound absorption coefficient	–
α_θ	Sound absorption coefficient at a given angle of incidence, θ	–
α_n	Sound absorption coefficient at normal incidence	–
δS	Small element of area	m^2
θ	Angle of incidence	
λ	Wavelength	
ρ	Density	kg m^{-3}
τ	Sound transmittance coefficient	–
	Transmissibility of a mounting	–

References

BS 359, Recommendations on preferred frequencies for acoustical measurements, 1963(86).

Beranek, L. L., Revised criteria for noise in buildings. *Noise Control*, **3,** 19–27, 1957.

Kosten, C. W. and Van Os, G. J., Community reaction criteria for external noises. *Proc. Conf. Control of Noise*, HMSO London, 1962.

ASHRAE Handbook, *Systems*, Chapter 35, 1980.

ASHRAE Handbook, *Heating, Ventilating and Air Conditioning Applications*, SI Edition, Chapter 42, Sections 42.2-42.4, 42.14, 1991.

BS 2750,Measurement of sound insulation in buildings and of building elements, 1987.

BS 661, Glossary of acoustical terms, 1969.

Cook, K. and Chrzanowski, P., Transmission of airborne noise through walls and floors. In Harris, C. M. (ed.), *Handbook of Noise Control*, McGraw-Hill, Sections 20.1–20.15, 1957.

BS 848, Fans for general purposes, Part 2. Methods of noise testing, 1985.

CIBSE Guide, Section B12, Sound control, 1986.

Sabine, H. J., *J Acoust Soc Am*, **12,** 53, 1940.

Rogers, W. L., *Trans. ASHVE*, 60, 41, 1954.

Blazier, W. E., Chiller noise: its impact on building design. ASHRAE semi-annual meeting, New Orleans, 1972.

BRE, BRS Digest No 153, Motorway noise and dwellings,1973a.

BRE, Designing offices against noise, BRE CP 6/73,1973b.

BRE, BRS Digest No 186, Part 2, Prediction of traffic noise, 1976.

8

Economics

8.1 Capital costs

Generalisation about capital costs is difficult because the options available to the designers of buildings and services are so many that an accurate correlation with design parameters is not easy. Nevertheless, the cost of mechanical services in £ m^{-2} of treated floor area has been extensively used for budget price indications and proved of some value with office blocks but of less worth in other applications. Alternative cost indicators, measured per unit of installed refrigeration capacity and per unit of supply airflow rate are also useful at times.

Whichever approach is adopted its accuracy is significantly affected by the following building features: total treated floor area; building shape; type of glazing and shading; proportion of glazing in the facade; illumination level and the heat dissipated by electric lights and business machines; population density; and, for office blocks, the width of the module. The *U*-values of the structural elements in the fabric are generally of less importance, except in the case of low rise-buildings having large plan areas, e.g. hypermarkets, where the thermal transmittance of the roof is significant. The thermal mass of the building structure also plays a part, notably in reducing the impact of solar gains through glass when it is large. The choice of system according to Swain *et al.* (1964) and Watson (1978) and the way it is designed are major influences on capital cost. It is evident that taking account of all the factors is a hindrance to the accurate preparation of indicative costs. However, if the total cost of a system is broken down to show the percentage contributions of its elements, the influence of design changes on the overall cost can be seen but, because of the many ways of grouping the component costs, an exact consensus of opinion cannot be obtained. Tables 8.1 and 8.2, based on Swain *et al.* (1964) and Watson (1978) illustrate such cost analyses. It should be noted, however, that it is always difficult to know exactly what comprises the cost of each component without access to the details.

The results of statistics collected for office blocks by Haden Young Ltd from 1959 to 1985 within the range of 1000–10 000 m^2 of treated floor area and with refrigeration loads between 110 and 150 W m^{-2} are given as comparative figures in Table 8.3.

The figures in Table 8.3 do not necessarily compare similar systems. For instance, a low velocity ducted system from a central plant, perhaps with simple zoned control, does not necessarily provide the same quality of air conditioning as do the other systems listed. Further, accommodating extensive low velocity duct runs poses serious problems in an office block, where space is at a premium. In the case of systems using terminal units, which are

Table 8.1 Percentage contribution of the components of a system to its total cost

Component	Approximate percentage of the total capital cost for various systems		
	Fan coil plus ducted air	Fan coil plus local air	Low velocity ducted air
Refrigeration plant and cooling towers	21	24	18
Air-handling plant and controls	7	2	18
Pumps	1	1	1
Pipe, etc.	6	6	2
Insulation on pipes and plant	7	8	1
Ducts, insulation, diffusers and grilles	27	14	51
Terminal units	22	34	–
Individual control on terminal units	inc.	inc.	
Air compressors and accessories	inc.	inc.	inc.
Boilers	3	4	3
Electrical wiring and switchgear	6	7	6
Total	100.0	100.0	100.0

These figures were determined by costing comparative designs for a hypothetical office building of about 14 200 m^2 treated floor area, according to JIHVE (1960).

usual in office air conditioning, the number of units per module is a major influence on the capital cost and also, of course, on the options for partition arrangement open to the tenant (Sections 2.6 and 2.11).

For applications other than offices, low velocity ducted systems are often a good solution and Table 8.4, based on similar statistical evidence, suggests comparative costs, the ratios again being related to unity for the fan coil system in Table 8.3.

Attempts have been made by Swain *et al.* (1964) and Watson (1978) to relate capital costs to treated floor area but the results have not always shown good correlation, mainly because of the weightings imposed by the many different features of the building and services design. The conclusion of an analysis by Watson (1978) of the costs for 20 different buildings was that an accurate correlation was impossible on this basis. Figure 8.1 shows the results of some attempted correlations, based on Swain *et al.* (1964).

8.2 Energy consumption

Energy consumption falls into two parts: the thermal energy used in burning fossil fuel for heating and the electrical energy used to drive the mechanical plant or, occasionally, used for heating. Making an accurate estimate is virtually impossible at the design stage because

Table 8.2 Percentage contribution of the components of a system to its total cost

Component	Approximate percentage of the total capital cost for a VAV system plus perimeter heating over a treated floor area of 6000 m^2	15 000 m^2
Refrigeration plant, cooling towers and pumps	8	6
Air-handling plant and silencers	15	14
Chilled water pipe and insulation	3	3
Cooling water pipe	2	1
LTHW pipe, insulation and pumps	5	4
Ductwork, insulation, grilles and diffusers	39	45
Terminal units	11	11
Automatic controls	5	5
Boilers, flues and gas installation	4	3
Space heating	3	3
Electrical supplies to control panels and plant	2	2
Water treatment	1	1
Fire detection	2	2
Totals	100%	100%

Table 8.3 Relative costs of various systems

	Approximate relative costs of various systems: Fan coil plus ducted air	VAV plus perimeter heating	Dual duct	Low velocity ducted
High	1.3	1.4	1.5	1.3
Average	1.0	1.0	1.2	1.2
Low	0.7	0.6	0.9	1.1

Table 8.4 Relative costs of systems for different applications

	Approximate relative costs of systems for various applications: Public rooms in hotels	Theatres, etc.	Hospitals	Restaurants	Factories
High	1.4	1.3	3.9	1.4	1.2
Average	1.0	1.0	2.6	1.0	0.8
Low	0.6	0.6	1.4	0.6	0.5

Note: The costs are related to a unit cost for the fan coil system in Table 8.3.

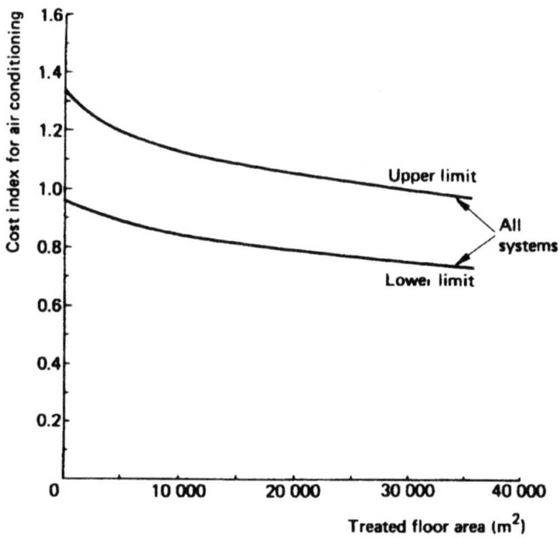

Figure 8.1 Approximate relative costs for various air conditioning systems against treated floor area. For example, a four-pipe fan coil system, with low velocity ducted air dealing with the latent gains, might lie between the curves, but nearer to the upper limit.

the actual consumption will depend greatly on the manner of system operation. Nevertheless, estimates are often necessary, even if for no other reason than to compare the performances of two different, competing systems on the same basis, at the conceptual stage. The CIBSE (1996a, b) propose a way of calculating the mean, annual power required by a building and its services, representing the total energy consumption divided by the total number of hours in a year (8760). This method goes to some lengths to assess the contribution of casual gains by solar radiation through windows and by electric lighting, both of which reduce the energy requirement for heating if the system is thermostatically controlled in an appropriate way.

THE ENERGY USED BY BURNING FUEL

The energy used by burning fuel depends on the type of system and the way it is operated and maintained. Any four-pipe system will take full credit for casual gains whereas other systems, such as VAV with perimeter heating, may not. Although more involved techniques are sometimes adopted, the traditional method, according to the CIBSE (1986a), is the use of degree days. For any given day when the average temperature is less than a chosen base temperature, the number of degree days is the difference between the base temperature and the mean temperature. Degree days are summed over the heating season to give a measure of the heating energy requirement.

Before estimating the consumption of fuel energy the design boiler power must be determined. It will be the greater of:

(a) The design fabric heat loss, including natural infiltration, plus a margin for warming up the building structure during the preheat period, in the case of intermittent operation or

(b) the design fabric heat loss, including natural infiltration, plus the fresh air load (because of the air introduced mechanically from outside), plus the HWS load.

Usually, (b) exceeds (a). The fabric heat loss plus the fresh air load, Q'_h, is calculated by

$$Q'_h = U'(t_r - t_o)/1000 \text{ (kW)} \tag{8.1}$$

where t_r and t_o are the inside and outside design temperatures, respectively and

$$U' = (\Sigma AU + nV/3) \text{ (W K}^{-1}\text{)} \tag{8.2}$$

where A is an element of area of the building envelope, U is its thermal transmittance coefficient, n is the number of fresh air changes per hour, natural and/or mechanical, and V is the volume of the treated space. Natural ventilation by infiltration should only be included for a distance of 6 m inwards from the outside wall.

The number of degree days to a base temperature of 15.5°C is compiled by the Department of the Environment (1993) for various parts of the UK and Table 8.5 lists some 20-year averages for the Thames Valley area.

In practice, degree days are usually quoted in the UK to a base temperature of 15.5°C, it having been argued in the past by Dufton (1934) and Faber and Kell (1945) that this is the highest outside temperature for which heating is needed. If a higher base temperature is to be used the number of degree days to the base 15.5°C may be increased by about 12% for each degree above 1 5.5°C by inference from CIBSE (1986a). A more involved method by the Department of Energy (1978) may be adopted if greater accuracy is thought worthwhile.

If the heating system operates under thermostatic control and the influence of casual gains is ignored then the average usage factor, F, over the year is given by

$$F' = D_{15.5} \times 24 \times f/(8760(t_r - t_o)) \tag{8.3}$$

wherein $D_{1.15}$ is the total number of degree days in the year to the base 15.5°C and f is a correction factor to cover the period of occupancy; the mode of operation, continuous or intermittent); the class of building structure; and the base temperature adopted according to the CIBSE (1986a). The mean annual boiler power is then:

$$\bar{Q}_h = FQ'_h \text{ (kW)} \tag{8.4}$$

If the boiler efficiency is η, expressed as a fraction, then the total fuel energy consumption over a year is

$$H = (\bar{Q}_h \times 8760 \times 3600)/(1\,000\,000\,\eta) - G \text{ (GJ)} \tag{8.5}$$

where G represents the casual gains

The ventilation air change rate, n, in Equation (8.2) refers to natural infiltration and any additional outside air introduced by a mechanical ventilation system. An air conditioning system handles a mixture of outside and recirculated air. A fixed fraction of fresh air may be used (as is sometimes the case with small installations) or the mixing proportions may be varied, under thermostatic control, to exploit the cooling capacity of cold air when convenient and so economise on the energy used by the mechanical refrigeration plant. Under such circumstances, provided that the recirculated air fraction is not too small, the outside air is heated by mixing with the warmer recirculated air to a temperature equal to the design off-coil temperature. There is then no additional fresh air heating load. However, if the proportion of fresh air is large, the mixed temperature may be less than the design off-coil temperature and there is a heating load. Furthermore, air supplied to the conditioned space at the off-coil temperature or the mixture temperature, plus a few degrees for fan power, may be too cool for the conditioned space and if the casual gains are not enough to provide the warmth needed, extra heating is required in the room when the VAV system is turned down to its minimum air supply rate. This could introduce a problem on start-up in cold winter weather.

The consumption of domestic hot water (abb HWS) should be included.

EXAMPLE 8.1

Estimate the design rate of heating and the probable annual thermal energy requirement for the hypothetical office block (Section 1.2), under steady-state conditions and in the absence of beneficial casual heat gains, assuming it is conditioned by a VAV system with perimeter heating designed to maintain a room temperature of 20°C for an outside temperature of −2°C. The LTHW flow temperature is compensated against outside air temperature. Assume a location in London, a heavy-weight structure, a 9-hour day, a 5-day week, intermittent operation and a gas-fired boiler plant. The average infiltration rate is 0.75 air changes per hour. Ignore heat flow through the ground floor slab.

It is given that there is no heat required for the mixture of outside and recirculated air and no additional heating load in the room. The energy used for HWS would normally be included but is not, in this example, because toilets are regarded as being in the two service blocks as a simplification.

Answer

$$\begin{aligned}\Sigma(AU) &= 0.6(86.4\times3.3\times12\times2)0.45 + (13.5\times3.3\times12\times2)0.45\\ &\quad + 0.4(86.4\times3.3\times12\times2)3.3 + (86.4\times13.5)0.45\\ &= 11886\ \mathrm{W\,K^{-1}}\end{aligned}$$

$$nV/3 = 0.75\times(86.4\times13.5\times2.6\times12)/3 = 9098\ \mathrm{W\,K^{-1}}$$

From Equation (8.2)

$$U' = 11\,886 + 9098 = 20\,984\ \mathrm{W\,K^{-1}}$$

From Equation (8.1)

$$Q'_h = 20\,984(20 + 2)/1000 = 462\ \mathrm{kW}$$

This covers fabric heat loss plus natural infiltration.

A reasonable assumption for the mean seasonal boiler efficiency is 78%, according to the CIBSE (1996a) and from the CIBSE (1986a) the following factors can be determined: 9-hour day, 1.005; 5-day week, 0.85; intermittent operation, 0.95. A variable air volume system as defined will ignore casual gains and attempt to maintain 20°C through its open loop, perimeter heating schedule. Therefore, 20°C should be selected as the base temperature for degree days. However, during the summer outside temperatures less than 20°C are most likely to occur during the night, when the system is off. Therefore, the number of degree days for the 9 months from September to May, inclusive should be taken. $D_{15.5}$= 2029 for this period (Table 8.5). The degree day correction factor is (1 +0.12(20 – 15.5)) = 1.54 and thus the overall correction factor, *f*, is 1.005 × 0.85 × 0.95 × 1.54 = 1.25. Then from Equation (8.3)

$$F = (2029 \times 24 \times 1.25)/(8760 \times (20 + 2)) = 0.316$$

and from Equation (8.4)

$$Q'_h = 0.316 \times 462 = 146 \text{ kW}$$

and from Equation (8.5) the total annual heat energy used is

$$H = (146 \times 8760 \times 3600)/(1\,000\,000 \times 0.78) = 5903 \text{ GJ}$$

The specific heat energy consumption, per unit of treated floor area is

$$5903/(86.4 \times 13.5 \times 12) = 0.42 \text{ GJ m}^{-2}$$

If a four-pipe fan coil system were used casual gains would reduce the heat energy consumption because of the mode of thermostatic control over the units, heating capacity being in sequence with cooling capacity. For casual gains, Table 1.2 gives diversity factors for people throughout the year, but as regards lighting it is reasonable to assume a factor of 1.0 for, say, the winter months, November–February, and to apply the tabulated diversity factors for the remaining months.

Reference CIBSE (1996a) gives a more detailed and accurate approach to the assessment of the heat gains from electric lighting but it is too complicated for inclusion in this text. For business machines the CIBSE (1992) publishes extensive data on the heat dissipation of various types of business machine, computing equipment and other ancillary systems, from which a practical evaluation of a diversity factor is possible. As an approximation, without reference to such data, a diversity factor of 0.7 might be assumed, although this is probably on the high side. The heating benefit from solar gain through windows is much more complicated and is particularly hard to assess because of the uncertainty about the benefit of solar and other gains and the possibility of undesirable overheating. One approach to casual solar gains, adopted by the CIBSE (1996a), uses the general principle that the solar gain is the mean gain over 24 hours plus the variation about the mean. Solar irradiation values are taken from data tabulated in the CIBSE Example Weather Year for Kew (1 October 1964 to 30 September 1965) and values for the solar gain through glazing and the relevant factors are from Table A9.15 in the CIBSE Guide (1986b), also given in Table A.5 in the Appendix. The following equation is proposed

Table 8.5 Average number of degree days over a typical period of twenty years in the Thames Valley

Month	Jan	Feb	Mar	Apr	May	Jun	Jul	Aug	Sep	Oct	Nov	Dec	Total
Number of degree days	349	304	285	199	113	49	24	26	55	132	256	336	2128

$$Q_{sg} = \Sigma(F_s A_g q_{sg}) H_p / (24 \times 1000) \tag{8.6}$$

where Q_{sg} is the mean monthly rate of solar gain through glazing (during the hours of occupancy) in kW, F_s is the blind correction factor from CIBSE (1986b), A_g is the area of glazing in m^2, for a particular type of glass and orientation, Q_{sg} is the mean monthly rate of solar gain for a particular type of glass and orientation over the CIBSE weather year (Table 8.6) in W m^{-2} and H_p is the occupied period in hours. For a particular month Q_{sg} is summed over all the relevant orientations. The answer is then summed over each month of the heating season, for which they are assumed to be useful. For an air conditioned building it is recommended that Table 8.6(b) is used for all building faces, except north, because the blinds will be drawn when significant solar radiation is incident on the windows. Thus the value of F_s to be used, from CIBSE Table A 9.15, is 0.77.

Table 8.6 Monthly average rate of solar gain per unit area (W m^{-2}) of glazing (6 mm single clear glass) for a lightweight building .
(a) Unshaded: occupied period 8 hours

Month	N	NE	E	SE	S	SW	W	NW
Jan	22.6	21.8	22.0	41.8	64.5	61.4	36.1	22.2
Feb	30.1	30.8	35.3	46.2	56.5	53.5	41.0	31.6
Mar	50.8	51.9	67.2	99.9	143.3	153.6	118.4	70.7
Apr	70.9	73.2	89.1	115.6	152.8	170.6	144.8	97.1
May	94.2	98.3	126.4	157.8	184.0	195.2	173.6	127.7
Jun	96.6	101.0	128.6	156.1	176.9	189.8	173.9	131.3
Jul	92.5	95.8	111.2	126.8	138.9	141.7	133.9	110.0
Aug	83.1	86.8	111.4	140.0	159.6	160.2	138.7	105.2
Sep	61.4	63.3	81.2	113.6	146.8	148.8	116.7	78.3
Oct	44.7	44.4	55.2	89.8	130.5	130.1	90.3	54.0
Nov	25.5	24.5	24.4	37.3	53.1	51.8	35.5	25.6
Dec	18.3	17.3	16.5	31.5	49.9	48.0	28.3	17.5

(b) Light slatted blinds; occupied period 8 hours

Month	N	NE	E	SE	S	SW	W	NW
Jan	14.9	14.3	14.4	27.4	42.3	40.3	23.7	14.6
Feb	19.7	20.2	23.1	30.3	37.1	35.1	26.9	20.7
Mar	33.0	33.8	43.9	65.3	93.8	100.7	77.6	46.1
Apr	46.0	47.5	57.8	75.2	99.8	111.5	94.5	63.1
May	60.7	63.2	81.3	102.0	119.7	127.0	112.6	82.4
Jun	62.0	64.6	82.5	100.7	115.0	123.4	112.6	84.5
Jul	59.9	61.9	71.8	82.1	90.3	94.1	86.9	71.2
Aug	53.8	56.0	71.9	90.7	104.0	104.4	90.2	68.1
Sep	39.8	41.0	52.8	74.1	96.0	97.4	76.3	51.0
Oct	29.2	29.2	36.1	58.8	85.5	85.3	59.2	35.3
Nov	16.7	16.1	16.0	24.5	34.8	34.0	23.3	16.8
Dec	12.1	11.4	10.8	20.7	32.8	31.5	18.6	11.6

Reproduced from the draft CIBSE Building Energy Code: Heated only buildings, by permission of the Chartered Institution of Building Services Engineers.

EXAMPLE 8.2

Repeat Example 8.1 assuming that a four-pipe fan coil system is used. Assume a population density of 9 m^2 per person with a sensible heat emission of 90 W each and a diversity factor of 0.75 (Table 1.2). Take 17 W m^{-2} as the heat liberated by the lighting, refer to Table 1.2 and use a diversity factor of (0.7 + 0.85)/2 = 0.775 for the eight months from March to October and 1.0 for the four months from November to February. Assume the emission from business machines is 20 W m^{-2} and the diversity factor is 0.7. Use Table 8.6(b) and refer to the CIBSE (1986a), as necessary to evaluate the casual gains from solar gain through windows. The occupied period is 9 hours a day, 5 days a week and there are eight statutory holidays.

Answer

The useful casual gains are:

People: [(86.4 × 13.5 × 12)/9] × [(90 × 0.75)/1000]	= 105.0 kW
Lights: [(86.4 × 13.5 × 12 × 17)/1000] × [(1 × 4/12) + (0.775 × 8/12)]	= 202.3 kW
Machines: (86.4 × 13.5 × 12 × 20 × 0.7)/1000	= 196.0 kW

The casual solar gains (q_{sg}) are determined from Table 8.6(b) for east and west orientations, for each month of a heating season from November to March, inclusive, and summed. From the CIBSE guide (1986b) the blind correction factor, F_s, for a lightweight building with internal shades drawn is 0.77. The total area of glass, A_g, on each of the two (east and west) facades of the hypothetical building (Section 1.2) is 0.4 × 86.4 × 3.3 × 12 = 1369 m^2.

Using Equation (8.6), the following tabulation is compiled:

Month	East	West	Total
November	0.77 × 1369 × 16.0 × 9/(24 × 1000) = 6.3 kW	0.77 × 1369 × 23.3 × 9/(24 × 1000) = 9.2 kW	15.5 kW
December	0.77 × 1369 × 10.8 × 9/(24 × 1000) = 4.3 kW	0.77 × 1369 × 18.6 × 9/(24 × 1000) = 7.4 kW	11.7 kW
January	0.77 × 1369 × 14.4 × 9/(24 × 1000) = 5.7 kW	0.77 × 1369 × 23.7 × 9/(24 × 1000) = 9.4 kW	15.1 kW
February	0.77 × 1369 × 23.1 × 9/(24 × 1000) = 9.1 kW	0.77 × 1369 × 26.9 × 9/(24 × 1000) = 10.6 kW	19.7 kW
March	0.77 × 1369 × 43.9 × 9/(24 × 1000) = 17.4 kW	0.77 × 1369 × 77.6 × 9/(24 × 1000) = 30.7 kW	48.1 kW
			Q_{sg} = 110.1 kW

Summarising, the total useful casual gains are:

People:	105
Lights	202
Machines	196
Sub total	503 kW
Solar	110
Total	613 kW

The period over which casual gains make a useful contribution to the heating requirements is questionable. However, in this example assume that it is over a period of $150 \times 9 = 1350$ hours. The useful gain is then: $1350 \times 613 \times 3600/10^6 = 2979$ GJ. Using the results of Example 8.1 and Equation (8.5) the total average annual thermal consumption is $H = 5903 - 2979 = 2924$ GJ. The specific thermal requirement is $2924/(86.4 \times 13.5 \times 12) = 0.21$ GJ m^{-2}.

THE ELECTRICAL ENERGY USED TO DRIVE MECHANICAL PLANT

The electrical energy used clearly depends on the type of system and the way it is designed. The energy consumed may be estimated by listing the items of plant that use electricity, their maximum absorbed powers and their utilisation factors. The products of such powers and factors are summed to yield the average annual electrical power likely to be required.

EXAMPLE 8.3

If the hypothetical office building (Section 1.2) is treated by a four-pipe fan coil system, provided with an auxiliary low velocity ducted supply of dehumidified air, estimate the likely annual consumption of electrical energy. Refer to Example 1.11 as necessary and assume the following:

Supply fan duty: 46.072 m^3 s^{-1} at 0.625 kPa fan total pressure, fan total efficiency 85%, motor and drive efficiency 90%, 37.6 kW absorbed power.
Extract fan duty: 46.071 m^3 s^{-1} at 0.20 kPa fan total pressure, 75% fan total efficiency, 90% motor and drive efficiency, 13.7 kW absorbed power.
Primary chilled water pump: 96.5 l s^{-1} at 200 kPa pump pressure, 65% pump efficiency, 90% motor and drive efficiency, 33.0 kW power absorbed.
Secondary chilled water pump: as primary chilled water pump (in this particular case, but not in general).
Cooling water pump: 105.0 l s^{-1} at 200 kPa, 65% pump efficiency, 90% motor and drive efficiency, 35.9 kW power absorbed.
Cooling tower fan power: 15 kW absorbed.
Installed boiler power: 1300 kW.
Forced draught fan power: 6 kW absorbed.
LTHW primary pump power: 3 kW absorbed.
LTHW secondary pump power: 3 kW absorbed (in this particular case but not in general).
HWS pump power: 1.5 kW absorbed.

Water chiller duty: 1616 kW of refrigeration (Example 1.11).
Refrigeration compressor motor power: 404 kW absorbed.

Answer

Utilisation factors express the equivalent number of hours that an item of plant runs at full load as a fraction of the total number of hours in a year (8760). Reasonable estimates can be made for most of the items, remembering that the total time of occupancy in the building is 2268 hours in a year. Thus a plant item that runs continuously at full load during the hours of occupancy has a utilisation factor of 2268/8760 = 0.26. In the case of items used during pre-heating for a building operated with an intermittent heating regime, an addition must be made to the hours of running. It is suggested that 1.5 hours per day be added to such plant operating hours during a heating season of 150 days per year. Hence the utilisation factor of the LTHW pumps is (2268 + 1.5 × 150)8760 = 0.28. A factor of 0.26 is used for the HWS pump. In the case of the forced draught fans, operation is intermittent, even during normal occupancy time. It is suggested that a lower figure of 0.2 be used for their utilisation

Table 8.7 Annual average electrical power in a hypothetical office block

Item	Maximum power (kW)	Utilisation factor –	Annual average power (kW)	Per cent (%)
Supply fan	37.6	0.26	9.8	9.9
Extract fan	13.7	0.26	3.6	3.6
Primary chilled water pump	33.0	0.26	8.6	8.7
Secondary chilled water pump	33.0	0.26	8.6	8.7
Cooling water pump	35.9	0.13	4.7	4.7
Cooling tower fans	15.0	0.13	2.0	2.0
Forced drught fans	6.0	0.2	1.2	1.2
LTHW primary pump	3.0	0.28	0.8	0.8
LTHW secondary pump	3.0	0.28	0.8	0.8
HWS pump	1.5	0.26	0.4	0.4
Refrigeration compressor motor	404.0	0.10	40.4	40.8
Fan coil unit fans (Example 1.11)	70.0	0.26	18.2	17.4
Totals			99.1 kW	100.0%

factor, without much error, because their contribution to the total power consumption for the building is comparatively small. The factors for the refrigeration compressor motors, cooling tower fans and cooling water pumps are always contentious, in the absence of firm recommendations. The CIBSE (1980) offered some guidance and Millbank *et al.* (1971) suggested 1000 equivalent full load running hours was appropriate for refrigeration plant but this appears to be too large. It is proposed here that 900 hours be used for air–water systems (fan coil and chilled ceiling). The corresponding utilisation factor is 900/8760 = 0.10 and this is adopted for the compressor motor. A larger factor should be taken for the cooling water pumps because they will run whenever the refrigeration plant is in use, which is likely to be for at least half the year, regardless of its actual duty, and hence a factor of 0.13 is adopted in this example. Although the cooling tower fans may be controlled to exploit natural draught to effect evaporative cooling in the tower when the outside wet-bulb is low enough, the extent of this is unpredictable and 0.13 is also used here for the fans. The primary and secondary chilled water pumps may run for the whole of the occupied period with a four-pipe fan coil system, because cooling capacity is necessary at the fan coil units at all times during the occupied hours. The refrigeration plant is off in the winter and free cooling is then used, in one form or another. Hence the factor taken for these two pumps is 0.26, in this example. It is emphasised that engineering judgement is needed when deciding the utilisation factors to be adopted.

The above considerations are summarised in Table 8.7.

The total annual electrical energy used is $99.1 \times 8760 \times 3600 \times 10^{-6} = 3125$ GJ or 0.22 GJ m^{-2} of treated floor area.

8.3 Electrical and thermal energy used by VAV systems

Electrical and thermal energy used by VAV systems is a function of sensible heat gain; fan total pressure; the proportional band of the pressure sensor used to vary the speed or change the blade pitch angle of the centrifugal or axial flow supply and extract fans (see Section 2.6 and Figures 2.13, 2.14 and 2.15), the turn-down of the VAV terminals; and the fan efficiency. Estimating the probable energy consumption of the fans depends in the first place on assessing the variation in sensible heat gains, not for summer design weather but for typical weather during the occupied hours for each month of the year. This estimation is able to be done by a computer and such a study for the hypothetical office building, modified to have U-values of 0.45, 0.25 and 3.3 W m^{-2} K^{-1} for the walls, roof and windows, respectively, to conform with the building regulations, yields the results shown in Figure 8.2. It was assumed for this that air handling plants fed both faces of the building from common ducts in order to take full advantage of the variation in the solar load. Account was taken of cloud cover according to the Meteorological Office (1953) and mean daily variations of outside temperature. Diversity factors of 0.75, 0.8 and 0.7 were applied to heat gains from people, lights and business machines, respectively. Table 8.8 gives calculated values of the typical sensible heat gains for each hour of the day and each month of the year, as fractions of the maximum sensible gains, occurring at 16.00–17.00 h sun time in July, for a modular pair having east and west facing windows in the hypothetical office block with 50% of its two, long, outer facades glazed. The fractions may then be interpreted as airflow factors.

The same air handling unit serves both modules and full account is taken of the diversity of solar gains through glass.

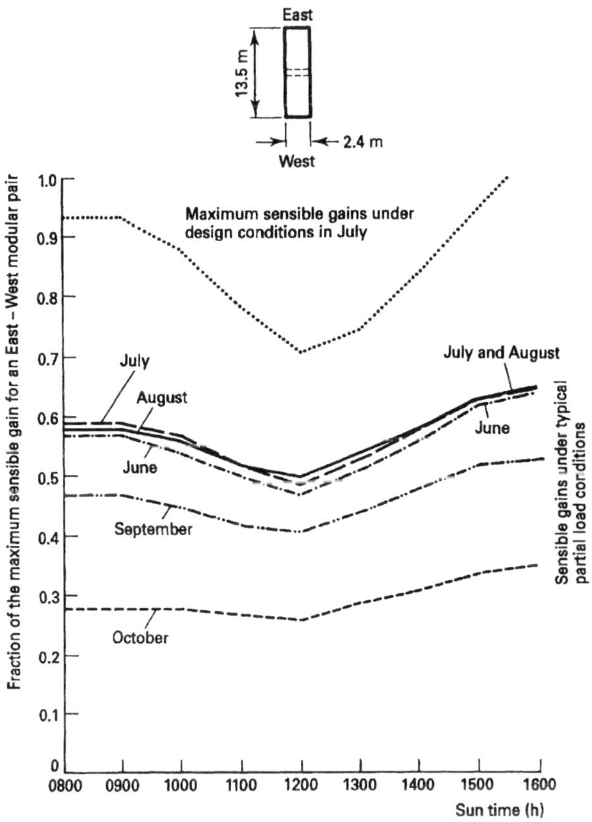

Figure 8.2 Sensible gains under typical conditions (taking account of diversity factors on lights, people and business machines, daily variations in outside temperature and cloud cover) compared with maximum design conditions. Other months show similar patterns. Note 40% of the outer facade is double glazed.

EXAMPLE 8.4

Determine the electrical energy consumption of a VAV system used to air condition the hypothetical office building, given that the maximum simultaneous sensible heat gains are 1032 kW (Example 1.11) and that four air-handling plants are to be used, delivering air at 13.5°C from common ducts to both glazed faces. Make use of the data in Table 8.8 and the fan curves in Figure 8.3. Take the fan total pressures of the supply and extract axial flow fans as 1.0 and 0.25 kPa, respectively. Assume the minimum allowable static pressure at a VAV terminal is 180 Pa and the proportional band of the pressure sensor regulating the blade pitch angles is 100 Pa. Assume also that the VAV units will turn down to a minimum of 25% of design airflow.

Table 8.8 Typical sensible beat gains for a modular pair having east and west-facing windows in an office block with a major building axis pointing north-south. See Section 1.2

	Typical heat gains expressed as a fraction of the maximum, gain occurring at about 1600 h sun time in July at latitude 51.7°								
	Sun time (h):								
Month	0800	0900	1000	1100	1200	1300	1400	1500	1600
Jan	0	0	0	0	0	0.02	0.02	0.03	0.03
Feb	0	0.07	0.06	0.05	0.05	0.08	0.09	0.11	0.11
Mar	0.22	0.22	0.21	0.19	0.17	0.21	0.24	0.28	0.28
Apr	0.36	0.36	0.34	0.31	0.28	0.32	0.36	0.41	0.43
May	0.47	0.48	0.45	0.41	0.38	0.42	0.47	0.52	0.54
Jun	0.57	0.57	0.54	0.50	0.47	0.51	0.56	0.62	0.64
Jul	0.59	0.59	0.57	0.52	0.50	0.54	0.58	0.63	0.65
Aug	0.58	0.58	0.56	0.52	0.49	0.53	0.58	0.63	0.65
Sep	0.47	0.47	0.45	0.42	0.41	0.44	0.18	0.52	0.53
Oct	0.28	0.28	0.28	0.27	0.26	0.29	0.31	0.34	0.35
Nov	0	0.10	0.10	0.11	0.11	0.14	0.15	0.16	0.16
Dec	0	0	0	0.01	0.02	0.04	0.04	0.05	0.05

The same air handling unit serves both modules and full account is taken of the diversity of solar gains through glass.

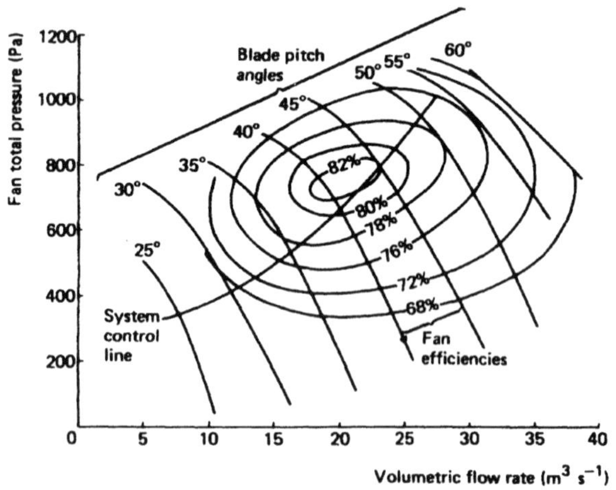

Figure 8.3 Characteristic curves for a typical axial flow fan with controllable blade pitch. The system control line is the characteristic curve for the system using a square law for the system resistance plus a linear relationship between the pressure sensed by the sensor regulating the blade pitch angle and the volume flowing.

Answer

Maximum total supply air quantity = [1032/(22 – 13.5)] + [(273 × 13.5)/358]
= 97.2 $m^3 s^{-1}$

Air quantity per plant = 24.3 ≃ 25.0 $m^3 s^{-1}$

If the static pressure sensing probe is assumed to be located in a position along the index run, probably two-thirds to three-quarters of the way from the fan, where it senses representative changes of pressure as the volume handled changes and. further, that the proportional band of the pressure sensor gives a linear relationship between volumetric flow rate and pressure sensed, a square law can be applied to the rest of the system and pressure/flow values for a system characteristic or control curve can be calculated. Thus, for a volumetric flow rate of 25%, the loss through the VAV terminal unit is 180 + 100 = 280 Pa but for 30% flow it is 280 [(30 – 25)/(100 – 25)] × (280 – 180) = 273 Pa. For the rest of the system, according to a square law, the pressure loss at 25% flow is $(25/100)^2$ × (1000 – 180) = 51 Pa and at 30% flow it is $(30/100)^2$ × (1000 – 180) = 74 Pa. Hence Table 8.9 is compiled It should be noted that in the duct feeding the index VAV unit the design air velocity is likely to be not more than about 2.5 m s^{-1} and that the velocity pressure is therefore trivial, static pressure being virtually the same as total pressure, particularly at reduced flow rates. As the static pressure measured at the sensor rises from 180 Pa, for the design flow, the blade pitch angles on the supply and extract fans are reduced in unison.

The final column in Table 8.9 is plotted in Figure 8.3 to form the pressure/volume control curve for the system as it throttles at the VAV unit terminals. From the intersections of the fan curves, at various blade pitch angles, with the system control line flow rates can be read off, $\dot{v}$, fan total pressures, p_{tF}, and efficiencies, η. The absorbed fan power at each flow rate, W_f, is then given by

$$W_f = \dot{v} p_{tF}/\eta \tag{8.7}$$

Table 8.9 Pressure/flow values for a system characteristic or control curve, Example 8.4

Flow rate		Total pressure loss (Pa)		
%	$m^3 s^{-1}$	Duct system	VAV terminal	System plus terminal
100	25.0	820	180	1000
90	22.5	664	193	857
80	20.0	525	207	732
70	17.5	402	220	622
60	15.0	295	233	528
50	12.5	205	247	452
40	10.0	131	260	391
30	7.5	74	273	347
25	6.25	51	280	331

With the simplifying assumption that the motor efficiency is constant at 90%, the electrical power absorbed at various flow rates can be calculated (Table 8.10).

Referring to Figure 8.3, the fan efficiency can be read for volumetric flow rates from 100% down to the minimum of 25%. The fan power may be calculated from Equation (8.7) and, using the value of 90% given for the efficiency of the motor and drive, the absorbed power against the volumetric airflow rate is established. Table 8.10 shows the results.

The supply and extract air quantities handled by the VAV system are proportional to the fractions of the sensible heat gains listed in Table 8.8. A given fraction from Table 8.8 can the be used to interpolate between values for the absorbed supply fan powers listed in Table 8.10 to yield the energy consumed In each hour in each month. Thus, at 15.00 h in July, the fraction from Table 8.8 is 0.63. Interpolating in Table 8.10 between 60–70% airflow rates gives an absorbed supply fan power of 12.8 kW. Over the 1-hour period centred at 15.00 h sun time in July the electrical energy used is $1 \times 12.8 \times 3600/1000 = 46.1$ MJ. On the other hand, since the minimum supply airflow rate from the VAV terminals is 25%, this is the airflow rate delivered at 15.00 h sun time in January, even though the fraction of the sensible heat gains is only 0.03 from Table 8.8. Hence the power absorbed is 4.2 kW (Table 8.10) and the energy used in the hour centred about 15.00 h sun time in January is $1 \times 4.2 \times 3600/1000 = 15.1$ MJ. Table 8.11 lists the results of using the above method to determine the total electrical energy consumption for a year (CIBSE Example Weather Year).

Since the extract fan is of the same type and controlled in the same way, in unison with the supply fan, one can assume, for simplicity, that the energy it uses is proportional to the design fan total pressure, i.e. that it equals $55.28 \times 0.25/1.0 = 13.82$ GJ, annually. The energy consumed by all fans is thus $4\,(55.28 + 13.82) = 276.4$ GJ.

The refrigeration load for the building when treated by a VAV system will be different from that calculated in Example 1.11, which was for a fan coil system. Some of the results of Example 1.11 can be used but the total cooling load, Q_{ref}, must be recalculated and is as follows, using the notation of Equation (1.13)

Table 8.10 Electrical power absorbed at various flow rates for the supply system plus a terminal

Flow rate %	Flow rate $m^3\,s^{-1}$	Fan total pressure (Pa)	Fan efficiency (%)	Fan power (kW)	Power absorbed (kW)
100	25.0	1000	78	32.1	35.6
90	22.5	857	82	23.5	26.1
80	20.0	732	80	18.3	20.3
70	17.5	622	78	14.0	15.5
60	15.0	528	75	10.6	11.7
50	12.5	452	68	8.3	9.2
40	10.0	391	65*	6.0	6.7
30	7.5	347	62*	4.2	4.7
25	6.25	331	55*	3.8	4.2

*These efficiencies are inferred from the fan powers quoted by the fan manufacturer.

Table 8.11 Electrical energy consumed for each hour of the day in each month of the year by each supply fan

Month	Energy consumed in each hour (MJ) Sun time (h) 08.00	09.00	10.00	11.00	12.00	13.00	14.00	15.00	16.00	Daily totals (MJ)	Average number of working days	Monthly totals (GJ)
January	15.1	15.1	15.1	15.1	15.1	15.1	15.1	15.1	15.1	135.9	20.3	2.76
Febrary	15.1	15.1	15.1	15.1	15.1	15.1	15.1	15.1	15.1	135.9	20.0	2.72
March	15.1	15.1	15.1	15.1	15.1	15.1	15.1	15.7	15.7	137.1	21.9	3.00
April	21.2	21.2	19.8	17.6	16.2	18.4	21.2	25.2	26.6	187.4	20.0	3.75
May	30.2	31.3	28.8	25.2	22.7	25.9	30.2	34.9	36.7	265.9	21.0	5.58
June	39.6	39.6	36.7	33.1	30.2	33.8	38.5	45.0	47.5	344.0	22.0	7.57
July	41.0	41.0	39.6	34.9	33.1	36.7	40.3	46.1	49.0	361.7	21.0	7.60
August	40.3	40.3	38.5	34.9	32.4	36.0	40.3	46.1	49.0	357.8	21.3	7.51
September	30.2	30.2	28.8	25.9	25.2	27.7	31.3	34.9	36.0	270.2	21.6	5.84
October	16.2	16.2	16.2	15.8	15.5	16.6	17.6	19.8	20.5	154.4	21.3	3.29
November	15.1	15.1	15.1	15.1	15.1	15.1	15.1	15.1	15.1	135.9	21.9	2.98
December	15.1	15.1	15.1	15.1	15.1	15.1	15.1	15.1	15.1	135.9	19.7	2.68
											252.0	55.28

$$
\begin{aligned}
&Q_s && = 1032\text{ kW}\\
&Q_l && = 92\text{ kW}\\
&Q_{fa} && = 281\text{ kW}\\
&Q_{sf} = (97.2 \times 1.0 \times 358)/(273 + 13.5) && = 121\text{ kW}\\
&Q_{sd} = (97.2 \times 1.5 \times 358)/(273 + 13.5) && = 182\text{ kW}\\
&Q_{ra} = (97.2 \times 0.25 \times 358)/(273 + 13.5) && = 30\text{ kW}\\
&\text{Total:} \quad Q_{ref} && = 1738\text{ kW}
\end{aligned}
$$

This corresponds to a specific refrigeration load of 124 W m^{-2}. The factor f_p is taken as 1.0 because there is considerably less chilled water piping with a VAV system than with a fan coil system.

Assuming a coefficient of performance of 4.0 under design operating conditions the power absorbed by the compressor is 1738/4 = 434.5 kW.

Meteorological records in the London area over a typical 10 year period for the frequency of occurrence of dry- and wet bulb temperatures show that, between the hours of 08.00 and 16.00, sun time, the dry-bulb exceeds 10°C for about 1712 hours and the wet-bulb is greater than 5°C for about 2226 hours, for 7 days a week throughout the year. It may be inferred from this that the refrigeration compressor for a VAV system will run for a shorter period of the year than will that for an air water system (fan coil or chilled ceiling), because it can give free cooling when the dry-bulb is less than 10°C, whereas the latter can only do so when the outside wet-bulb is less than about 5°C. It is therefore reasonable, for the case of a VAV system, to multiply the utilisation factor of 0.10 for the fan coil system in Example 8.3 by 1712/2226, namely, 0.77. However, in practice, there is a tendency for a refrigeration plant to be shut off in the winter months to economise on the electrical demand charges, and this would bring the operational hours of the refrigeration plant in the two types of system closer together. It is therefore suggested that the factor be increased from 0.77 to 0.8 which, when applied to the utilisation factor of 0.1 adopted for the compressor of the fan coil system in Example 8.3, becomes a utilisation factor of 0.08 for the compressor in the VAV system for this example. It is reasonable to use a factor of 0.13 for the primary chilled water pump since there is no need for a supply of chilled water in the winter months. The dry-bulb/wet-bulb frequency factor of 0.8 is not applied since it is argued that the chilled water pump would not be switched off for small periods in marginal weather when the refrigeration plant did not have to run because the dry-bulb was below 10°C for a short period of time. Similarly, the cooling water pump and cooling tower fans have utilisation factors of 0.13. The maximum powers of the chilled and cooling water pumps, and the fans, have to be increased in the ratio of 1738/1616 = 1.075, because the refrigeration load for the VAV system is greater than that for the fan coil system. Since there are no fan coil units there will be no secondary chilled water or LTHW pumps dedicated to their use.

The annual average electrical power required for the building with a VAV system installed is then calculated and summarised in Table 8.12.

The annual electrical energy consumption for the items in Table 8.12 is $48.93 \times 8760 \times 3600 \times 10^{-6} = 1543$ GJ. To this must be added the annual energy consumption of the fans, determined earlier in this example, namely, 276.4 GJ. Hence the total annual electrical energy consumption for the hypothetical building, when treated by a VAV system is 1543 + 276.4 = 1819 GJ which, when referred to the total treated floor area, is 0.13 GJ m^{-2}. This is to be compared with 0.22 GJ m^{-2} obtained in the answer to Example 8.3 for a fan coil system. It is clear that the values chosen for the utilisation factors exercise a significant effect.

A variable air volume system with perimeter heating will use more thermal energy than a four-pipe fan coil system. The minimum turn-down of the VAV system is 0.25, yet for much of the winter the airflow factor is less than this and in some cases no cooling air is wanted at all because there is a net heat loss, on average. The VAV system is thus overcooling at times and this must be cancelled by heating.

Table 8.12 The annual average electrical power required for the hypothetical building with a VAV system installed, excluding the supply and extract VAV fans

Item	Maximum power (kW)	Utilisation factor	Annual average power (kW)
Primary chilled water pump	35.5	0.13	4.62
Refrigeration compressor	434.5	0.08	34.76
Cooling water pump	38.6	0.13	5.02
Cooling tower fans	16.1	0.13	2.09
Forced draught fans	6.0	0.2	1.2
LTHW primary pump	3.0	0.28	0.84
HWS pump	1.5	0.26	0.39
Total			48.93 kW

Taking the difference between 0.25 and the factors in Table 8.6 that are less, the mean airflow that must be warmed from 13.5°C to 20°C can be determined over a 9-hour day in each month of the winter as the following calculations show.

January: $[(0.25 \times 5 + 0.23 + 0.23 + 0.22 + 0.22)/9] \times 97.2 = 23.22\ m^3\ s^{-1}$

February: $[(0.25 \times 1 + 0.18 + 0.19 + 0.20 + 0.20 + 0.17 + 0.16 + 0.14 + 0.14)/9] \times 97.2 = 17.60\ m^3\ s^{-1}$

March: $[(0.03 + 0.03 + 0.04 + 0.06 + 0.08 + 0.04 + 0.01)/9] \times 97.2 = 31.32\ m^3\ s^{-1}$

November: $[(0.25 \times 1 + 0.15 + 0.15 + 0.14 + 0.14 + 0.11 + 0.10 + 0.09 + 0.09)/9] \times 97.1 = 13.18\ m^3\ s^{-1}$

December: $[(0.25 \times 3 + 0.24 + 0.23 + 0.21 + 0.21 + 0.20 + 0.20)/9] \times 97.2 = 22.03\ m^3\ s^{-1}$

Taking the average number of working days in each of the above five months as 20.3, 20, 21.9, 21.9 and 19.7, respectively, the average airflow rate over the five-month period is:

$$[(23.22 \times 20.3) + (17.6 \times 20) + (31.32 \times 21.9) + (13.18 \times 21.9) + (22.03 \times 19.7)]/(20.3 + 20 + 21.9 + 21.9 + 19.7) = 2231.9/103.8 = 21.5\ m^3\ s^{-1}$$

Using Equation (2.3) the average rate of heat input to the airflow is $[21.5 \times (20 - 13.5) \times 358]/(273 + 13.5) = 174.6$ kW over the 103.8-day period and the thermal energy supplied for this purpose is $(174.6 \times 103.8 \times 9 \times 3600)/106 = 587.2$ GJ. Assuming an annual average boiler efficiency of 78%, as before, this becomes 753 GJ annually.

The total rate of heating required for a VAV system provided with perimeter heating compensated against outside air temperature (Example 8.1) was 462 kW. The annual thermal energy requirement was determined as 5903 GJ, corresponding to 0.42 GJ m^{-2}, These figures must be revised upwards to 5903 + 753 = 6656 GJ, corresponding to 0.48 GJ m^{-2}, to account for the heat needed to warm the minimum supply air rate in the winter months.

This is summarised in Table 8.13 and the energy used in the building is converted into primary energy as contained in fossil fuels like coal, oil and gas, by adopting fuel factors used in the Building Energy Code (1980) based on work done by the Building Research Establishment. These factors reflect the losses in producing energy as a usable medium from its fossil form and in distributing it to the building. Factors of 1.07 and 3.82 are taken for natural gas and electricity, respectively.

Table 8.13 Comparative annual average energy consumptions for two systems

	Average annual energy used ($GJ\ m^{-2}$ floor area)					
	Energy used in the building			Primary energy used		
System	Heat	Electricity	Total	Heat	Electricity	Total
Four-pipe, fan coil:	0.21	0.22	0.43	0.22	0.84	1.06
VAV + perimeter heating	0.48	0.13	0.61	0.65	0.50	1.15

8.4 Economic appraisal

To obtain a realistic view of the engineering economics of a system all the financial outgoings and receipts must be accounted over its estimated life. The technique currently most favoured for this is the present value method. This consists of establishing the discounted cash flow (DCF) throughout the life of the plant by converting all future expenditure and income into an equivalent sum of money at present day values, allowing for inflation and for the interest that must be paid on borrowed capital. Money lent cannot be used otherwise by the lender and the interest received after a period of time is a reward for his or her postponement of its expenditure. To convert a sum of money received 1 year from now, for example, into an equivalent sum today it must be divided by a discount factor determined from the interest rate paid. Thus an interest rate, or discount rate, of 10% per annum corresponds to a discount factor of 1.1 and so £1100 received in one year's time is equivalent to £1000 received today. Inflation is a different process that erodes the value of money and an inflation factor, corresponding to an assumed annual percentage for a future inflation rate, must be further applied to obtain a true picture of present value as Carsberg and Hope (1976) shows. The product of the discount and inflation factors is termed the discount money factor. A sum of money received or spent in 12 month's time, when divided by the money discount factor, gives the equivalent present value of the sum. When estimating the discounted cash flow over the life of a system it is possible to apply different inflation factors to different elements in the analysis. For example wages, or energy prices, could be assumed to inflate more rapidly than the general inflation rate for the national economy.

EXAMPLE 8.5

A plate heat exchanger (static recuperator) is to be considered for transferring heat between the incoming fresh air and the outgoing vitiated air in an air-conditioning system. The total capital expenditure initially required is £7721 and it is estimated that the annual saving by energy conservation will be £799, at present day prices. Do a discounted cash flow analysis to establish the present value and so show if the project is worthwhile. Take a discount rate of 10% to reflect the passage of time and a general inflation rate also of 10% but assume that the cost of energy will double, in real terms, over the next 20 years. The life of the plant is considered to be 20 years and there are no maintenance costs.

Answer

The calculations involved are all based on the formula for compound interest:

$$S = P(1 + r)^n \tag{8.8}$$

where S is the sum of money received after n years if an annual interest rate of r, as a fraction, is charged on the principal sum, P.

Equation (8.8) can be used to determine the annual rate of increase in energy costs if they are to double in 20 years because $S/P = 2$ and so $r = (2^{1/20} - 1) = 0.0353$, or 3.53% per annum. The specific inflation factor to apply to energy costs is the general inflation factor, 1.1, times the real energy inflation factor, 1.0353, which is, therefore, 1.139. The money discount factor, to be used to convert future money receipts into present day values, is the general inflation factor, 1.1, times the discount rate, 1.1, which is, therefore, 1.21. The calculations of cash flow are summarised in Table 8.14.

Table 8.14 Cash flow calculations

	At present-day prices						
Year	Cost of plant (£)	Energy saving (£)	Specific energy inflation factor	Money cash flow* (£)	Money discount factor	Present value† (£)	Progressive sum of the present values (£)
0	−7721.0			−7721.0	1.0	−7721.0	−7721.0
1		+799	1.139	+910.1	1.21	+752.1	−6968.9
2		+799	1.139^2	+1036.6	1.21^2	+708.0	−6260.9
3		+799	1.139^3	+1180.6	1.21^3	+666.4	−5594.5
.							
.							
.							
15		+799	1.139^{15}	+5628.6	1.21^{15}	+322.6	−77.3
16		+799	1.139^{16}	+6411.0	1.21^{16}	+303.6	+226.3
.							
.							
.							
20		+799	1.139^{20}	+10790.0	1.21^{20}	+238.4	+1272.9

*Energy saving at present prices × specific energy inflation factor.
†Money cash flow/money discount factor.

The net present value at the end of the life of the plant is £1272.90, from Table 8.14, and the project is clearly cost-effective. It is worth noting that where a company is entitled to capital allowances these should be accounted since the tax saved constitutes an income throughout the life of the plant and the cost effectiveness is improved.

A much cruder technique, that frequently provides a misleading answer and tends to underestimate the benefits of a proposed investment under inflation, is the calculation of the pay-back period. This simply divides the initial capital expenditure for a project by the estimated annual saving, at present day prices, to give the period in years needed for the return of the outlay. Thus, in Example 8.5 the simple pay-back period is £7721/£799 per annum = 9.66 years. Discount rates and general inflation are ignored and no attempt is made to assess the worth of the energy saved after the pay-back period has expired, up to the end of the life of the plant. The crude technique of the simple pay back period should only be used for the most obvious, preliminary indications and should always be backed up by a discounted cash flow analysis, including the most realistic assumptions possible for future inflation. An essential feature of this method is an allowance for the life of the plant. Table 8.15 lists some suggested lives for items of plant.

Not all the parts of a piece of plant age at the same rate. For example, in the case of cooler coils, the framework, which is usually of galvanised steel, invariably corrodes away before the fins and tubes because of the extreme, relative positions of copper and zinc in the electrochemical scale.

The annual cost of a system is the annual sum of money that is equivalent to all expenditure and income and it is determined using similar principles to those in the calculation of present value, covering owning (capital) and operating costs over the estimated life of the system. The present value (PV), of £1 received annually over a period of n years and

Table 8.15 Some suggested life-times for plant according to Hassan (1969), Eyers (1967) and Day (1976)

Item	Lifetime (years)
HPHW (high pressure hot water) and steam boilers:	
Shell	15–25
Water tube	25–30
M and LPHW (medium and low pressure hot water) boilers:	
Shell	15–20
Cast iron sectional	20–25
Steel sectional	5–15
Boiler plant ancillaries:	
Oil tanks	40
Combustion controls	15–20
Gas and oil burners	15–25
Fans	15–25
Heating equipment:	
Radiators	15–20
Convectors	15–20
Closed piping systems	25–60
Pumps in closed piping systems	20–25
Air-conditioning equipment:	
Vapour compression refrigeration plant	15–30
Cooling towers	10–25
Fans	20–30
Filters, excluding filter medium	15–25
Cooling coils copper fins, tinned	20–30
Cooling coils – aluminium fins	15–20
Heater batteries	20–30
Packaged air-handling units	8–15
Galvanised steel ductwork	30–60
Fan coil units	10–25
VAV units	15–30
Double duct units	15–30

discounted at a rate r as a fraction is given by

$$PV = [1 - (1 + r)^{-n}]/r \qquad (8.9)$$

and values for the solution of this are tabulated elsewhere by CIBSE (1986a).

EXAMPLE 8.6

Calculate the annual sum needed to repay a loan of £7721 over 20 years when the discount rate is 10% per annum.

Answer
From Equation (8.8)

$$PV = [1 - (1 + 0.1)^{-20}]/0.1 = £8.514 \text{ per } £.$$

Since the present value of the sum borrowed is £7721 the annual repayment must be £7721/£8.514 per £ = £906.86.

Conversely, using Equation (8.8) and the method adopted in Example 8.5, Table 8.16 shows the PV of £906.86 paid annually over 20 years with a discount rate of 10%.

In the past, the sum that must be paid annually to cover interest and the repayment of capital (sinking fund) was calculated by the reciprocal of Equation (8.9).

Maintenance costs, to include repairs, routine lubrication and cleaning, replacements, labour, tools, power, water and overheads, have often been disregarded or underestimated.

The owning and operating costs of air-conditioning systems are often compared by determining their annual costs.

EXAMPLE 8.7

Estimate the owning and operating costs, i.e. annual costs, of a four-pipe fan coil system and a VAV system with compensated perimeter heating, for air conditioning the hypothetical office building (Section 1.2) in the year 1996. Take a discount rate of 10% per annum and assume energy prices to be flat rates of 8.5p per kWh (£23.61 per GJ) for electricity, and 1.7p per kWh (£4.72 per GJ) for natural gas.

Answer
Using Table 8.15 different lifetimes might be inferred for each system but it is not uncommon to use 20 years, regardless of the system type, and this is taken here as the life for both systems. (It is to be noted that Table 8.15 indicates that fan coil units have a shorter life than VAV units and this could bias the assumption of 20 years for both systems. However obsolescence plays a part and this could redress any imbalance.) Approximate capital costs in 1996 over a treated floor area of 14 000 m^2 are taken as £225 per m^2 for the VAV system

Table 8.16 Cash flow calculation

Year	Cash flow (£)	Discount factor	Present value (PV) (£)
1	906.86	1.1	824.42
2	906.86	1.1^2	749.47
.			
.			
.			
.			
19	906.86	1.1^{19}	148.28
20	906.86	1.1^{20}	134.80
Total present value			7721.00

Table 8.17 Mean monthly outside air temperatures from the CIBSE Example Weather Year for Kew (01.10.64–30.09.65)

	Mean monthly outside air temperature (°C) averaged over the stated occupied period (h)	
Month	8	24
January	5.0	4.4
February	4.7	4.1
March	7.9	6.4
April	10.7	9.0
May	14.8	12.9
June	17.3	15.5
July	17.1	15.5
August	18.3	16.4
September	15.0	13.1
October	11.1	9.4
November	9.5	8.7
December	5.3	4.8

and £250 per m^2 for the fan coil system, corresponding to totals of £3 150 000 and £3 500 000, respectively. Maintenance costs are particularly difficult to assess but a figure of £15 per m^2 is assumed here for each system, although it is probable that different systems would have different costs for maintenance. Reference to Table 8.13 gives the annual average energy used on site. From Equation (8.9) and Example 8.6 the annual repayment to cover interest and the money borrowed, for a lifetime of 20 years and a discount rate of 10%, is £8.514 per pound. The annual costs for 1996 are therefore:

System:	Four-pipe fan coil		VAV	
Capitalised annual cost (£ m^{-2})	250/8.514	= 29.36	225/8.514	= 26.43
Maintenance (£ m^{-2})		15.00		15.00
Natural gas (£ m^{-2})	4.72 × 0.21	= 0.99	4.72 × 0.48	= 2.27
Electricity (£ m^{-2})	23.61 × 0.22	= 5.19	23.61 × 0.13	= 3.07
Annual total (£ m^{-2})		= 50.54		= 46.77

The method adopted in Examples 8.1 and 8.2 for determining the heat needed to warm the fresh air introduced by the VAV and fan coil systems is open to question. Both systems use a mixture of fresh and recirculated air and it is customary to vary the proportions as the seasons change, under automatic control, so as to economise on the use of mechanical refrigeration. Thus minimum fresh air is used in summer as long as the enthalpy of the outside air exceeds that of the recirculated air. When the outside enthalpy is less than that of the recirculated air it is pointless using recirculated air and 100% fresh air is handled, because this is more economical in refrigeration usage. In the case of an all-air system, such as VAV, the required off-coil dry-bulb temperature can be obtained without refrigeration, by mixing colder fresh air with warmer-recirculated air. In the case of air–water systems, such as the four-pipe fan coil, fresh/recirculated air mixing dampers may be operated to get the required secondary chilled water flow temperature for the terminal units by free cooling:

colder fresh air flows over the cooler coil in the air handling unit and chills the water flowing within the coil. There are also other methods of free cooling and hence it becomes difficult to generalise on how much fresh air will be used in a typical year and how much thermal energy is likely to be needed for the fresh air handled. One approach is to consider that minimum fresh air will be used at all times and that this will only require heat to warm it from the mean monthly outside dry-bulb, during the hours of plant operation, to the design temperature of the air leaving the cooler coil in the air handling plant.

EXAMPLE 8.8

Re-work Examples 8.1 and 8.2, modifying the thermal requirement for fresh air by using the minimum fresh airflow rate warmed from the mean monthly outside air temperature (Table 8.17) over an occupied day of eight hours to a design off-coil temperature of 11°C.

Answer

Refer to Table 8.17 for mean outside air temperatures in each month during an 8-hour period of occupancy. $\Sigma(AU)$ stays at 11 886 W K^{-1} and n is 2.69 per hour (infiltration at 0.75 air changes per hour plus 1.4 l s^{-1} m^{-2} minimum fresh air). Thus $nV/3$ is 32631 W K^{-1} and $(\Sigma(AU) + nV/3)$ is (11 886 + 32 631)/1000 = 44.517 kW. From Equation (8.1) we can write:

$$Q_{sh} = [\Sigma(AU) + nV/3](t_w - t_{om})/1000 \text{ kW}$$

where Q_{sh} in kW is the mean monthly rate of heat input for the minimum fresh air, t_w is the design off-coil temperature in °C and t_{om} is the mean monthly outside air temperature in °C for an 8-hour occupied period from Table 8.17. The following calculations are tabulated:

Month	$(\Sigma(AU) + nV/3)/10^3$ (kW K^{-1})	$(t_w - t_o)$ (K)	Q_{sh} (kW)
January	44.517	(11 – 5.0)	267.1
February	44.517	(11 – 4.7)	235.9
March	44.517	(11 – 7.9)	138.0
April	44.517	(11 – 10.7)	13.4
May	44.517	(11 – 14.8)	–
June	44.517	(11 – 17.3)	–
July	44.517	(11 – 17.1)	–
August	44.517	(11 18.3)	–
September	44.517	(11 – 15.0)	–
October	44.517	(11 – 11.1)	–
November	44.517	(11 – 9.5)	66.8
December	44.517	(11 – 5.3)	253.7
Total (Q_{sh})			974.9 kW

The total fresh air thermal requirement for each system is the sum of the product $8Q_{mh}$ over 12 months, multiplied by 9/8, because Table 8.17 is for an 8-hour day and the example is for a 9-hour day. The result is 8 × 974.9 × 9/8 = 8774 kWh, or 28.59 GJ.

Thus, for Example 8.1 (VAV plus perimeter heating) the additional fresh air load would increase the annual thermal consumption from 5903 to 5932 GJ. For Example 8.2 (four-pipe fan coil) the annual consumption would rise to 2953 GJ. There would be a small increase in the specific consumptions in GJ m^{-2} given in the table of owning and operating costs in Example 8.7.

It has been shown by Pullinger (1978) that these figures represent less than 3% of the average staff salaries in an office and, allowing for overheads, less than about 1.5% of the total costs of such employment. A major element in the economic appraisal of Example 8.7 is capital cost and this can vary greatly since it depends on many factors (Section 8.1). Furthermore, because of the way in which charges are usually made for electrical energy, involving the demand for maximum power as well as the cost of the energy actually used, the load factor exercises a significant influence. Generalisations on comparative annual costs should, therefore, only be made with caution.

There is another useful concept, termed internal rate of return. This reflects the fact that the present value of a proposal becomes zero at the end of its chosen life according to the discount rate adopted for the calculation. Such a discount rate is termed the internal rate of return and the present value of a project will be positive at the end of its life, for any discount rate, that is less than the internal rate of return. The cost effectiveness of different proposals can be ranked in the order of their internal rates of return. Projects with a larger internal rate of return will be more cost-effective than those with smaller internal rates of return. This is because the internal rate of return corresponds to the discount rate, at which the capital expenditure is earning money on the original investment.

An approximate formula for determining the internal rate of return, i, is as follows:

$$i \simeq [(1/t_{spb}) - (t_{spb})/L^2)]100 \qquad (8.10)$$

where t_{spb} is the simple pay-back period, in years, and L is the chosen life of the installation, in years. Equation (8.9) is appropriate if the life does not exceed four years, otherwise the following exact equation should be used:

$$i = (1/t_{spb})\,[1 - 1/(1 + i)^L] \qquad (8.11)$$

with the same notation as in Equation (8.9).

Using Equation (8.11) requires an iterative approach, for lives exceeding 4 years.

EXAMPLE 8.9

Establish the cost effectiveness of three optional proposals for saving energy, by calculating their internal rates of return:

(a) Capital cost £4500, annual saving £2250, life 3.8 years.
(b) Capital cost £5000, annual saving £3000, life 3.9 years.
(c) Capital cost £5500, annual saving £3250, life 4.0 years.

Answer

Using Equation (8.10) compile the following:

Proposal	(a)	(b)	(c)
Capital expenditure	£4500	£5000	£5500
Money saved annually	£2550	£3000	£3250
Simple pay-back period (years)	1.76	1.67	1.69
Life (years)	3.8	3.9	4.0
i	44.6%	48.9%	48.6%

Proposal (b) appears to be the most cost effective since it has the largest internal rate of return but it would be reasonable to repeat the calculations, using Equation (8.11), because the comparative results are so close. The approximate values of i, obtained by Equation (8.10) and tabulated above, are inserted in the right-hand side of Equation (8.11) and give a more accurate value

$$\text{Proposal (a): } i = (1/1.76)[1 - 1/(1.446)^{3.8}]100 = 42.83\%$$
$$\text{Proposal (b): } i = (1/1.67)[1 - 1/(1.489)^{3.9}]100 = 47.20\%$$
$$\text{Proposal (c): } i = (1/1.69)[1 - 1/(1.486)^{4.0}]100 = 47.04\%$$

Proposal (b) is marginally better than (c). Further iterative calculations could be done to obtain a closer verification, if thought worthwhile.

EXERCISES

1 Repeat Example 8.1 assuming that the U-values are 0.5 and 0.3 W m^{-2} K^{-1} for the walls and roof, respectively and that only 30% of the two long, outer facades are glazed. (*Answer* 385 kW, 4933 GJ, 0.35 GJ m^{-2})

2 Repeat Example 8.2 but assuming the same U-values and glazing as above. (*Answer* 1954 GJ, 0.14 GJ m^{-2})

3 Repeat Example 8.5 but assume that energy costs will increase, in real terms, by a factor of five over the next 20 years. (*Answer* Net present value = £592)

Notation

Symbols	Description	Unit
A	Element of area in the building envelop	m^2
A_g	Area of glazing for a given type of glass and orientation	m^2
$D_{15.5}$	Number of degree days in the year to a base temperature of 15.5°C	°C days
f	Correction factor to cover period of building occupancy, mode of system operation, class of building structure	–
F	Annual average usage factor	–
F_s	Blind correction factor	–
G	Heating benefit from casual gains	GJ

H	Total annual average consumption of thermal energy	GJ
H_p	Occupied period	h
i	Internal rate of return	%
L	Life of an installation	years
n	Number of fresh air changes per hour (natural and/or mechanical) or a period of time	h^{-1} or years
p_{tF}	Fan total pressure	kPa
P	Principal sum of money	£
q_{sg}	Mean monthly solar heat gain through glass for a given type of glass and orientation	$W\ m^{-2}$
Q'_h	Design fabric heat loss rate plus fresh air load	kW
Q_h	Mean annual boiler power	kW
Q_{mh}	Mean monthly rate of heat input for minimum fresh air	kW
Q_{sg}	Mean monthly solar gain through glass	kW
r	Annual rate of interest as a fraction	-
S	Sum of money	£
t_o	Outside design temperature	°C
t_{om}	Mean monthly outside temperature for an 8-hour occupied period	°C
t_r	Design room temperature	°C
t_{spb}	Simple pay back period	years
t_w	Off-coil temperature	°C
U	Thermal transmittance coefficient of an element in the building envelope	$W\ m^{-2}\ K^{-1}$
U'	$\Sigma(AU) + nV/3$	$W\ m^{-2}\ K^{-1}$
$\dot{v}$	Volumetric airflow rate	$m^3\ s^{-1}$
$\dot{v}_t$	Volumetric airflow rate at a temperature t	$m^3\ s^{-1}$
V	Volume of the treated space	m^3
W_f	Absorbed fan power	kW
η	Boiler or fan efficiency, as a fraction	–

References

Carsberg, B. and Hope, A. *Business Investment Decisions under Inflation, Theory and Practice*. The Institution of Chartered Accountants in England and Wales, London, 1976.

CIBSE Guide, Building Energy – Code, Part 2, Calculation of energy demands and targets for the design of new buildings and services, Section (a) Heated and naturally ventilated buildings. CIBSE, 1980.

CIBSE Guide, Section B18, Owning and operating costs. CIBSE, 1986a.

CIBSE Guide, Section A9, Estimation of plant capacity. CIBSE, 1986b.

CIBSE Guide, Information technology and buildings. AM7, CIBSE, 1992.

CIBSE Guide, Building energy code, heated-only buildings. CIBSE (to be published), 1996a.

CIBSE Guide, Building energy code: air conditioned buildings. CIBSE (to be published), 1996b.

Day, P., Capital investment appraisal for mechanical and electrical services in commercial buildings. *The Heating and Ventilating Engineer*, pp. 4–7, February 1976.

Department of Energy, Fuel Note No 7, Department of Energy, HMSO, London, 1978.

Department of the Environment, Degree days. Fuel Efficiency Office, Department of the Environment, 1986a.

Dufton, A. F., Degree days. *JIHVE*, **2,** 1934.
Eyers, J., Engineering insurance, inspection and breakdown of heating and ventilation plant. *JIHVE*, **35,** 163, September 1967.
Faber, O. and Kell, J. R., *Heating and Air Conditioning of Buildings*. The Architectural Press, 1945.
Hassan, A., Reliability of HVAC equipment – introduction and literature survey. HVRA Laboratory Report 56, 1969.
JIHVE, Symposium of high velocity air conditioning. *JIHVE*, **28,** 225, 333, 1960.
Meteorological Office, *Average of bright sunshine for Great Britain and Northern Ireland, 1921–1950*. HMSO, 1953.
Millbank, N. O., Dowdall, J. P. and Slater, A., Investigation of maintenance and energy costs for services in office buildings. *JIHVE*, **39,** 145, October 1971.
Pullinger, F. A., Low energy air conditioning – a challenge for the industry. CIBS/RICS/ Institute of Refrigeration Conference, Nottingham University, pp. 20–21 April 1978.
Swain, C. P., Thornley, D. L. and Wensley, R. The choice of air conditioning systems. *JIHVE*, **32,** 7, 307, 1964.
Watson, R. B., Capital and running costs of air conditioning systems. CIBS/RICS/Institute of Refrigeration Conference on Air Conditioning and Energy Conservation, Nottingham University, 20–21 April, 1978.

9
Energy Conservation

9.1 Building design

Before looking at energy conservation in systems it is worth examining the influence of the building itself on the cooling load and energy consumption. A study by Jones (1974) in the design of buildings to minimise energy use, based on the hypothetical office block (Section 1.2), concluded that the following principles appear to apply:

(1) In the UK there is no economic justification for double glazing in office blocks but, for continuously occupied buildings, such as hospitals, there may be. This conclusion would be different for a country with a colder winter and/or higher energy costs.
(2) The influence of the mass of the walls on the cooling load is not significant. Floor slabs have a marginal effect, heavier floors slightly reducing the cooling load contribution from solar radiation through glass. Carpeting and furniture tend to insulate the floor slab from the solar heat gain and so reduce its effective mass.
(3) The mass of the roof slab is only significant for low-rise structures but should be heavy, rather than light, acoustic considerations and over-flying aircraft being relevant.
(4) *U*-values for walls and roofs should be 0.4 and 0.2 W m^{-2} K^{-1}, respectively, or better according to Jones (1975), due consideration being paid to problems of condensation. (It is to be noted that the building regulations may require a smaller *U* value for walls.)
(5) The major building axis should point east–west, particularly when there is a lot of glazing.
(6) The amount of glass and its *U*-value dominates the energy flux through the building envelope and also affects the need for artificial lighting. Regarding human preference, it is likely that from 25% to 30% of the outer facade ought to be glazed. Some benefit accrues from the penetration of natural daylight and solar warmth as the Building Energy Code (1980) points out.
(7) Large buildings are more economical than small ones, in terms of cooling load per unit area of usable floor according to the Building Energy Code (1980).
(8) In the UK climate, the heating and air-conditioning systems in buildings should be operated intermittently, rather than continuously, when the structure is lightweight as Billington *et al.* (1964) and Harrison (1956) point out (Figure 9.1).

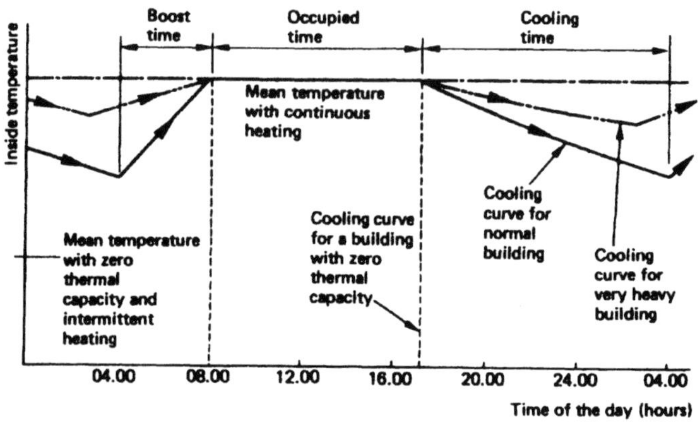

Figure 9.1 Intermittent heating.

9.2 Energy conservation techniques in systems

RECIRCULATING AIR

The well-established method of recirculating air minimises refrigeration load and running costs. It is one of the simplest and most effective techniques. The proportions of fresh and recirculated air are automatically mixed to reduce the fresh air component of the cooling load in summer and to switch off the refrigeration plant in winter when the outside air is cool enough to permit so-called free cooling, i.e. at about 5°C wet-bulb and 10°C dry-bulb for air–water and all-air systems, respectively, in the UK. Assuming normal office hours and a 5-day week, the outside dry-bulb is less than 10°C in the London area for roughly 38% and the wet-bulb is below 5°C for about 21% of the occupied period. These figures suggest that significant benefit can result from free cooling with an air–water system and, in fact, the heat gains to be dealt with in marginal weather are less than the design summer loads and the outside wet bulb that permits the refrigeration plant to be switched off can be a degree or so higher than 5°C. In any case, mixing control effects an energy saving in running cost during mid-seasons by using 100% fresh air when the outside wet-bulb is less than the room wet-bulb, the enthalpy of the fresh air being then lower than the enthalpy of mixed air. An analysis of the meteorological data for London (Heathrow) Airport shows that, during the occupied hours assumed, the outside wet-bulb exceeds the room wet-bulb, assumed as 16°C screen, for about 170 hours a year, i.e. for about 7.5% of a total working period of 2268 hours. This means that mixing is effective in saving energy for 100 −(7.5 + 38) = 54.5% and 100 − (7.5+21) = 71.5% of the time the refrigeration plant runs for all-air and air–water systems. respectively.

RUN-AROUND COILS

Run-around coils have also been used to a limited extent, for many years and they too provide an effective way of conserving energy, heat being transferred between the outgoing, vitiated air and the incoming fresh air. Figure 9.2 shows the principle for a winter case with a plant handling 100% fresh air, e.g. an operating theatre. Holmes and Hamilton (1978) showed that, assuming eight-row coils with 320 fins m^{-1} and face velocities of

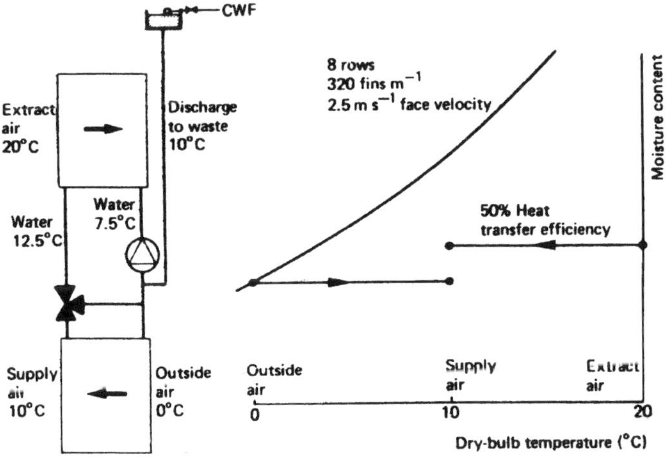

Figure 9.2 Run-around coils.

2.5 m s^{-1}, an approach efficiency of heat transfer of about 50% is possible. The advantage of this method over some of the others is that the two coils need not be close together since the link between them is pipework. There is also no possibility of cross-contamination between the two airstreams.

THERMAL WHEELS

These are regenerative heat exchangers and Figure 9.3 illustrates the principle. A slowly rotating drum or wheel (0.1–10 rpm) crosses the incoming airstream of cold, fresh air in winter and also the outgoing airstream of warm air being exhausted from the building. A small segment, used as a purge, separates the two major sections and a special seal wipes the face of the wheel to prevent the cross flow of air during rotation, although problems have sometimes occurred with seal failures. The commonest type of wheel has a spirally

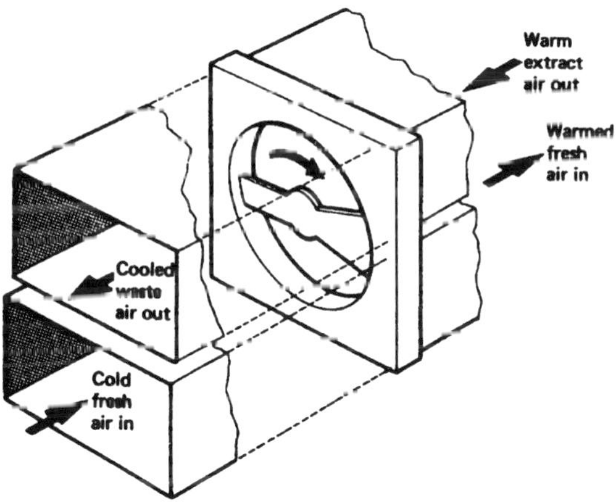

Figure 9.3 Thermal wheel.

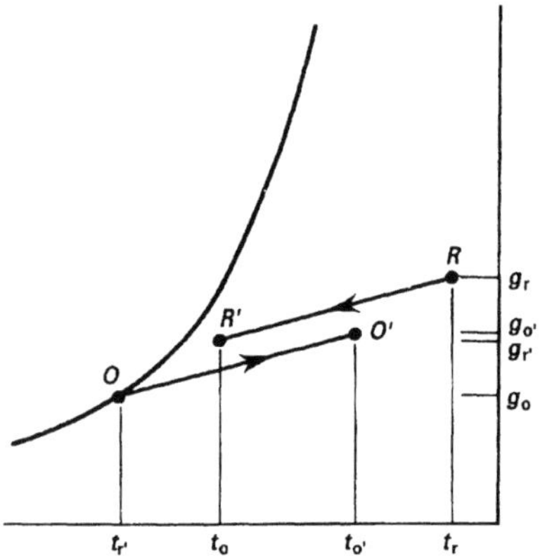

Figure 9.4 Psychrometry for the performance of thermal wheels. Sensible efficiency = $(t_r - t_{r'})/(t_r - t_0) \times 100$. Latent efficiency = $(g_r - g_{r'})/(g_r - g_0) \times 100$. Sensible and latent efficiencies are not necessarily equal.

wound filling made of alternate layers of corrugated and flat sheet aluminium or stainless steel. The diameter of the corrugations is small and the material used is coated with a hygroscopic material, impregnated with a desiccant. In winter the exhaust airstream from the building is warmer and more humid than the incoming cold, fresh air. The freshly desiccated material in the segment passing across the exhaust airstream cools and dehumidifies it, removing both sensible and latent heat. The segment then rotates over the fresh airstream, and the desiccant gives up the sensible and latent heat, recently acquired from the exhaust air, as it warms and humidifies the fresh airflow (Figure 9.4). In summer, the incoming fresh air is cooled and dehumidified at the expense of the exhaust air, which is warmed and humidified prior to discharge from the building. The surfaces of the wheel filling are dry during normal operation and this discourages the presence of bacteria and the growth of mould. Rotation is continuous at between 0.1 and 10 rpm and hence sensible and latent heat are steadily transferred between the two airstreams. Capacity control is usually by modulating the speed of the rotor between the limits mentioned, varying the speed of the geared-down electric driving motor to achieve this. Constant speed drives are also sometimes used and motor powers to drive the rotor are small. As the rotor passes through the purge sector, between the incoming and outgoing airstreams, fresh air is blown through the upper half of the sector and drawn out through the lower half. Air carryover between the two major sectors is negligible, provided that the fan connections are arranged properly: to achieve the desired air pressure balance between the two major sectors it is essential that both the supply and extract fans are on the same side of the wheel. Locating the supply fan on one side and the extract fan on the other is wrong and will not give the necessary pressure balance between the two main airstreams. The corrugations can pass dust particles of size up to 0.4 mm and the manufacturers claim that a pre-filter is not needed for conventional applications, although one would be necessary for hospitals, pharmaceutical laboratories or the like. Volumetric capacities are in the range from 450 $l\ s^{-1}$ to over 30 $m^3\ s^{-1}$, face velocities are 1.0–5.0 $m\ s^{-1}$, and air pressure drops are from 300 to 450 Pa, depending on the type of wheel and the face velocity. Temperatures up to 350°C can be accommodated if the correct material is used for the fill. Total heat transfer efficiencies approaching 90%

are claimed. When handling very humid air, wheels with hygroscopic fills have occasionally suffered difficulties with waterlogging.

Another type of wheel uses a rotor fill that comprises knitted aluminium or stainless steel wire. This type of wheel does not appear to have a fill impregnated with a hygroscopic or desiccant substance and so heat exchange is likely to be mostly sensible. However, it is possible for latent heat exchange to occur, provided that the incoming outside airstream is colder than the dew-point of the exhaust air. Efficiencies up to 84% are claimed by the manufacturers for sensible heat transfer, and up to 66% for moisture removal when latent cooling takes place, under favourable circumstances. Severe corrosion has occurred in the past when the metallic fill was wrong for the application and it is essential to provide a drain from the wheel.

A further form of thermal wheel has a rotor fill made in the form of a solid honeycomb of inorganic silica gel, which is very effective for dehumidification and is totally incombustible.

STATIC RECUPERATORS

Static recuperators are air-to-air plate heat exchangers and the principle is shown in Figure 9.5. The metal casing is airtight and lagged. The individual heat transfer plates within are usually made of aluminium, with protective coatings when necessary to cope with corrosion in aggressive atmospheres. Volumetric airflow rates are from 140 $l\ s^{-1}$ to 17 $m^3\ s^{-1}$ with pressure drops from about 100 to 450 Pa, depending on the face velocity. Sensible heat transfer efficiency approaching 90% is claimed but 60–75% appears typical. Latent cooling is also possible when the incoming cold air has a temperature lower than the dew-point of the air discharged to waste. A drain point is essential. Occasional cleaning is required, usually by brushing or by jets of compressed air, and inspection ports are often provided. In installations where the air is contaminated with greasy or sticky particles or droplets, cleaning should be done by washing with hot water and a suitable detergent. Piping can be provided to facilitate this.

As with run-around coils and thermal wheels, heat transfer efficiency is expressed in terms of the approach between the entering and leaving air states. In Figure 9.2 heat transfer cools the incoming airstream from 20 to 10°C, and warms the outgoing airflow from 0 to 10°C. The sensible heat transfer efficiency, or temperature approach, is 50%.

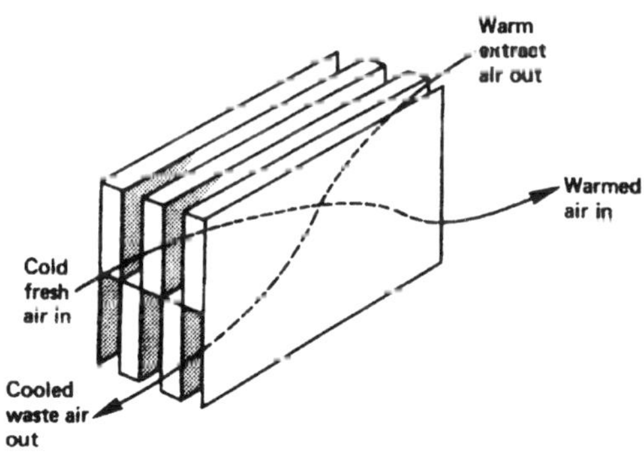

Figure 9.5 Static recuperator.

HEAT PIPES

A heat pipe is an externally finned tube, closed at both ends, containing a volatile fluid that exists in both its liquid and vapour phases. If cold air or fluid flows over the external finning at one end of the pipe it condenses the vapour within to a liquid. The liquid then flows by capillary action, or under the influence of gravity, to the other end of the pipe where warm air or fluid, flowing over the external finning, causes the volatile liquid within to evaporate and return to the original end of the pipe. In the case of horizontal tubes, a capillary action is promoted by a wick, which is often merely a concentric groove cut in the pipe wall. Capacity is self-regulating to some extent but control is possible by tilting the tubes in the evaporating half of the coil above the horizontal. Alternatively, motorised face and by-pass dampers may be used. In Figure 9.6 a plan is shown of a battery of horizontal heat pipes. The small dashed arrows represent heat flowing from the warm airstream, through the tubes by the mechanism described above, into the cold airstream.

Another version of heat pipe uses vertical tubes that return the condensed liquid from the upper to the lower half of the assembly by gravity, without the aid of a wick. Control is then only possible by motorised face-and-by-pass dampers.

There are no cross-connections between individual tubes which are assembled into batteries, commonly comprising three to seven rows of tubes with 600 fins m^{-1} (14 fins per inch). Face velocities are from 1.5 to 4.0 m s^{-1} with air pressure drops from about 75 to 750 Pa. The practical range of temperatures appears to be from −50 to +50°C. When transferring heat between moving airstreams thermal efficiencies are up to 50%. The very large heat transfer within the tubes, from one end to the other, cannot be matched by the heat flow between the finned tube and the moving airstreams, which is of the order of 80 W m^{-2} K^{-1} Hence it seems that efficiencies beyond 70% are not possible.

HEAT PUMPING

Heat can only be rejected from a condenser if there is a cooling load at the evaporator. Also,

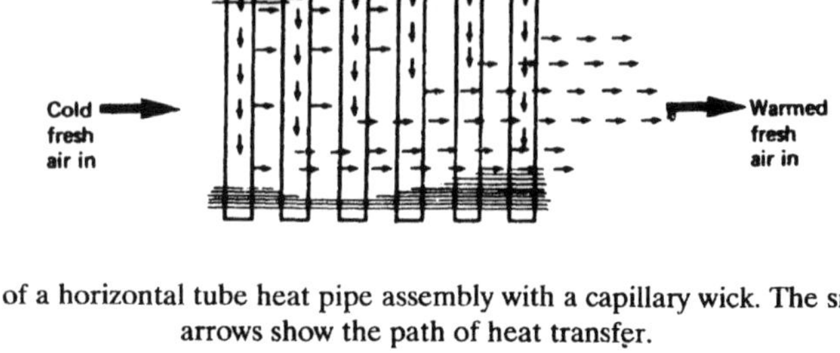

Figure 9.6 Plan of a horizontal tube heat pipe assembly with a capillary wick. The small dashed line arrows show the path of heat transfer.

heat pumping is only worth while if the rejected heat can be usefully deployed, i.e. if there is a fabric heat loss or a ventilation heating load. With all-air systems there is generally a cooling load when the outside temperature exceeds 10°C dry-bulb and a need for heating when it is below 20°C dry-bulb. In the London area, for the office occupancy assumed earlier, the outside temperature lies between these values for about 55% of the working time. For air–water systems the outside limits are 5°C wet-bulb and 20°C dry bulb and the temperature lies in this region for about 70% of the working time. With smaller size installations, having cooling loads up to about 100 kW of refrigeration, fixed proportions of minimum fresh and maximum recirculated air are used throughout the year because this gives the lowest capital cost and, although the running cost is high, probably the lowest owning and operating cost over the life of the system. The conditions for useful heat pumping are then met for about 90% of the working time. In the above terms, the heat pumping looks good for office blocks. The economic situation is not so favourable in many cases, however, because of the high extra capital expenditure involved in reclaiming the condenser heat and the relatively long pay-back period compared with the life of the plant. Nevertheless, if electricity is the only source of energy available on a site, heat pumping is the most economical way of heating, its coefficient of performance under design summer conditions of about 5:1 making it a very cheap way of using electricity for heating (cheaper than using fossil fuels directly), if the low-grade heat produced at the condenser can be used.

There are several limitations on the use of heat pumps that it is well to consider:

(1) A positive displacement compressor, i.e. screw or reciprocating, should be chosen in preference to a centrifugal. This is because there is no risk of surge with such machines. A centrifugal machine tends to surge under the very conditions when heat pumping is most needed, i.e. when the outside air temperature is low and the cooling load is correspondingly low but high condensing temperatures are needed to give as much heat as possible.

(2) Screw compressors can operate at high condensing temperatures but these are restricted by lubrication and motor cooling problems. The open screw is much preferred because the hermetic model sometimes has its stator windings cooled by discharge gas and cannot therefore work at such high condensing temperatures.

(3) The compression ratio should be kept as low as possible if the desirably high coefficients of performance are to be obtained.

(4) The lowest possible cooling water leaving temperature ought to be adopted for the condenser. Although higher values have been used, a recommended upper limit is 40°C.

(5) Similarly, the lowest possible entering water temperature is desirable on the condenser and a suggested maximum is 35°C.

(6) Even if these suggestions are followed it is prudent to select the heater batteries to work with lower temperatures, say 38°C flow and 33°C return.

(7) The minimum design temperature difference between the chilled water and the water leaving the condenser is 14 K.

(8) Of the five methods, i.e. double bundle condenser, auxiliary condenser, open cooling tower with a plate heat exchanger, open cooling tower with cleanable tubes at the heater batteries, closed cooling tower, the most economical and practical is generally the open cooling tower with dirty water pumped from it to one side of a plate heat exchanger and clean water pumped from the condenser through its clean side, via the heater batteries.

(9) Hot gas air heater batteries can be used, in conjunction with conventional air cooled condensers and liquid receivers. Condensing and evaporating pressures must be controlled and a good deal of care must be exercised in the design, installation and commissioning of the system. In particular, control valves must be carefully sized.

(10) A coefficient of performance having a value of five is only appropriate when the plant is working under design conditions in the summer when heating is not likely to be needed. For winter operation, when heating will be wanted, the heat pumping coefficient of performance will be significantly less than five and the price of electricity to effect heat pumping will be much less attractive in comparison with gas or oil for conventional heating. For example, if the heat source in winter is the outside air, its design temperature might be –2°C and an evaporating temperature of –5°C or less required to achieve the necessary heat flow. A high condensing temperature would be needed to heat the building. It follows that the evaporating pressure will be low and the condensing pressure high, causing a high compression ratio and reducing the heat pumping coefficient of performance to perhaps as low as 2.0. Frost will form on the outer surfaces of the evaporator coil and impede the heat flow from the outside air into the evaporating refrigerant. The evaporating temperature and pressure will drop, reducing the compression ratio still further and increasing the compression ratio. The effect is cumulative and heat pumping must be stopped while de-frosting is carried out. Meanwhile, heating must still be provided for the building and this is likely to be by the use of direct electricity at the full tariff. The situation will recur until the cold weather has passed. It is evident that the seasonal coefficient of performance for heat pumping will be a good deal less than the value of five, considered earlier. A value of 2.0 or 2.5 is suggested as an annual average value for the coefficient of heat pumping. There may still be a case for heat pumping but this must be carefully considered. On the other hand, for some applications, such as swimming pools (see Section 3.12), the case is very good for heat pumping because the load is comparatively constant, with a high coefficient of performance for most of the year.

To compare some of these methods it was assumed that the hypothetical office block (Section 1.2) was maintained at 20°C with the same moisture content inside as outside. A ventilation allowance of 1.3 $l\ s^{-1}\ m^{-2}$ was adopted and different methods of heat exchange between the incoming fresh air and the outgoing vitiated air were evaluated. Although a discounted cash flow analysis would have been better it was felt that within the limitations of the simplifying assumptions made the cruder test of the pay back period (Section 8.4) was enough to show the trend. Meteorological data from London (Heathrow) Airport were used and analysed to yield the values of outside temperatures and their durations throughout the occupied working year. The total number of working hours for which the outside temperature was less than 20°C was estimated to be about 2085, or 90% of the working time. It was then possible to determine the heat energy requirement of the ventilation air. Because there is an air pressure drop across air-to-air heat exchangers, and a water pressure drop too for the run-around coils, a significant waste of electrical energy is involved. Such losses were debited to the energy saved. The results obtained varied a little because of the options possible when selecting a heat exchanger from a commercial range of sizes, and getting a selection at the best end of the commercial range is vital to making a good case for heat reclaim. Table 9.1 summarises the results.

The significance of the pay back periods in Table 9.1 can only be seen in relation to the estimated lives of the plant items in question (see Table 8.15). A life of 20 years can be

Table 9.1 Comparison of methods of energy conservation in air-conditioning systems

Method	Face velocity ($m\ s^{-1}$)	Heat transfer efficiency (%)	Simple pay back period (years)
Run-around coils	2.5–3.75	50–47	9–12
Non-metallic thermal wheels	3.3	70+	13
Metallic thermal wheels	3.15–3.66	82–79	8
Static recuperators	3.33	66	13

assumed for the run-around coils and more than this for the associated pump and pipework, so a pay back in 9–12 years should give a positive present value over the life of the plant and make the method economically viable. Thermal wheels are more difficult to assess but it is possible that their life is a little shorter than that of run-around coils, because of the moving parts. Assuming a life-time of 15 years, a successful economic case is still obtained with a pay back of 8–13 years. Static recuperators could be regarded as similar to ductwork, with a life of more than 30 years but for these and the metallic wheels the economic case is less easy to establish because of the risk of corrosion giving a reduced life, particularly if condensate drainage is inadequate and maintenance poor.

It is to be noted that better cases can often be made with industrial applications, where the air temperatures are likely to be higher and the energy saved consequently greater. Heat pumps may also be used in industry with a more powerful economic justification than in commercial applications. It is possible to lift condenser cooling water to a leaving temperature of from 60 to 110°C by pumping heat from a relatively low-grade source of fluid at between 27–77°C. Notwithstanding what has been said earlier, the centrifugal compressor has been successfully used for this because its selection is possible in an area of performance remote from the surge condition, with the particular combinations of temperature and pressure involved.

9.3 System operation

The most potent way of conserving energy is by operating the system in a rational manner and in this respect there are two notable steps that should be considered:

(1) Intermittent heating – as has been indisputably demonstrated theoretically by Billington *et al.* (1964) and Harrison (1956), and verified in practice, switching off the heating system at night saves energy unless the building is of very heavy construction when the cooling down and heating up curves are so shallow that the system might just as well run continuously (Figure 9.1).

(2) Reduced distribution – it is also possible to save large amounts of energy by operating the prime movers, i.e. fans and pumps, at reduced speeds during the occupied period, if some variation in noise and air movement is acceptable to the occupants and if a temporary reduction in the rate of the fresh air supply is tolerable. It has already been seen that systems with automatic mixing of the fresh and recirculated air quantities only operate with minimum fresh air for about 7.5% of the working time and so, on average, will get much more than the minimum fresh air allowance, even if fan operation at reduced speed is adopted. The argument is that except at conditions of summer and

winter design loads, when the plant must run continuously at design speed if conditions are to be maintained, it is possible to reduce the speed of fans for part of the occupied period in a year, allowing the heat energy or cooling capacity stored in the building structure to make good the temporary deficit, room temperatures falling or rising during the off period but remaining at acceptable values, although some individual thermostatic control is likely to be lost. For a large building, with multiple air-handling units, the fans can be reduced in speed in a planned sequence, certain areas of the building being given priority if necessary. The principle of fan switching, allowing the thermal inertia of the room to keep the temperature within reasonable limits, is not new. For many years fan coil units and room air conditioners have been controlled in this way (see Example 2.5). Jones (1976) claims that 15–20% savings in energy consumption are possible if automatic, programmed switching is adopted for a commercial building. If fan or motor speeds are reduced, energy will only be saved if the method of speed reduction is efficient. Invertors, which vary the electrical frequency supplied to the driving motor are efficient. Other methods, notably using eddy current control, are not efficient and some may even use more electrical power at partial duty than full load.

Notation

Symbols	Description	Unit
g_o	Moisture content of air at state *O*	kg kg^{-1} dry air
$g_{o'}$	Moisture content of air at state *O′*	kg kg^{-1} dry air
g_r	Moisture content of air at state *R*	kg kg^{-1} dry air
$g_{r'}$	Moisture content of air at state *R′*	kg kg^{-1} dry air
t_o	Dry-bulb temperature of air at state *O*	°C
$t_{o'}$	Dry-bulb temperature of air at state *O′*	°C
t_r	Dry-bulb temperature of air at state *R*	°C
$t_{r'}$	Dry-bulb temperature of air at state *R′*	°C

References

Jones, W. P., *Designing Air Conditioned Buildings to Minimise Energy Use, Conference on Integrated Environment in Building Design*. University of Nottingham, Applied Science Publishers, 1974.

Jones, W. P., Built form and energy needs. Joint Conference, Department of Energy, Institute of Fuel, IES, CIBS, London, November 1975.

Building Energy Code. 1980: Part 2, Calculation of energy demands and targets for the design of new buildings and services, Section 1a, Heated and naturally ventilated buildings, CIBS, 1980.

Building Energy Code 1. 1999: Energy demands and targets for heated and ventilated buildings, CIBSE, 1999.

Building Energy Code 2. 1999: Energy demands for air conditioned buildings, CIBSE, 1999.

Billington, N. S. *et al.*, Intermittent Heating, HVRA Laboratory Report No 26, November 1964.

Harrison, E., The intermittent heating of buildings. *JIHVE*, **24**, 145–96, 1956.

Holmes, M. and Hamilton, G., Heat recovery with run-around coils. BSRIA Technical Note TN2/78, 1978.

Jones, W. P., Energy recovery in ventilation systems. Conference on Airflow and Building Design, University of Sheffield, 5–6 January 1976.

Appendix

Table A.1 Solar gains through internally shaded glass (reproduced by kind permission of Haden Young Ltd)
(a) For room surface density of 500 kg m^{-2}

		Solar air-conditioning loads (W m^{-2})												
		Sun time (hours)												
Month	Exposure	06.00	07.00	08.00	09.00	10.00	11.00	12.00	13.00	14.00	15.00	16.00	17.00	18.00
June 21	N	3	9	22	22	22	25	25	25	25	28	28	28	28
	NE	117	145	139	113	73	60	47	47	41	38	38	32	25
	E	106	183	205	199	164	110	76	69	63	57	50	44	38
	SE	6	73	123	158	173	170	148	113	76	63	54	47	38
	S	6	6	41	69	95	113	126	129	123	107	82	47	41
	SW	16	19	19	19	25	57	98	132	161	170	154	120	63
	W	22	25	26	26	28	28	28	57	113	170	208	214	189
	NW	16	19	22	22	25	25	25	25	38	82	126	161	157
July 23	N	3	9	10	19	19	22	22	22	22	22	25	25	25
and	NE	107	136	130	104	69	54	44	44	38	35	35	28	25
May 21	E	127	183	205	199	164	111	76	69	63	57	50	44	38
	SE	9	79	130	171	186	180	158	123	82	66	60	50	41
	S	9	9	44	79	107	130	142	145	142	120	91	54	44
	SW	19	22	22	22	28	66	111	155	183	196	177	139	85
	W	22	25	25	28	28	28	28	57	114	171	208	215	189
	NW	16	19	19	19	22	22	22	22	35	76	117	148	145
August 24	N	0	6	16	16	16	16	16	19	19	19	19	19	19
and	NE	88	111	104	85	54	44	38	35	32	28	28	25	19
April 20	E	123	177	199	193	158	107	73	66	60	54	47	44	38
	SE	9	85	142	186	205	199	174	133	88	73	63	54	44
	S	9	9	60	101	139	167	186	189	183	158	120	69	60
	SW	22	25	25	25	32	73	120	167	202	212	196	152	79
	W	19	25	25	28	28	28	28	54	111	164	202	208	183
	NW	13	16	16	16	19	19	19	19	28	63	95	117	117
September 22	N	0	6	9	13	13	13	13	13	13	13	13	13	13
and	NE	54	66	63	50	35	28	22	22	19	19	16	16	13
March 22	E	107	155	174	167	139	95	63	60	54	47	44	38	32
	SE	9	88	148	193	212	205	180	139	91	76	66	57	47
	S	13	13	66	117	158	193	215	218	212	180	139	79	66
	SW	22	25	25	25	32	76	127	174	208	221	202	158	82
	W	19	22	22	25	25	25	25	47	98	145	177	183	161
	NW	6	9	9	9	13	13	13	13	19	38	57	73	73
October 23	N	0	3	6	6	6	6	6	9	9	9	9	9	9
and	NE	28	35	32	25	16	13	13	9	9	9	9	9	9
February 20	E	82	117	133	127	107	73	47	44	41	38	32	28	25
	SE	9	85	142	186	205	199	174	133	88	73	63	54	44
	S	13	13	73	123	167	205	228	231	224	189	145	85	73
	SW	22	25	25	25	32	73	123	167	202	212	196	152	79
	W	16	16	16	19	19	19	19	38	73	111	136	139	123
	NW	3	3	6	6	6	6	6	6	9	19	28	38	35

Table A.1 (Continued)
(b) For room surface density of 150 kg m^{-2}

		Solar air-conditioning loads (W m^{-2})												
		Sun time (hours)												
Month	Exposure	06.00	07.00	08.00	09.00	10.00	11.00	12.00	13.00	14.00	15.00	16.00	17.00	18.00
June 21	N	0	6	22	25	28	28	28	28	28	32	32	32	32
	NE	136	186	177	142	88	60	47	41	38	32	28	28	16
	E	145	221	253	249	202	133	79	60	50	44	35	28	22
	SE	0	79	148	196	221	212	180	130	79	54	44	35	22
	S	19	38	79	114	139	155	158	143	101	88	44	28	19
	SW	6	9	16	19	22	60	123	177	212	224	205	158	69
	W	9	12	19	22	25	25	25	60	133	205	256	263	234
	NW	6	13	16	19	22	22	25	25	41	95	158	196	193
July 23 and May 21	N	0	6	19	22	25	25	25	25	25	25	25	28	28
	NE	127	174	164	133	82	54	44	38	35	28	28	25	16
	E	145	221	253	249	202	133	79	60	50	44	35	28	22
	SE	0	82	158	208	234	224	193	139	82	57	47	35	25
	S	22	44	88	130	158	177	180	167	114	101	50	32	22
	SW	9	13	16	19	25	63	130	186	224	237	218	167	73
	W	9	13	19	22	25	25	25	60	133	205	256	268	234
	NW	6	13	16	19	19	19	22	22	38	88	142	183	180
August 24 and April 20	N	0	6	16	19	19	19	19	19	19	22	22	22	22
	NE	101	139	132	104	68	44	35	32	28	22	22	19	13
	E	147	215	246	243	196	130	76	57	50	44	35	28	22
	SE	0	91	174	228	256	246	208	152	91	60	50	41	28
	S	25	57	114	167	205	231	234	218	146	133	63	44	28
	SW	9	13	19	22	28	69	142	205	246	262	240	183	79
	W	9	13	19	22	25	25	25	57	130	199	249	259	228
	NW	6	9	13	16	16	16	19	19	32	73	114	145	145
September 22 and March 22	N	0	3	13	13	13	13	16	16	16	16	16	16	16
	NE	63	85	82	66	41	28	22	19	16	16	13	13	6
	E	123	186	215	212	171	114	66	50	44	38	28	25	19
	SE	0	95	180	237	265	256	218	158	95	68	54	41	28
	S	32	63	133	193	237	262	268	249	171	152	78	50	35
	SW	9	13	19	22	28	73	148	212	256	271	249	189	32
	W	9	9	16	19	22	22	22	50	114	174	218	228	199
	NW	3	6	6	9	9	9	13	13	19	44	69	88	88
October 23 and February 20	N	0	3	6	6	9	9	9	9	9	9	9	9	9
	NE	32	44	41	32	19	13	9	9	9	6	6	6	3
	E	95	142	164	161	130	85	50	38	32	28	22	19	16
	SE	0	91	174	228	256	246	212	152	91	60	50	41	27
	S	32	69	139	205	249	278	284	265	180	161	79	50	35
	SW	9	13	19	22	28	69	142	199	246	262	240	183	79
	W	6	9	13	16	16	16	16	38	85	133	164	174	152
	NW	3	3	3	3	6	6	6	6	9	22	35	44	44

Table A.1 shows solar air-conditioning loads through windows in the UK only, for room surface densities of 500 kg m^{-2} and 150 kg m^{-2}, expressed in W m^{-2}. Values are for single plate or float glass and, where correction is necessary for other types of glazing, should be multiplied by the factors given in Table A.2. The area to be used is the opening in the wall for metal-framed windows and the area of the glass for wooden-framed windows. A haze factor of 0.9 has been allowed. It is assumed that shades are not provided on the windows facing north and that all other exposures have blinds that will be raised when the windows are not in direct sunlight. Scattered radiation is included and the storage effect of the building taken into account. Air-to-air transmission is excluded.

Table A.2 Factors for use with Table A.1 (reproduced by kind permission of Haden Young Ltd)

		Blind position		
Outer pane	Inner pane	Internal	Between the panes	No blind
Clear sheet (4 mm)		1.00		
Clear plate or float		1.00		
Heat-absorbing bronze 49/66		0.96		1.43
Heat-absorbing green 75/60		0.92		1.30
Heat-absorbing grey 41/60		0.92		1.30
Heat-reflecting bronze 11/28				0.61
Heat-reflecting gold 15/22				0.50
Heat-reflecting gold 39/34				0.72
Clear plate or float	Clear plate or float	1.00	0.49	
Heat-absorbing bronze 49/66	Clear plate or float	0.80	0.47	1.17
Heat-absorbing green 75/60	Clear plate or float	0.73	0.45	1.03
Heat-absorbing grey 41/60	Clear plate or float	0.73	0.45	1.03

The factors given in Table A.2 are approximate and in each case are the ratio of the shading coefficient of the particular glass to that of clear glass with internal Venetian blinds, taken as 0.53. Thus heat-absorbing bronze 49/66, which has a shading coefficient of 0.76, will have a factor to apply to Table A.l of 0.76/0.53 or 1.43. Shading coefficients cannot be applied as factors directly to the solar loads given in Table A.3 for bare or externally shaded glass. This is because the shading coefficient is compounded of an instantaneous convective gain that is not susceptible to the building storage effect and a direct radiation gain that is susceptible to the building storage gain, whereas the values in Table A.3 refer to heat gains that occur largely by direct radiation through bare glass and, therefore, are heavily influenced by building storage effect. The procedure of relating the shading coefficient to that of clear glass with venetian blinds, for which the heat gain does have a significant instantaneous convective component, and referring it to Table A.1 is, therefore, proposed as an acceptable compromise. All the factors except the first refer to 6 mm plate or float glass. The figures quoted after the entries for the heat-absorbing and heat-reflecting glasses give the light and total heat transmittances, respectively. For types of glass having heat transmittances other than those listed, interpolation may be used to yield a good approximate factor to refer to Table A.1. For a more exact approach see Example 1.4.

Table A.3 Solar gains through bare glass (reproduced by kind permission of Haden Young Ltd)
(a) For room surface density of 500 kg m^{-2}

		Solar air-conditioning loads (W m^{-2})												
		Sun time (hours)												
Month	Exposure	06.00	07.00	08.00	09.00	10.00	11.00	12.00	13.00	14.00	15.00	16.00	17.00	18.00
June 21	N	6	19	25	28	32	35	38	38	41	41	44	44	32
	NE	82	136	164	171	158	148	117	104	95	91	82	73	69
	E	91	164	220	259	259	237	205	177	158	142	130	114	101
	SE	22	57	107	164	205	225	237	218	189	167	145	127	111
	S	22	19	38	63	95	123	152	171	183	180	167	142	117
	SW	41	41	38	41	41	66	101	145	196	234	246	237	205
	W	50	50	50	50	50	50	57	69	107	171	228	271	287
	NW	35	41	41	41	41	41	41	41	47	82	123	174	199
	Horizontal	28	63	104	155	206	259	310	354	391	420	436	443	434
July 23	N	6	16	22	25	28	32	32	35	38	38	8	38	28
and	NE	76	127	155	158	145	136	111	98	88	85	76	69	63
May 21	E	88	164	224	259	259	259	237	174	158	142	130	114	101
	SE	25	60	114	174	218	243	253	234	202	180	155	133	120
	S	25	22	44	73	111	142	177	199	212	208	196	164	136
	SW	44	44	41	44	44	69	111	155	208	253	262	253	218
	W	50	50	50	50	50	50	57	66	107	167	224	268	287
	NW	32	38	38	38	38	38	38	38	44	76	117	161	186
	Horizontal	9	32	70	114	164	219	269	314	350	380	391	391	376
August 24	N	3	13	16	19	22	22	25	25	28	28	32	32	22
and	NE	63	101	123	127	117	111	88	79	73	69	63	54	50
April 20	E	83	158	218	249	249	231	196	171	152	136	127	111	98
	SE	28	104	171	224	262	262	234	205	183	161	142	127	114
	S	35	28	57	95	142	186	228	256	278	271	253	215	177
	SW	47	47	44	47	47	76	120	167	228	271	287	278	240
	W	50	50	50	50	50	50	54	66	104	164	218	262	278
	NW	25	28	28	28	28	28	28	28	35	63	95	130	148
	Horizontal	0	9	32	70	114	161	209	253	285	306	316	310	285
September 22	N	3	9	11	13	16	16	19	19	19	22	22	22	16
and	NE	38	63	76	79	73	69	54	47	44	41	38	35	32
March 22	E	76	139	189	218	218	199	171	148	133	120	111	95	85
	SE	28	66	130	196	248	275	287	267	231	202	174	152	136
	S	38	32	66	111	164	212	262	294	316	310	290	246	202
	SW	50	50	47	50	50	79	123	174	237	281	297	287	246
	W	44	44	44	44	44	44	47	57	91	112	139	223	243
	NW	16	19	19	19	19	19	19	19	22	38	57	79	91
	Horizontal	0	0	9	32	63	104	145	180	209	225	225	209	190

Table A.3 (Continued)
(b) For room surface density of 150 kg m^{-2}

		Solar air-conditioning loads (W m^{-2})												
		Sun time (hours)												
Month	Exposure	06.00	07.00	08.00	09.00	10.00	11.00	12.00	13.00	14.00	15.00	16.00	17.00	18.00
June 21	N	0	25	35	41	44	47	38	41	54	54	54	54	28
	NE	136	243	281	265	199	142	114	91	79	69	60	50	38
	E	152	281	378	413	384	300	215	152	123	101	85	66	50
	SE	0	82	186	275	335	357	335	278	205	148	107	82	66
	S	0	0	36	91	155	205	240	262	259	240	196	136	88
	SW	9	13	22	28	38	57	158	246	316	363	363	316	215
	W	13	15	28	35	41	41	44	72	164	278	378	428	422
	NW	9	16	22	32	35	38	44	44	57	117	208	281	316
	Horizontal	63	130	209	291	375	455	521	572	608	620	606	585	535
July 23 and May 21	N	0	22	32	38	38	41	35	35	44	47	47	47	25
	NE	127	224	262	246	186	133	104	85	73	63	57	47	35
	E	152	281	374	409	381	297	215	152	123	101	85	66	50
	SE	0	88	196	290	354	378	354	294	215	158	114	88	69
	S	0	0	41	104	174	234	275	300	297	275	221	155	101
	SW	9	16	25	28	38	60	167	259	335	381	381	335	228
	W	13	16	28	35	38	38	44	79	164	275	374	425	419
	NW	9	16	19	28	32	38	41	41	54	107	193	262	294
	Horizontal	19	70	139	225	310	395	468	525	560	571	560	521	459
August 24 and April 20	N	0	19	25	28	32	32	28	28	35	38	38	38	19
	NE	101	189	212	199	148	107	85	69	57	50	44	38	28
	E	145	271	363	397	368	287	205	145	120	98	82	66	50
	SE	0	91	215	319	387	416	387	325	237	174	123	98	76
	S	0	0	57	139	228	303	357	387	384	357	290	199	133
	SW	9	16	28	32	44	65	183	287	366	419	419	366	249
	W	9	16	28	32	38	38	44	76	158	265	363	413	406
	NW	6	13	16	22	25	28	32	32	41	88	155	212	237
	Horizontal	0	19	70	142	221	304	375	433	468	477	455	408	335
September 22 and March 22	N	0	13	19	22	22	25	19	19	25	25	28	28	16
	NE	63	111	130	120	91	66	50	41	35	32	28	25	19
	E	130	237	319	347	322	253	180	130	104	85	73	57	44
	SE	0	101	224	332	403	432	403	335	246	180	130	101	79
	S	0	0	66	161	265	354	413	450	447	413	338	231	155
	SW	13	16	28	35	44	66	189	297	381	425	435	381	259
	W	9	16	25	28	35	35	38	66	139	234	319	360	357
	NW	3	9	9	13	16	19	19	19	25	54	95	130	145
	Horizontal	0	0	19	66	133	202	268	323	350	350	326	269	211

Table A.3 shows solar air-conditioning loads through unshaded windows in the UK only, for room surface densities of 500 and 150 kg m^{-3}, expressed in W m^{-2}. Values are for single plate or float glass and should be multiplied by the factors given in Table A.4 where correction is necessary for external shaded or other treatment. The area to be used is the opening in the wall for metal-framed windows and the area of the glass for wooden-framed windows. A haze factor of 0.9 has been allowed and scattered radiation is included, together with the storage effect of the building. Air-to-air transmission is excluded.

Table A.4 Factors for use with Table A.3 (reproduced by kind permission of Haden Young Ltd)

			Type of outside shading		
Outer pane	Inner pane	No shade	Light Venetian blind	Light awning	Dark awning
4 mm clear sheet		1.0	0.15	0.20	0.25
6 mm clear plate or float		0.97	0.14	0.19	0.24
4 mm clear sheet	4 mm clear sheet		0.14	0.18	0.22
6 mm clear plate or float	6 mm clear plate or float	0.85	0.12	0.16	0.20
4 mm clear sheet, painted light colour		0.28			
4 mm clear sheet, painted medium colour		0.39			

Table A.5 Cooling load due to solar gain through vertical glazing (10 h plant operation) – W m^{-2}. For constant dry resultant temperature, lightweight building, intermittent blinds, 51.7° N latitude (reproduced from CIBSE Guide A by permission of the Chartered Institution of Building Services Engineers)

Date	Climatic constants	Orien-tation	Sun time											Orien-tation
			0800	0900	1000	1100	1200	1300	1400	1500	1600	1700	1800	
June 21	I=0.66	N	81	96	114	128	138	142	137	126	111	95	122	N
		NE	223	169	81	144	151	154	150	139	124	106	87	NE
	k_c=1.96	E	328	306	254	184	88	156	151	140	125	107	89	E
		SE	252	280	282	257	207	143	69	125	110	93	74	SE
	k_r=0.20	S	102	124	179	220	238	230	197	145	89	39	61	S
		SW	69	87	105	184	183	239	273	281	261	217	152	SW
	C=0.14	W	78	97	114	129	139	230	227	284	314	313	274	W
		NW	79	98	115	130	140	143	141	217	204	240	243	NW
July 23 and May 22	I=0.89	N	85	107	128	145	157	161	156	143	125	104	86	N
		NE	205	161	80	155	167	170	165	152	135	114	92	NE
	k_c=1.33	E	306	292	250	189	90	172	167	154	136	115	93	E
		SE	247	277	283	264	219	159	73	140	122	101	79	SE
	k_r=0.20	S	75	209	194	234	252	244	212	161	59	112	68	S
		SW	74	96	117	211	196	247	275	278	256	210	144	SW
	C=0.18	W	80	102	123	140	152	248	222	270	293	285	239	W
		NW	81	103	123	141	152	156	152	219	186	213	207	NW
August 24 and April 22	I=0.89	N	59	82	102	119	131	134	129	117	100	79	58	N
		NE	159	118	57	126	137	141	136	124	106	85	63	NE
	k_c=1.34	E	277	270	229	166	71	143	138	126	108	87	65	E
		SE	246	283	292	274	228	163	64	121	103	82	59	SE
	k_r=0.20	S	118	153	212	254	272	264	231	177	111	36	52	S
		SW	56	78	99	205	204	257	285	286	259	203	120	SW
	C=0.23	W	59	81	102	119	130	222	207	254	273	253	177	W
		NW	59	81	102	119	130	134	129	177	148	169	141	NW
September 22 and March 21	I=0.97	N	32	55	76	93	104	107	102	90	73	51	29	N
		NE	205	39	89	94	105	109	104	92	74	53	30	NE
	k_c=161	E	427	260	220	147	52	109	104	92	75	53	31	E
		SE	402	310	330	313	262	184	59	110	74	53	30	SE
	k_r=0.20	S	137	189	262	312	333	324	285	219	137	36	48	S
		SW	31	53	78	224	237	298	328	322	274	175	37	SW
	C=0.27	W	31	54	75	92	103	192	201	252	258	192	43	W
		NW	32	54	75	92	103	107	102	93	156	106	28	NW
October 23 and Febrary 21	I=1.18	N	10	32	53	70	81	85	80	68	50	28	8	N
		NE	19	61	54	71	82	85	81	69	51	29	9	NE
	k_c=1.31	E	136	183	170	115	37	84	80	67	49	28	7	E
		SE	148	242	283	280	241	175	50	106	50	28	8	SE
	k_r=0.20	S	78	164	241	291	313	303	264	197	108	23	10	S
		SW	10	32	58	225	221	271	289	265	183	41	21	SW
	C=0.31	W	9	31	52	70	81	155	159	188	158	35	21	W
		NW	10	32	54	71	82	86	81	69	55	58	14	NW

Table A.5 (Continued)

Date	Climatic constants	Orientation	Sun time 0800	0900	1000	1100	1200	1300	1400	1500	1600	1700	1800	Orientation
November 21	I=1.29	N	4	9	28	44	55	58	54	42	25	7	4	N
and		NE	5	11	28	44	55	58	54	42	25	7	4	NE
January 21	k_c=1.34	E	7	121	120	38	103	59	55	43	26	8	5	E
		SE	7	165	226	242	216	158	42	92	27	8	5	SE
	k_r=0.20	S	6	122	198	259	284	273	226	143	30	25	5	S
		SW	5	10	36	203	198	237	236	176	43	39	5	SW
	C=0.35	W	5	9	29	45	56	65	192	117	28	33	5	W
		NW	4	9	28	44	55	58	54	42	25	9	4	NW
December 21	I=1.18	N	3	4	16	31	40	43	39	29	14	3	3	N
		NE	3	4	16	31	40	43	39	29	14	3	3	NE
	k_c=1.70	E	5	12	177	32	85	45	41	31	16	5	5	E
		SE	6	15	314	225	206	150	39	82	19	6	6	SE
	k_r=0.20	S	5	13	267	245	273	261	206	109	26	5	5	S
		SW	4	5	25	186	187	222	211	129	34	4	4	SW
	C=0.40	W	3	4	17	31	41	50	160	81	20	3	3	W
		NW	3	4	16	31	40	43	39	29	14	3	3	NW

Correction factors for tabulated values

Types of glass (outside pane for double glazing)	Building weight	Single glazing: Light slatted blind Open	Single glazing: Light slatted blind Closed	Single glazing: Linen roller blind	Double glazing, internal shade: Light slatted blind Open	Double glazing, internal shade: Light slatted blind Closed	Double glazing, internal shade: Linen roller blind	Double glazing, mid-pane shade: Light slatted blind Open	Double glazing, mid-pane shade: Light slatted blind Closed	Double glazing, mid-pane shade: Linen roller blind
Clear 6mm	Light	1.00	0.77	0.66	0.95	0.74	0.65	0.58	0.39	0.42
	Heavy	0.97	0.77	0.63	0.94	0.76	0.64	0.56	0.40	0.40
BTG 6mm	Light	0.86	0.77	0.72	0.66	0.55	0.51	0.45	0.36	0.38
	Heavy	0.85	0.77	0.71	0.66	0.57	0.51	0.44	0.37	0.37
BTG 10mm	Light	0.78	0.73	0.70	0.54	0.47	0.45	0.38	0.34	0.36
	Heavy	0.77	0.73	0.70	0.53	0.48	0.45	0.37	0.34	0.34
Reflecting	Light	0.64	0.57	0.54	0.48	0.41	0.38	0.33	0.27	0.29
	Heavy	0.62	0.57	0.53	0.47	0.41	0.38	0.32	0.27	0.28
Strongly reflecting	Light	0.36	0.34	0.32	0.23	0.21	0.21	0.17	0.16	0.16
	Heavy	0.35	0.34	0.32	0.23	0.21	0.21	0.17	0.16	0.16
Additional factor for air point control	Light	0.91	0.91	0.91	0.91	0.91	0.91	0.80	0.80	0.80
	Heavy	0.83	0.83	0.83	0.90	0.90	0.90	0.78	0.78	0.78

Table A.6 Approximate time lags for building structures

Density ($kg\ m^{-3}$)	Approximate time lags (hours) Thickness (mm) 100	150	200	250	300	350	400
≤ 1200	2.7	4.9	6.8	9.0	11.2	13.3	15.2
1200–1800	2.7	4.9	6.4	8.0	9.6	11.2	12.7
1800–2400	2.7	4.6	5.8	7.3	8.6	9.9	11.2

For more accurate details refer to the CIBSE Guide, Section A3, Thermal properties of building structures (1980).

Table A.7 Approximate decrement factors for building structures

Insulation	Approximate decrement factors Thickness (mm) 100	150	200	250	300	350	400
None	0.87	0.67	0.48	0.33	0.22	0.15	0.11
On the inside surface	0.67	0.47	0.33	0.22	0.16	0.11	0.08
On the outside surface	0.51	0.34	0.22	0.15	0.10	0.07	0.04

For more accurate details refer to the CIBSE Guide, Section A3, Thermal properties of building structures (1980).

Table A.8 Equivalent temperature differences (reproduced by kind permission of Haden Young Ltd)

Exposure	Surface density ($kg\ m^{-2}$)	Equivalent temperature difference (K) Sun time (hours) 06.00	07.00	08.00	09.00	10.00	11.00	12.00	13.00	14.00	15.00	16.00	17.00	18.00
N	100	−3.9	−3.9	−4.5	−3.9	−3.4	−1.7	0	2.2	3.3	4.5	5.6	5.0	4.5
	300	−3.9	−3.9	−4.5	−3.9	−3.4	−2.8	−2.2	−0.6	1.1	2.2	3.4	3.9	4.5
	500	−1.7	−1.7	−2.2	−2.2	−2.2	−2.2	−2.2	−1.7	−1.1	−0.6	0	0.6	0.6
NE	100	0.6	5.6	8.9	9.5	10.0	7.8	5.0	5.0	4.5	5.0	5.6	5.6	5.6
	300	−2.8	−3.4	−3.4	0	10.0	8.9	7.8	5.6	3.4	3.9	4.5	5.0	5.6
	500	0	−0.6	0	0	0	2.8	6.1	5.6	5.0	4.5	3.4	3.9	4.5
E	100	−1.7	7.2	14.5	16.1	17.8	17.2	15.6	8.9	4.5	5.0	5.6	5.6	5.6
	300	−2.8	−2.8	−2.2	9.5	14.4	15.0	15.0	8.4	5.6	5.0	4.5	5.0	5.6
	500	0.6	0.6	1.1	2.2	5.6	8.9	11.1	11.7	11.1	8.9	7.8	6.7	5.6
SE	100	4.5	1.7	6.7	10.0	14.4	15.0	15.6	13.9	12.2	8.9	6.7	6.1	5.6
	300	−1.7	−1.7	−2.2	6.1	10.6	13.3	15.6	13.9	13.3	10.6	8.4	6.7	5.6
	500	2.2	2.2	1.7	1.7	1.7	5.0	7.8	8.4	8.9	9.5	8.9	7.8	6.1
S	100	−2.2	−2.8	−4.5	−0.6	1.7	9.5	15.6	18.3	20.0	18.3	16.1	11.1	7.8
	300	−2.2	−3.9	−4.5	−3.9	−3.4	3.9	8.4	13.9	16.7	16.7	17.2	14.4	11.1
	500	1.1	1.1	−0.6	−0.6	−0.6	0.6	1.1	4.5	7.2	10.0	10.6	11.7	11.7
SW	100	−3.4	−4.5	−4.5	−3.4	−2.2	0	1.1	9.4	13.3	18.3	22.2	22.8	23.4
	300	−0.6	−1.1	−1.7	−2.2	−2.2	−1.7	−1.1	2.2	5.0	12.2	17.2	19.4	20.0
	500	2.2	1.1	1.7	1.1	0.6	1.1	1.1	1.7	2.8	5.0	6.1	9.5	11.7
W	100	−3.4	−3.9	−4.5	−3.4	−2.2	−0.6	1.1	5.6	8.9	15.6	20.0	22.8	24.4
	300	−1.1	−1.7	−2.2	−2.2	−2.2	−1.1	0	1.7	3.4	8.4	12.2	16.7	20.0
	500	1.7	1.7	1.1	1.1	1.1	1.1	1.1	1.7	2.2	3.4	4.5	7.2	8.9
NW	100	−3.9	−4.5	−4.5	−3.4	−2.2	−0.6	1.1	3.4	4.5	8.4	10.6	15.0	18.3
	300	−3.4	−3.9	−4.5	−3.9	−3.4	−2.2	−1.1	1.1	2.2	3.4	4.5	8.9	13.3
	500	−0.6	0	0	0	0	0	0	0	0	0.6	1.1	2.8	4.5
Horizontal	50	−4.5	−5.6	−6.1	−5.0	−2.8	1.1	5.6	10.6	14.4	17.8	20.0	21.6	21.1
in	100	−2.2	−2.8	−3.3	−2.8	−1.1	2.2	6.1	10.0	13.3	16.7	18.3	20.0	20.0
sunshine	200	−0.6	−1.1	−1.1	−1.1	0.6	2.8	6.1	9.4	12.2	15.0	17.2	18.3	18.9
	300	2.2	1.7	0.6	1.1	0.6	3.3	5.6	8.9	11.7	13.9	15.5	17.2	18.3
	400	4.5	3.9	3.3	3.3	3.9	4.5	5.6	8.9	11.1	12.2	13.9	15.5	17.2

Table A.8 gives equivalent temperature differences for some typical walls and roofs at a latitude of 51.5°N in the UK, based on a 6 K difference between outside air temperature and room air temperature at 15.00 h sun time. For other temperature differences at 15.00 h, add or subtract the appropriate correction in each case. Diurnal variation of air temperature as well as fluctuation in solar radiation has been taken into account.

Table A.9 Sol-air temperatures at Kew. These are based on measurements taken at Kew during the ten-year period 1959–68, the weather for two consecutive days being averaged. The solar absorptance values assumed are 0.9 for dark surfaces and 0.5 for light surfaces.
Reproduced from CIBSE Guide A by permission of the Chartered Institution of Building Services Engineers

(e) July 23

Sun time	Air temp. t_{ao}/(°C)	Sol-air temperature, t_{av}/(°C)																	
		Horizontal		North		North-East		East		South-East		South		South-West		West		North-West	
		Dark	Light	Dark	Light	Dark	Light	Dark	Light	Dark	Light	Dark	Light	Dark	Light	Dark	Light	Dark	Light
00	16.0	12.5	12.5	15.0	15.0	15.0	15.0	15.0	15.0	15.0	15.0	15.0	15.0	15.0	15.0	15.0	15.0	15.0	15.0
01	14.5	11.5	11.5	14.0	14.0	14.0	14.0	14.0	14.0	14.0	14.0	14.0	14.0	14.0	14.0	14.0	14.0	14.0	14.0
02	13.5	10.5	10.5	13.0	13.0	13.0	13.0	13.0	13.0	13.0	13.0	13.0	13.0	13.0	13.0	13.0	13.0	13.0	13.0
03	13.0	10.0	10.0	12.0	12.0	12.0	12.0	12.0	12.0	12.0	12.0	12.0	12.0	12.0	12.0	12.0	12.0	12.0	12.0
04	13.0	9.5	9.5	12.0	12.0	12.0	12.0	12.0	12.0	12.0	12.0	12.0	12.0	12.0	12.0	12.0	12.0	12.0	12.0
05	13.0	12.0	11.0	17.5	15.0	22.5	18.0	22.5	18.0	17.0	15.0	13.5	13.0	13.5	13.0	13.5	13.0	13.0	13.0
06	13.5	17.5	14.5	19.5	16.5	31.0	23.0	33.5	24.5	25.5	20.0	15.5	14.5	15.5	14.5	15.5	14.5	15.5	14.0
07	14.5	24.5	18.5	18.0	16.0	34.0	25.0	40.5	28.5	33.5	24.5	18.0	16.0	18.0	16.0	18.0	16.0	18.0	16.0
08	16.0	31.5	23.0	20.5	18.0	33.5	25.5	43.5	31.0	40.0	29.0	25.5	20.5	21.0	18.5	21.0	18.5	20.5	18.9
09	17.5	38.5	27.5	23.0	20.0	31.0	24.5	44.0	31.5	44.5	32.0	33.5	26.0	24.0	20.5	24.0	20.5	23.5	20.5
10	19.0	44.5	31.5	26.0	22.5	27.5	23.5	41.5	31.0	46.5	34.0	40.0	30.5	27.0	23.0	27.0	23.0	26.5	22.5
11	20.5	49.0	34.5	28.5	24.5	29.0	24.5	37.5	29.5	46.0	34.5	45.5	34.0	35.0	28.0	29.5	25.0	29.0	24.5
12	21.5	51.5	36.5	30.0	26.0	30.5	26.5	31.5	26.5	43.0	33.0	48.0	36.0	43.0	33.0	31.5	26.5	30.5	26.5
13	23.0	51.5	37.5	31.0	27.0	31.5	27.5	32.0	27.5	38.0	31.0	48.0	36.5	49.0	37.0	40.0	32.0	31.5	27.5
14	24.0	49.5	36.5	31.0	27.5	31.5	28.0	32.0	28.0	32.0	28.0	45.0	35.5	51.5	39.0	46.5	36.0	32.5	28.5
15	24.5	45.5	35.0	30.5	27.5	31.0	27.5	31.5	28.0	31.5	28.0	40.5	33.0	51.5	39.0	51.0	39.0	38.5	32.0
16	24.5	40.5	32.0	29.0	26.5	29.5	27.0	30.0	27.0	30.0	27.0	34.0	29.5	49.0	37.5	52.5	39.5	42.5	34.0
17	24.5	34.5	28.5	28.0	26.0	27.5	26.0	28.0	26.0	28.0	26.0	28.0	26.0	43.5	34.5	50.5	38.5	45.0	35.0
18	24.0	28.0	24.5	29.5	26.5	25.5	24.5	25.5	24.5	25.5	24.5	25.5	24.5	35.5	30.0	44.0	34.5	41.5	33.0

Table A.9 (Continued)

Sun time	Air temp. t_{ao} (°C)	Sol-air temperature, t_{av}/(°C)																	
		Horizontal		North		North-East		East		South-East		South		South-West		West		North-West	
		Dark	Light	Dark	Light	Dark	Light	Dark	Light	Dark	Light	Dark	Light	Dark	Light	Dark	Light	Dark	Light
19	23.0	22.0	21.0	27.0	25.0	23.0	22.5	23.0	22.5	23.0	22.5	23.0	22.5	26.5	24.5	32.5	27.5	32.5	28.0
20	21.5	18.5	18.5	21.0	21.0	21.0	21.0	21.0	21.0	21.0	21.0	21.0	21.0	21.0	21.0	21.0	21.0	21.0	21.0
21	20.5	17.0	17.0	19.5	19.5	19.5	19.5	19.5	19.5	19.5	19.5	19.5	19.5	19.5	19.5	19.5	19.5	19.5	19.5
22	19.0	15.5	15.5	18.0	18.0	18.0	18.0	18.0	18.0	18.0	18.0	18.0	18.0	18.0	18.0	18.0	18.0	18.0	18.0
23	17.5	14.0	14.0	16.5	16.5	16.5	16.5	16.5	16.5	16.5	16.5	16.5	16.5	16.5	16.5	16.5	16.5	16.5	16.5
Mean	19.0	27.5	22.0	22.0	20.0	24.0	21.5	26.5	22.5	27.0	23.0	26.0	22.5	27.0	23.0	26.5	22.5	24.0	21.5

Note I_l (horizontal surface) = 79 W/m^2 I_l (vertical surface) = 18 W/m^2.

Table A.9 (Continued)

(f) August 22

Sun time	Air temp. $t_{ao}/(°C)$	Sol-air temperature, $t_{av}/(°C)$																	
		Horizontal		North		North-East		East		South-East		South		South-West		West		North-West	
		Dark	Light	Dark	Light	Dark	Light	Dark	Light	Dark	Light	Dark	Light	Dark	Light	Dark	Light	Dark	Light
00	15.0	12.0	12.0	14.0	14.0	14.0	14.0	14.0	14.0	14.0	14.0	14.0	14.0	14.0	14.0	14.0	14.0	14.0	14.0
01	14.0	11.0	11.0	13.0	13.0	13.0	13.0	13.0	13.0	13.0	13.0	13.0	13.0	13.0	13.0	13.0	13.0	13.0	13.0
02	13.0	10.0	10.0	12.0	12.0	12.0	12.0	12.0	12.0	12.0	12.0	12.0	12.0	12.0	12.0	12.0	12.0	12.0	12.0
03	12.5	9.5	9.5	11.5	11.5	11.5	11.5	11.5	11.5	11.5	11.5	11.5	11.5	11.5	11.5	11.5	11.5	11.5	11.5
04	12.0	9.0	9.0	11.0	11.0	11.0	11.0	11.0	11.0	11.0	11.0	11.0	11.0	11.0	11.0	11.0	11.0	11.0	11.0
05	12.0	9.0	8.5	11.0	11.0	11.5	11.0	11.5	11.0	11.0	11.0	11.0	11.0	11.0	11.0	11.0	11.0	11.0	11.0
06	12.0	12.5	11.0	13.5	12.5	21.0	16.5	23.0	17.5	19.0	15.5	12.5	12.0	12.5	12.0	12.5	12.0	12.5	12.0
07	12.5	18.5	14.5	14.0	13.0	25.0	19.0	31.0	22.5	27.0	20.0	15.5	14.0	14.5	13.0	14.5	13.0	14.0	13.0
08	13.5	25.5	19.0	16.0	14.5	25.5	19.5	34.5	25.0	33.0	24.0	22.0	18.0	17.0	15.0	17.0	15.0	16.5	14.5
09	14.5	32.0	23.0	18.5	16.5	23.5	19.0	35.5	26.0	37.5	27.0	29.0	22.0	19.5	17.0	19.5	17.0	19.0	16.5
10	16.0	38.0	27.0	21.0	18.5	21.5	18.5	33.5	25.5	39.5	28.5	35.0	26.0	22.5	19.0	21.5	19.0	21.5	18.5
11	17.5	42.5	30.0	23.0	20.0	23.5	20.5	30.5	24.0	39.5	29.0	39.5	29.0	30.5	24.0	24.0	20.5	23.5	20.5
12	18.5	44.5	32.0	24.5	21.5	25.0	22.0	25.5	22.0	37.0	28.5	41.5	31.0	37.0	28.5	25.5	22.0	25.0	22.0
13	20.0	45.0	32.5	25.5	22.5	26.0	23.0	26.5	23.0	33.0	26.5	42.0	31.5	42.0	31.5	33.0	26.5	26.0	23.0
14	21.0	43.0	31.5	26.0	23.0	26.0	23.5	26.5	23.5	27.5	24.0	39.5	31.0	44.5	33.5	38.5	30.5	26.0	23.5
15	21.5	39.0	30.0	25.5	23.0	25.5	23.5	26.0	23.5	26.0	23.5	35.5	29.0	44.0	33.5	42.0	32.5	30.0	26.0
16	21.5	33.5	27.0	24.5	22.5	24.5	23.0	25.0	23.0	25.0	23.0	30.5	26.0	41.5	32.0	43.0	33.0	33.5	28.0
17	21.5	27.5	23.5	23.0	22.0	23.0	22.0	23.0	22.0	23.0	22.0	24.5	22.5	36.0	29.0	40.0	31.5	34.0	28.0
18	21.0	21.5	20.0	22.5	21.5	21.5	21.0	21.5	21.0	21.5	21.0	21.5	21.0	28.0	24.5	32.0	26.5	30.0	25.5

Table A.9 (Continued)

Sun time	Air temp. t_{ao}/(°C)	Sol-air temperature, t_{av}/(°C)																	
		Horizontal		North		North-East		East		South-East		South		South-West		West		North-West	
		Dark	Light	Dark	Light	Dark	Light	Dark	Light	Dark	Light	Dark	Light	Dark	Light	Dark	Light	Dark	Light
19	20.5	17.5	17.5	19.5	19.5	19.5	19.5	19.5	19.5	19.5	19.5	19.5	19.5	19.5	19.5	20.0	19.5	20.0	19.5
20	19.5	16.5	16.5	18.5	18.5	18.5	18.5	18.5	18.5	18.5	18.5	18.5	18.5	18.5	18.5	18.5	18.5	18.5	18.5
21	18.5	15.5	15.5	17.5	17.5	17.5	17.5	17.5	17.5	17.5	17.5	17.5	17.5	17.5	17.5	17.5	17.5	17.5	17.5
22	17.0	14.0	14.0	16.5	16.5	16.5	16.5	16.5	16.5	16.5	16.5	16.5	16.5	16.5	16.5	16.5	16.5	16.5	16.5
23	16.0	13.0	13.0	15.0	15.0	15.0	15.0	15.0	15.0	15.0	15.0	15.0	15.0	15.0	15.0	15.0	15.0	15.0	15.0
Mean	16.5	23.5	19.0	18.0	17.0	19.5	18.0	22.0	19.0	23.0	19.5	23.0	19.5	23.0	19.5	22.0	19.0	19.5	18.0

Note I_l (horizontal surface) = 79 W/m² I_l (vertical surface) = 18 W/m².

Table A.10 Approximate climatic refrigeration loads for office blocks in the UK, assuming a lightweight building (150 kg m^{-2}) with a heavyweight roof (300 kg m^{-2})

Amount of glass (%)	Number of storeys in building 1			2			4			8			30		
	Sun time (hours)	Month	Load (Wm^{-2})	Sun time (hours)	Month	Load (Wm^{-2})	Sun time (hours)	Month	Load (Wm^{-2})	Sun time (hours)	Month	Load (Wm^{-2})	Sun time (hours)	Month	Load (Wm^{-2})
Major building axis E–W:															
0	17.00	July	46	16.00	August	36	16.00	August	32	16.00	August	29	16.00	August	28
25	15.00	August	56	15.00	August	47	13.00	August	44	13.00	August	42	13.00	August	42
50	13.00	August	69	13.00	August	65	13.00	August	62	13.00	August	61	13.00	August	60
75	13.00	August	88	13.00	August	83	13.00	August	81	13.00	August	80	12.00	August	79
Major building axis N–S:															
0	18.00	July	49	18.00	August	38	18.00	August	33	18.00	August	30	18.00	August	29
25	17.00	July	70	17.00	July	59	17.00	July	54	17.00	July	52	17.00	July	50
50	17.00	July	92	17.00	July	82	17.00	July	77	17.00	July	74	17.00	July	72
75	17.00	July	115	17.00	July	104	17.00	July	99	17.00	July	96	16.00	July	95
Major building axis NE–SW:															
0	18.00	July	47	18.00	August	36	18.00	August	31	18.00	August	28	18.00	August	26
25	17.00	July	62	17.00	July	52	17.00	July	47	17.00	July	44	17.00	July	42
50	17.00	July	80	17.00	July	69	17.00	July	64	17.00	July	62	17.00	July	60
75	17.00	July	97	17.00	July	87	17.00	July	82	17.00	July	79	11.00	August	78
Major building axis NW–SE:															
0	18.00	July	50	18.00	August	39	18.00	August	33	17.00	August	31	17.00	August	29
25	16.00	August	65	16.00	August	56	16.00	August	52	16.00	August	50	16.00	August	48
50	15.00	August	85	15.00	August	77	15.00	August	73	15.00	August	72	15.00	August	70
75	15.00	August	107	15.00	August	100	15.00	August	96	15.00	August	94	15.00	August	92

Table A.11 Meteorological data for Kew: 5°C 28′ N, 0°19′ W, 5 m above sea level, for period 1931–1960

Item	Temperature (°C)											
	Jan	Feb	Mar	April	May	June	July	Aug	Sept	Oct	Nov	Dec
Mean daily maximum	6.3	6.9	10.1	13.3	16.7	20.3	21.8	21.4	18.5	14.2	10.1	7.3
Mean daily minimum	2.2	2.2	3.3	5.5	8.2	11.6	13.5	13.2	11.3	7.9	5.3	3.5
Diurnal range	4.1	4.7	6.8	7.8	8.5	8.7	8.3	8.2	7.2	6.3	4.8	3.8
Mean monthly maximum	11.7	12.1	15.5	18.7	23.3	25.9	26.9	26.2	23.4	18.7	14.4	12.2
Mean monthly minimum	−4.3	−3.6	−2.3	0.1	2.7	6.9	9.3	8.5	5.4	0.4	−1.4	−3.2
Absolute maximum	14.3	16.1	21.4	23.5	30.2	32.7	33.8	33.1	29.8	25.6	19.0	15.1
Absolute minimum	−9.5	−9.4	−7.7	−2.1	−1.0	4.8	7.0	6.2	3.0	−3.6	−5.0	−7.0

Based on information from Met O. 856c. Meteorological Office. Tables of temperature, relative humidity, precipitation and sunshine. Part III, Europe and the Azores. HMSO. London. 1972 (reproduced by permission of HMSO. London).

The specific refrigeration loads listed in Table A.10 are based on the following assumptions: values (W m^{-2} K^{-1}): glass, 5.6; wall, 1.7; roof, 1.1: Surface densities (kg m^{-2}): walls, 300; roof, 300; floor slabs, 150. The results are from a computer study (see Chapter 1) for a hypothetical building (very similar to that in Figure 1.1) with various amounts of single glazing on its two long faces hut none on its short faces, fitted with internal Venetian blinds on all faces except north. The plan of the building is 86.4 m × 13.5 m with a floor-to-floor height of 3.3 m and a floor-to-ceiling height of 2.6 m. No natural infiltration is assumed. The loads are the maximum values for the whole building divided by the plan area × the number of storeys, and they include gains through the windows, walls and roof, plus the fresh air loads arising from the supply of 1.5 s^{-1} m^{-2} of floor area. They can, therefore, be converted into actual total refrigeration loads by adding the appropriate allowances for lighting, people, business machines, fan power, duct gains, etc. A sinusoidal variation in outside air temperature (Equation (1.2)) based on meteorological data from Kew (see Chapter 1) is used and the inside state is taken as 22°C dry-bulb and 50% saturation in all cases. The solar loads through glass are calculated using the data in Tables A.1–A.4, inclusive, based on the Carrier method, and equivalent temperature differences are adopted for the gains through walls and roofs.

The tabulated results can be regarded as typical of a modern, office block with carpeted floors. A heavyweight building (slab surface density 500 kg m^{-2}) gives results about 5% less for the higher rise structures with 75% and 50% glazing, but the times and months of the peak load are not greatly different. The same is true of buildings with lightweight roofs (50 kg m^{-2}), except for single-storey constructions where, although the time and month of the maximum is virtually unchanged, the maximum load tends to be about 10% greater when there is no glazing.

Derivation of Equation (2.3)

$$\text{Sensible heat gain} = \text{mass flow rate of supply air} \times \text{specific heat capacity} \times \text{temperature rise} = \dot{m} \times c\,(t_r - t_s)$$

where t_r and t_s are the room and supply air temperatures, respectively.

$$\text{Volumetric flow rate of air at temperature } t = \dot{v}_t = \dot{m}/\rho_t$$

where ρ_t is the density of air at temperature t, and also

$$\rho_t = \rho_o\,(273 + t_o)/(273 + t)$$

where ρ_o and t_o are a standard density and temperature, respectively.
Therefore

$$\text{Sensible heat gain} = [\dot{v}_t \times \rho_o \times (273 + t_o)/(273 + t)] \times c(t_r - t_s)$$

If $\rho_o = 1.191$ kg m^{-3} at 20°C dry-bulb and 50% saturation and c = 1.026 kJ kg^{-1} K^{-1} then

$$\dot{v}_t = [(\text{sensible heat gain})/(t_r - t_s)] \times [(273 + t)/358]$$

If the sensible heat gain is in W, $\dot{v}_t$ is in l s^{-1}. If the sensible heat gain is in kW, $\dot{v}_t$ is in m^3 s^{-1}

Derivation of Equation (2.4)

$$\text{Latent heat gain} = \text{mass flow rate of supply air} \times \text{moisture pick-up in kg kg}^{-1}\text{ air} \times \text{latent heat of evaporation in kJ kg}^{-1}\text{ water} = \dot{m} \times h_{fg}(g_r - g_s)$$

where g_r and g_s are the room and supply air moisture contents, respectively. As with Equation (2.3), $\dot{m} = \dot{v}_t \times \rho_o(273 + t_o)/(273 + t)$, therefore

$$\text{Latent heat gain} = [\dot{v}_t \times \rho_o \times (273 + t_o)/(273 + t)] \times h_{fg}(g_r - g_s)$$

If $\rho_o = 1.191$ kg m^{-3} at 20°C dry-bulb and 50% saturation and $h_{fg} = 2454$ kJ kg^{-1} at 20°C, then

$$\dot{v}_t = [(\text{sensible heat gain})/(g_r - g_s)] \times [(273 + t)/856]$$

where $\dot{v}_t$ is in m^3 s^{-1}, the latent heat gain is kW and $(g_r - g_s)$ **is in g kg^{-1} dry air.**

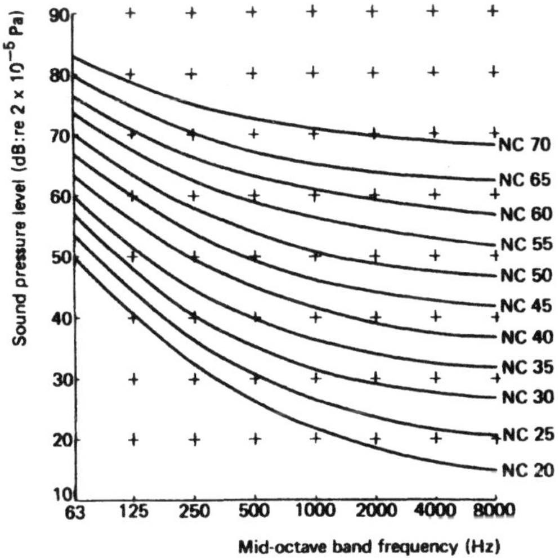

Figure A.1 Noise criteria curves (after Beranek; in Chapter 7).

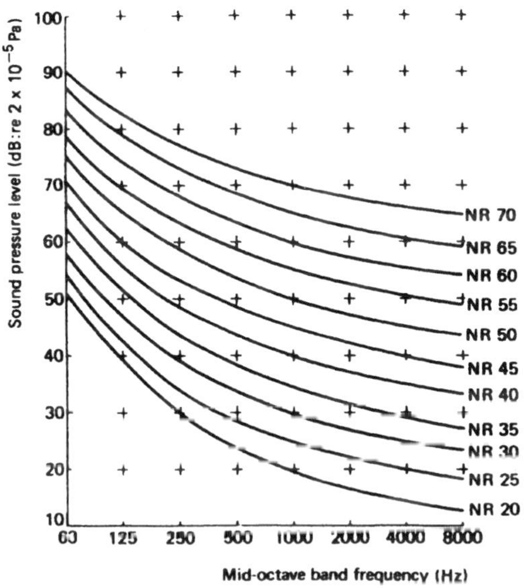

Figure A.2 Noise rating curves (after Kosten and van Os; in Chapter 7)

Index

Manufactured by Amazon.ca
Bolton, ON

28651183R00221